Mutualistic Interactions between Flowering Plants and Animals

Editors
Palatty Allesh Sinu
KR Shivanna

First published 2026
by CRC Press
4 Park Square, Milton Park, Abingdon, Oxon, OX14 4RN

and by CRC Press
2385 NW Executive Center Drive, Suite 320, Boca Raton FL 33431

CRC Press is an imprint of Informa UK Limited

British Library Cataloguing-in-Publication Data
A catalogue record for this book is available from the British Library

ISBN: 9781032826455 (hbk)
ISBN: 9781032826479 (pbk)
ISBN: 9781003505471 (ebk)

DOI: 10.1201/9781003505471

Typeset in Georgia
by Manipal Universal Press

MANIPAL
UNIVERSAL PRESS

Mutualistic Interactions between Flowering Plants and Animals

Editors
Palatty Allesh Sinu
KR Shivanna

MANIPAL
UNIVERSAL PRESS

CRC Press
Taylor & Francis Group
Boca Raton London New York

CRC Press is an imprint of the
Taylor & Francis Group, an **informa** business

Dedicated to
Late Professor TC Narendran
and
Professor KS Manilal
for their
invaluable contributions in documenting the
diversity of insects and flowering plants

Preface

"There is a mutual aid in many species, with services being repaid with good behaviour or in kind, and mutualism can well take its place beside commensalism"

(Pierre van Beneden 1876 in Animal Parasites and Messmates)

Flowering plants have established close association with the animals particularly the insects right from the beginning of their origin over 100 million years ago. This interaction has been considered to have acted as a driving force for diversification of both the partners. The flowering plants have achieved dominance amongst the land plants and insects became dominant amongst the animals. The plant-animal interactions are both mutualistic and antagonistic. Two of the most vital areas of reproduction in flowering plants that depend on mutualistic interactions are pollination, a pre-requisite for fruit and seed set, and seed dispersal, an essential requirement for successful regeneration and colonization of the wild species. These interactions are similar to biological barter; plants provide rewards in the form of food for animals in exchange for their services in carrying out pollination and seed dispersal. Herbivory, particularly of seeds and seedlings, is a major antagonistic interaction which limits seedling recruitment. Both the mutualistic and the antagonistic interactions, to a large extent, determine the reproductive success of the flowering plants and also the sustenance of animlas that depend on plant resources. Thus, plant-animal interaction is ubiquitous, although the type of interaction is highly variable.

Although habitat degradation, overexploitation, and climate change are the proximate causes for erosion of biodiversity, the ultimate causes that drive plant species to vulnerable category and eventual extinction are the negative effects of the proximate causes on reproductive success of plant species. They impose major constrains at different stages of reproduction; when reproductive constraints are prolonged, the species are pushed to vulnerable category and eventual extinction. Apart from the sustenance of biodiversity, plant-animal interactions affect the productivity of our crop plants. Development of fruits and seeds depends on adequate pollination which is mediated, in a majority of our crop plants, through the activities of the insects. Therefore, the effective conservation and sustainable utilization of our plant and animal diversity, the management of our forests and sustained productivity of our agri-horticultural crops require a thorough understanding of plant-animal interactions.

Because of their importance both in fundamental and applied areas of biology, extensive studies have been carried out on the interactions between the flowering plants and the animals, and enormous amount of literature has accumulated in this area in the Western countries. However, such studies in Indian subcontinent are rather limited. One of the main reasons for this lack of interest amongst our researchers is that the plant-animal interactions are hardly taught in our undergraduate and post-graduate courses. This lacuna is probably due to the fact that the curricula of these courses are framed largely with the mindset of Botany and Zoology, rather than Biology. This has resulted in lack of awareness amongst our researchers on the importance of studies on plant-animal interactions. The studies on the plants and animals in isolation are totally inadequate in understanding their ecology and evolution, and functioning of natural communities. Such studies have become more important in recent years in the light of massive erosion of our biodiversity in natural habitats and also the leveling of productivity of our crop plants as a result of anthropogenic activities.

The need is to increase our quantum of teaching on plant-animal interactions at the undergraduate and the post-graduate levels, and to encourage research on these areas in the country. In spite of extensive literature available on the subject, there is a need for a book to cover the basic aspects of the interaction between the flowering plants and the animals. We conducted a workshop essentially for young research students covering a number of areas of mutualistic interactions. The proceedings of the workshop are presented in this book. The authors of the review chapters are all experts in their respective fields. The volume includes the chapters on various aspects of the mutualistic plant-animal interactions particularly the two most important areas – pollination and seed dispersal. One of the chapters covers an interesting aspect of the plant-animal interaction in which the plants, upon attack by herbivores, recruit other insects that parasitize on caterpillars of herbivorous insects. The book caters to post-graduate students, teachers and researchers in the areas of biology, botany, zoology, ecology, agri-horticulture, forestry, and conservation biology.

We thank all the authors who have sent their manuscripts promptly. We also thank all the research scholars Asha G, Sangeetha Varma, Prabitha Mohan, Shibil VK, Rajesh TP, Manoj K, Prashanth Ballullaya U, Nayak K in the Department of Animal Science for their help in proof-reading of the entire manuscript. We are grateful to the Central University of Kerala and the British Ecological Society, London, for their financial support for conducting the workshop on the concepts and practices in ecology of plant-animal interactions which form the basis for this book. We also express our gratitude to Professor G Gopakumar, Vice Chancellor, Central University, Kerala, for his encouragement and support for the publication of this book. KRS is thankful to the Indian National Science Academy, New Delhi, for awarding the positions of INSA Senior Scientist (2003-08) and INSA Honarary Scientist (2009-to date).We sincerely thank Dr Neeta Inamdar, Professor and Head, Manipal Centre for European Studies and the Chief Editor of Manipal Universal Press, Ms Savita Sastri, Acquisition Editor, Mr Arvind N, Executive-Design and Production and other members of the Editorial Board of MUP for their co-operation in publishing this book.

Date: May 6, 2015

Palatty Allesh Sinu
KR Shivanna

Content

Contributors

G Asha, Department of Animal Science, Central University of Kerala, Padannakad 671 314, Kasaragod, Kerala, India

U Prashanth Ballullaya, Department of Animal Science, Central University of Kerala, Padannakad 671 314, Kasaragod, Kerala, India

Chandan Barman, Department of Botany, University of Delhi, Delhi 110 007, India

Parthiba Basu, Centre for Pollination Studies, Calcutta University, 35, Ballygunge Circular Road, Kolkata 700 019, India

Priyadarshini Chakrabarti, Centre for Pollination Studies, Calcutta University, 35, Ballygunge Circular Road, Kolkata 700 019, India

Pradnya V Deshpande, Chemistry group, Agharkar Research Institute, G G Agarkar Road, Pune 411 004, India

Sujata A Deshpande, Department of Zoology, St Xavier's College, 5 Mahapalika Marg, Mumbai 400 001, India

T Jobiraj, Department of Zoology, Government College, Kodanchery 673 580, Kerla, India

Giby Kuriakose, Assistant Professor, Department of Botany, Sacred Heart College, Thevara, Kochi 682013,Kerala, India

K Manoj, Department of Animal Science, Central University of Kerala, Padannakad 671 314, Kasaragod, Kerala, India

TN Manohara, Rain Forest Research Institute, PB No 136, Deovan, Jorhat, 785 001, Assam, India

Prabitha Mohan, Department of Animal Science, Central University of Kerala, Padannakad 671 314, Kasaragod, Kerala, India

Divya Mohanty, Department of Botany, University of Delhi, Delhi 110007, India

M Nagarajan, Department of Genomic Science, Central University of Kerala, 671 314, Kasaragod, Kerala, India

Anjankumar Naik, Department of Entomology, University of Agricultural Sciences, Dharwad 580 005, India

Dattatraya G Naik, Chemistry group, Agharkar Research Institute, G GAgarkar Road, Pune 411 004, India

AK Nimisha, Department of Genomic Science, School of Biological Sciences, Central University of Kerala, Kasaragod 671 314, Kerala, India

K Nimisha, Department of Genomic Science, School of Biological Sciences, Central University of Kerala, Kasaragod 671 314, Kerala, India

AG Pandurangan, Plant Systematics and Evolutionary Science Division, Department of Conservation Science, Jawaharlal Nehru Tropical Botanical Garden and Research Institute, Karimancode PO, Palode, Thiruvananthapuram 695 562, Kerala, India

TP Rajesh, Department of Animal Science, Central University of Kerala, Padannakad 671 314, Kasaragod, Kerala, India

AJ Solomon Raju, Department of Environmental Sciences, Andhra University, Visakhapatnam 530 003, India

VK Shibil, Department of Animal Science, Central University of Kerala, Padannakad 671 314, Kasaragod, Kerala, India

Vineet Kumar Singh, Department of Botany, University of Delhi, Delhi 110 007, India

Palatty Allesh Sinu, Department of Animal Science, Central University of Kerala, Padannakad PO, Kasaragod 671 314, Kerala, India

Deepu Sivadas, Plant Systematics and Evolutionary Science Division, Jawaharlal Nehru Tropical Botani Garden and Research Institute, Karimancode PO, Palode, Thiruvananthapuram 695 562, Kerala, India

KR Shivanna, Ashoka Trustfor Research in Ecology and the Environment,Royal Enclave, Srirampura, Jakkur Post, Bengaluru 560 064, India

Rajesh Tandon, Department of Botany, University of Delhi, Delhi 110 007, India

Sangeetha Varma, Department of Animal Science, Central University of Kerala, Padannakad 671 314, Kasaragod, Kerala, India

Shashidhar Viraktamath, Department of Entomology, University of Agricultural Sciences, Dharward 580 005, Karnataka, India

Chapter 1

General Introduction

K R Shivanna and Palatty Allesh Sinu

Land plants have interacted with animals since their origin in a variety of ways (see Herrera and Pellmyr 2002, Bronstein et al. 2006, Schaefer and Ruxton 2011). This interaction became more intimate and diverse in flowering plants (Labandeira 2002). Plant-animal interactions are both mutualistic and antagonistic, and occur during vegetative as well as reproductive phases. Thus, plant-animal interactions are ubiquitous and studies on plants and animals in isolation are inadequate to understand their ecology and evolution, and functioning of communities.

One of the critical phases of flowering plants that depend on plant-animal interactions is sexual reproduction, the basis of genetic recombination. Sexual reproduction is more complex in flowering plants when compared to other groups of plants and involves a series of sequential events (**Figure 1.1**). In cultivated plants, reproductive success is assessed on the basis of the extent of seed set. However, in wild species, it is assessed on the basis of recruitment of new young adults which is essential for the sustenance of populations and their colonization in new areas. Two of the most important events of sexual reproduction that depend on mutual interactions with animals are pollination, a pre-requisite for seed set, and seed dispersal, an essential requirement for effective colonization of the species. These mutualistic interactions are similar to biological barter; plants provide rewards in the form of food for animals in exchange for their services in carrying out pollination and seed dispersal. In fact, reproductive success of flowering plants is determined, to a large extent, on these mutualistic interactions. To understand these mutualistic interactions, it is essential to become familiar with some of terminologies used to describe the structural and functional aspects of the flower. Some of these terminologies are explained in **Box 1.1**. Familiarity with these terminologies helps to understand floral diversity of flowering plants in relation to their reproductive function.

DOI: 10.1201/9781003505471-1

Initiation and development of flowers

↓

Development of functional pollen grains and ovules

↓

Effective pollination

↓

Pollen-pistil interaction and fertilization

↓

Development of fruits and seeds

↓

Seed dispersal

↓

Seedling establishment and their growth into young adults

Fig. 1.1. Sequential events in sexual reproduction of flowering plants.

Box 1.1. Terminologies used to describe some structural and functional features of the flower

Types of flowers

Cleistogamous flowers: Flowers do not open and thus the stamens and the pistil remain enclosed within the flower bud; the anthers and the stigma lie close to each other. Pollen grains germinate inside the anther or after coming in contact with the stigma following anther dehiscence; pollen tubes enter the stigma and grow through the pistil. Thus, pollination in cleistogamous flowers is exclusively autogamous (within the flower).

Chasmogamous flowers: Flowers open and permit out-cross pollination mediated through pollinating agents.

Actinomorphic flowers: Flowers are radially symmetrical; can be divided into two identical halves by cutting vertical section in any plane passing through the centre of the flower.

Zygomorphic flowers: Flowers are bilaterally symmetrical; can be divided into two identical halves by cutting vertical section only in one plane passing through the centre.

Male and female phases of flowers in space and time

Herkogamy: The position/level of the anthers and the stigma in the flower are spatially separated.
Dichogamy: Onset of male (pollen shedding) and female (stigma receptivity) phases of the flower are temporally separated. The period between the onset of the two phases may range from a few hours to a few days.
Protogynous: The stigma becomes receptive before pollen grains are shed.
Protandrous: Pollen grains are shed before the stigma becomes receptive

Sexuality of flowers

Bisexual: Produce functional pollen grains as well as ovules
Male: Produce functional pollen grains but not ovules
Female: Produce functional ovules but not pollen grains

Sexuality of plants and populations
Hermophrodite: Each plant produces only bisexual flowers
Monoecious: Each plant produces male and female flowers
Andromonoecious: Each plant produces male and bisexual flowers
Gynomonoecious: Each plant produces female and bisexual flowers
Polygamomonoecious: Each plant produces male, female and bisexual flowers (more common condition: in addition to male and female flowers, some plants produce a small proportion of bisexual flowers also)
Dioecious: Each plant produces either male or female flowers
Androdioecious: Each plant produces either male or bisexual flowers
Gynodioecious: Each plant produces either female or bisexual flowers
Polygamodioecious: Each plant produces male or female or bisexual flowers [more common condition: in addition to male and female plants, some plants produce a small proportion of flowers of the opposite sex (males producing some female flowers or females producing some male flowers) and/or bisexual flowers].

Types of pollination
Autogamy: Transfer of pollen grains from the anther to the stigma of the same flower.
Geitonogamy: Transfer of pollen grains from the anther to the stigma of another flower of the same plant or of the same clone.
Xenogamy: Transfer of pollen grains from the anther to the stigma of another plant of non-clonal origin.

Breeding systems
Obligate inbreeding: Pollination is exclusively autogamous. Cleistogamous flowers and chasmogamous flowers of some species may show exclusive autogamy when pollinators are absent.
Mixed breeding: Both autogamy/geitonogamy and xenogamy occur, although the proportion of each type may vary depending on the structure of the flower and pollination environment.
Obligate outbreeding: Exclusively xenogamous. Autogamy and geitonogamy do not result in fertilization and seed set. Strictly self-incompatible (inability of self-pollen to effect fertilization) and dioecious species are obligate outbreeders. However, many of the self-incompatible species are leaky; self-pollination in such species results in some seed set. Even some dioecious species produce a small proportion of bisexual flowers permitting some inbreeding through auto-/geitonogamy.
Double fertilization: A unique feature of flowering plants involves two fusion events. Of the two sperm cells released by the pollen tube in the embryo sac (female gametophyte), one fuses with the egg to give rise to the zygote and the other fuses with the secondary nucleus (fused polar nuclei) of the central cell to give rise to the primary endosperm nucleus. The zygote develops into the embryo and the central cell gives rise to the endosperm, a nutritive tissue for the developing embryo.

Pollination, deposition of pollen grains from the anther to the stigma, is the basis of gene flow between individuals and populations. As plants are sedentary, they cannot achieve pollen transfer on their own; they have to make use of other agencies to perform this function. Majority of flowering plants (nearly 90%) make use of animals for pollination and the remaining species make use of wind or water. Pollination mediated through animals (biotic pollination) is considered primary as it is prevalent in all basal angiosperms; wind and water pollinations are considered to be derived (see Faegri and van der Pijl 1979, Pellmyr 2002, Schaefer and Ruxton 2011). Available evidences indicate that insects are the primary pollinators of primitive angiosperms (Faegri and van der Pijl 1979) including early branched angiosperms (basal angiosperms) (Thien et al. 2009) and other pollination syndromes seem to have evolved later. Biotic pollinators exert selection on floral traits and promote their evolutionary divergence amongst plant populations leading to reproductive isolation (Bronstein et al. 2006, Johnson 2006) and eventual speciation. This avenue of reproductive isolation is not available for species with abiotic pollination. This concept is in agreement with the reports on the phylogenetic analyses on species diversity indicating association of biotic pollination with more rapid diversification when compared to abiotic pollination (Dodd et al. 1999). Thus, biotic pollination mediated by insects acted as a powerful evolutionary force for the diversification of flowering plants as well as insects. Flowering plants, within a short period after their origin, became dominant amongst the land plants with the largest number of species. Although insects originated much earlier than flowering plants, their initial diversification was slow. However, after the origin of flowering plants, their diversification increased dramatically as the result of the evolution of pollination mutualism; they soon became dominant with the highest number of species amongst animals (*see* Pellmyr 2002, Price 2002).

Another dimension to pollination mutualism is its role in sustaining plant and animal diversity. As pollination is the basic requirement for optimal seed set and eventual recruitment of new adults, biotic pollination has become highly vulnerable in recent decades due to anthropogenic disturbances. A number of studies have highlighted the decline of pollinators particularly of insects and plant species in recent decades (Biesmeijer et al. 2006, Potts et al. 2010). More importantly, the decline is more intense amongst specialized (foraging only a limited number of plant species) and sedentary insects when compared to general (foraging a larger number of plant species) and migratory insects (Biesmeijer et al. 2006). Amongst plant species, the decline is more evident in species that depended on specialized pollinators than those that depended on generalized pollinators, indicating a parallel link between the decline of plant species and their pollinators. This decline of pollinator diversity and their density has raised serious concern on the sustainability of pollination services and thus biodiversity in the coming decades (*see* Potts et al. 2010, Cameron et al. 2011, Tylianakis 2013).

Because of its importance in fundamental and applied areas, extensive studies have been carried out on pollination ecology since the late eighteenth century (*see* Pellmyr 2002, Willmer 2011, Patiny 2012). The literature available in this area is probably more than that in any other area of reproductive biology. Two of the essential features required to achieve pollination through animals are:

a) Plants have to produce suitable rewards and advertise the presence of these rewards to attract animals to visit flowers of conspecific plants in a sustainable way, and

b) Plants have to position the anthers and the stigma in such a way that they come in contact with the body of visitors to bring about pollen transfer.

There are several chapters in the present volume on pollination. The chapter by K R Shivanna gives an overview on these essential features of biotic pollination. As fruits and seeds are the economic products of most of our crop plants, effective pollination is essential for optimal yield of crop species. Historically, native pollinators present on and around agricultural landscapes used to bring about crop pollination. In recent decades because of the decline of the density and diversity of pollinators due to a number of reasons, native pollinators are no more able to perform adequate pollination services of crop plants. This chapter also elaborates the reasons for pollination limitation and how managed pollinators are being used routinely in Western countries to sustain pollination services in agri-horticultural crops. It also highlights the importance of native wild bees in pollination services even in the presence of managed pollinators and recent attempts to increase and sustain native wild pollinators in Western countries by providing suitable nesting sites combined with eco-friendly agricultural practices. The chapter also highlights the inability of Indian farmers to use this eco-friendly technology of managed pollinators. This is largely due to lack of awareness amongst Indian farmers, insufficient data available on crop pollinators and pollination, and limitation of research backup on the management of pollinators.

Insects, particularly bees, form the major group of pollinators. They have evolved a number of morphological and behavioural adaptations to perform pollination services. The chapter by Sinu et al gives a comprehensive account on the origin, evolution, classification, natural history and foraging adaptations of bees relevant to pollination. Although these details form important components of pollination efficacy, they are not readily available in pollination literature. He also elaborates floral constancy which is important in preventing pollination with heterospecific pollen.

All animals that visit the flower are not pollinators. Some of them forage the rewards without bringing about pollination. These are called pollen and/or nectar robbers. The general concept has been that these robbers affect the pollination efficiency negatively thus reducing the fitness of the plant. However, there are many evidences in recent years to indicate that these robbers increase the fitness of plants in several species by encouraging the movement of pollinators between plants and thus increasing the extent of cross-pollination. The chapter on plant-pollinator-robber interaction by Rajesh Tandon and his associates analyzes our understanding of the dynamics of plant-pollinator-robber interactions and evolutionary significance of nectar robbing.

Orchids form one of the most diverse and specialized groups of flowering plants. They show great variation in floral morphology and mode of pollination. Pollination of orchids is generally considered as highly specialized, as most of them are pollinated by species-specific pollinators. The chapter by Giby Kuriokose elaborates all these aspects of orchid reproductive biology. One of the fascinating features particularly prevalent in orchids is deceptive pollination in which they get pollination services of animals without offering any rewards. In fact one third of orchid species (nearly 10,000 species) are known to be deceptive. This chapter also elaborates different forms of deception in orchids including sexual deception.

As pointed out earlier, keeping managed pollinators particularly honey bees amidst crop plants is a routine practice in Western countries to sustain pollination. One of the problems of maintaining managed pollinators is to keep their foraging to the target crop during its flowering period. If there are any other species flowering amidst or around the crop plants, and if they are more attractive and offer better rewards, the bees desert the crop species and start visiting the competing species. One of the potential approaches to retain them on the target crop is to spray the crop plants with bee attractants. These attractants are essentially based on bee foods (sugars and proteins), bee pheromones and plant-based chemicals. Two of the chapters deal on this aspect. The chapter by P V Deshpande and D G Naik describes in detail the pheromones isolated by Nosonov gland, their chemistry and efficacy in attracting bees. They also describe development of plant based attractants including the details of the work done in their laboratory on leaf extract of *Swertia densifolia* and the efficacy of plant based attractants. The chapter also includes the efficacy of commercially available honey bee attractants. In another related chapter S Viraktamath and A Naik describe the results of field studies on application of bee attractants, both commercial and indigenous, carried out by Viraktamath's group on a range of crop species. They report significant pollination enhancement by some bee attractants in some of the crop species.

One of the major causes for the decline of pollinators, both native and managed, in recent years is the extensive use of pesticides and herbicides on agricultural crops. Apart from inducing mortality of the foragers, they cause sub-lethal effects on the bees that pose threats to the health and survival of the entire colony. P Basu and P Chakrabarty describe sub-lethal effects of pesticides on pollinators, particularly honey bees. They show that long term and repeated exposure to pesticides incapacitate honey bees by impairing a number of physiological and behavioural aspects. More importantly, they highlight the risk associated with carrying pesticide residues to the hive through contaminated pollen and nectar to feed the brood, thereby imposing additional threats to the population. They emphasize the need to formulate strict pesticide regulations to monitor indiscriminate use of pesticides to sustain pollinators in agricultural landscapes. One of the potential approaches to reduce pesticide damage to bees, at least partially, is to prevent their visits to crop plants when they are sprayed with pesticides. Many studies have been carried out to study the potential of bee repellents. In one of the chapters P V Deshpande and D G Naik analyze the effects of different synthetic and plant-based substances as bee repellents.

Following pollination, pollen grains germinate on the stigma and the resulting pollen tubes grow through the tissues of the stigma and style and discharge the two male gametes in the embryo sac. One of them fuses with the egg to give rise to the zygote and the other with the central cell (double fretilization, a unique feature of flowering plants); the former develops into the embryo and the latter gives rise to the endosperm, a nutritive tissue for the developing embryo. The ovule develops into the seed and the ovary develops into the fruit. The details that occur from pollination until the pollen tubes enter the embryo sac (the female gametophyte) are generally referred to as pollen-pistil interaction and this is unique to flowering plants. Although pollen-pistil interaction does not involve plant-animal interaction, successful completion of pollen-pistil interaction is essential for fruit and seed development. One of the important functions of pollen-pistil interaction is pollen recognition. If pollen grains are of the right type (compatible), the pistil facilitates post-pollination events but if pollen grains are of the wrong type (incompatible), the pistil inhibits pollen germination or pollen tube growth before entering the embryo sac. Because of its importance, there has been tremendous progress in this area in recent years by integrating the techniques in molecular biology and genetics. The chapter on pollen-pistil interaction and fertilization (K R Shivanna) summarizes these recent advances.

Seed dispersal is critical for colonization of new habitats (Corlett 1998). In a majority of species seed dispersal is mediated by animals, largely vertebrates and some are dispersed by wind or water. The chapter by Sinu et al presents theoretical concepts of seed dispersal, its importance in recruitment of species, and elaborates on biotic mode of seed dispersal highlighting recent advances. It also highlights the concerns on the sustenance of this important eco-service in the light of extensive defaunation of animal dispersers on the recruitment of species in our forests. Another chapter by S A Raju on this aspect describes in detail different types of seed dispersal by biotic as well as abiotic agents. It also covers seed predation, an important hazard for seedling establishment. The chapter also describes many case studies, carried out in Raju's laboratory, on reproductive biology of a range of plant species, largely endemic and endangered tree species growing in the Eastern Ghats. Such studies are very valuable for effective conservation of these species.

Two chapters present case studies on seed dispersal. One of them by T N Manohara describes his work on wasp dispersal, which is rare, in an economically important and critically endangered plant, *Aquilaria malaccensis,* commonly known as agarwood plant. Another case study by D Sivadas and A G Pandurangan describes an antagonistic and mutualistic relationship (seed predation and dispersal) between ants and seeds of *Lagerstroemia speciosa,* a tree species in a moist deciduous forest in Southern Western Ghats.

An interesting paper by Sujata A Deshpande, gives an account of tri-trophic interaction, which can also be considered as a combination of antagonistic and mutualistic interaction. It gives a brief account on the defence mechanism that plants have evolved against insect herbivory and describe hyper-parasitism in which plants, upon attack by insect herbivores, recruit other insects that parasitize on caterpillars of herbivorous insects.

Plant-animal interactions are diverse and very complex. Traditional methods being used to study this complex interactions are rather cumbersome and take too much of time and efforts; often they may not reveal some of the finer aspects of these interactions. In recent years genomics has shown great promise in studies on plant-animal interactions and is bound to be used extensively in the coming years. The chapter by Nagarajan et al. highlights the potential of genomics in studies on plant animal interactions particularly plant-pollinator and predator-prey interactions including herbivory and frugivory.

One of the major problems of pollination biologists has been the identification of insect pollinators. The volume provides keys to Indian genera of Apidae (T Jobiraj), which would help pollination biologists in identifying the bees at least up to the genus. It is hoped that the volume provides base-line data on mutual interactions between flowering plants and animals. It facilitates not only in teaching but also in creating interest in researchers to take up studies on these critical aspects of reproduction, so important for the management of our biodiversity.

References

Biesmeijer JC, Roberts SPM, Reemer M et al. 2006. Parallel declines in pollinators and insect-pollinated plants in Britain and the Netherlands. Science 313: 351–354.

Bronstein JL, Alarcon R, Geber M 2006. Tansley review: The evolution of plant–insect mutualisms. New Phytol 172: 412–428.

Cameron SA, Lozier JD, Strange JP et al. 2011. Patterns of widespread decline in North-American bumble bees. Proc Natl Acad Sci USA 108: 662–667.

Corlett RT 1998. Frugivory and seed dispersal by vertebrates in the Oriental (Indomalayan) region. Biol Rev 73: 413-448.

Dodd ME, Silvertown J, Chase MW 1999. Phylogenetic analysis of trait evolution and species diversity variation among angiosperm families. Evolution 53: 732–744.

Faegri K, van der pijl L 1979. The principles of pollination ecology. Pergamon Press, Oxford.

Herrera CM, Pellmyr O 2002. Plant-animal interactions: An evolutionary approach. Blackwell Science Ltd, Malden, USA.

Johnson SD 2006. Pollination driven speciation in plants. In: Harder LD, Barrett SCH (Eds), Ecology and evolution of flowers. Oxford Univ Press, New York, pp 295–310.

Labandeira CC 2011. The history of associations between plants and animals. In: Herrera CM, Pellmyr O (Eds), Plant-animal interactions: An evolutionary approach. Blackwell Science Ltd, Malden, USA, pp 26–74.

Patiny S (Ed) 2012. Evolution of plant-pollinator relationships. The Systematics Association Special Vol 81, Cambridge Univ Press, Cambridge.

Pellmyr O 2002. Pollination by animals. In: Herrera CM, Pellmyr O (Eds) Plant-animal interactions: An evolutionary approach. Blackwell Science Ltd, Malden, USA, pp 157–184.

Potts SG, Biesmeijer JC, Kremen C et al. 2010. Global pollinator declines: trends, impacts and drivers. Trends Ecol Evol 25: 345–353.

Price PW 2002. Species interactions and the evolution of biodiversity. In: Herrera CM, Pellmyr O (Eds) Plant-animal interactions: An evolutionary approach. Blackwell Science Ltd, Malden, USA, pp 3–25.

Schaefer HM, Ruxton GD 2011. Plant-animal communication. Oxford Univ Press, New York Chapter 6, Floral communication and pollination, pp 116–146.

Thien LB, Bernhardt P, Devall MS et al. 2009. Pollination biology of basal angiosperms. (ANITA grade). Am J Bot 96: 166–182.

Tylianakis JM 2013. The global plight of pollinators. Science 339: 1532–1533.

Willmer P 2011. Pollination and floral ecology. Princeton Univ Press, New Jersey.

Chapter 2

Biotic Pollination: Fundamental and Applied Aspects

K R Shivanna
shivanna@atree.org

Abstract

Pollination, transfer of pollen grains from the anther to the stigma, is a key event in sexual reproduction of flowering plants. It is a pre-requisite for seed set and thus plays an important role in fundamental and applied areas of plant reproduction. About 90 per cent of the flowering plants use a range of animals, particularly insects for pollination services (biotic pollination). Pollination is largely mutualistic and both the partners are benefitted. Plants offer rewards in the form of pollen and nectar and animals visit the flowers to get the rewards and incidentally bring about pollination. In some species the flowers are open and any visitor can harvest the rewards; such flowers are generalists and a range of animals visit the flowers. However, in a number of species the rewards are hidden; such flowers show specialization to different degrees, and depending on the extent of specialization, the number of visitors that can harvest the rewards is proportionately reduced. In some species the specialization is extreme and each plant species can be pollinated by just one animal species as a result of co-evolution. Pollination limitation/failure often acts as a driving force to push the species to vulnerable category. Thus, information on pollination ecology is important for effective conservation of vulnerable species.

As pollination is a prerequisite for fruit and seed set, pollination limitation which has become common in recent years as the result of anthropogenic activities, reduces the yield of crop plants. A number of approaches especially the use of managed pollinators (largely honey bees and bumble bees), both in field grown and greenhouse grown crops are available to sustain pollination and are being used extensively in developed countries since long. Recent trend is on integrated use of natural pollinators, by maintaining eco-friendly habitat on and around crop fields, and managed pollinators in a judicious way to sustain

DOI: 10.1201/9781003505471-2

pollination services. However this eco-friendly technology of using managed pollinators is yet to be applied effectively in developing countries including India.

2.1 Introduction

Sexual reproduction in flowering plants is more complex when compared to that in the lower groups of plants. The flower is the seat of sexual reproduction. It shows great variation in the structure, sexuality and breeding system. Pollination is the transfer of pollen grains from the anthers to the stigma. As the plants are stationary, they cannot transfer pollen grains on their own; they have to depend on external agents to achieve pollination. About 90 per cent of the flowering plant species depend on animals for pollination (biotic pollination, Ollerton et al. 2011). A small proportion of them depend on wind as the pollinating agent and a very small number of species use water for pollen transfer. Biotic pollination is the ancestral condition and both wind and water pollinations are the derived conditions (Faegri and van der Pijl 1971, Pellmyr 2002). Wind pollination is considered to have evolved from animal pollinated species as a means of reproductive assurance in pollinator scarce environments (see Friedman and Barrett 2009). Water-mediated pollination is considered to have evolved as a result of movement of flowering plants to aquatic habitat secondarily. In this chapter the discussion is confined to biotic pollination.

Pollination is one of the most important interactions between flowering plants and animals. Pollination is the only means of gene flow between conspecific individuals within and between populations, and thus it is the basis of genetic recombination. Pollination is responsible for generating genetic variation in the population which is essential for maintaining evolutionary potential of the species. Also, pollination is a pre-requisite for seed set. Pollination efficacy determines, to a large extent reproductive success of flowering plants. Adequate pollination is necessary not only for the stability of the species in their natural habitat but also for the productivity of our crop plants. This chapter elaborates briefly the fundamental and applied aspects of pollination. There is enormous literature accumulated over the years on all aspects of pollination biology. For more details reader may refer to Bronstein et al. (2006), Currie (1997), Faegri and van der Pijl (1971), James and Pitts-Singer (2002), Mitchell et al. (2009), Patiny (2012), Pellmyr (2002), Roubik (1995), Shivanna (2014a, 2015), Waser and Ollerton (2006), Willmer (2011).

2.2 Essential Features of Biotic Pollination

Flowering plants were the last group of plants to evolve about 100-110 million years ago. However, they became the most dominant vegetation on the earth. Evolution of insects was much earlier to that of flowering plants, about 250 million years ago. But their rate of diversification was slow. Following the origin of flowering plants and their consequent association with insects for pollination services, there was a rapid diversification of both insects as well as flowering plants. Origin of the flowers which increased their efficiency for attracting the insect pollinators leading to enhanced cross-pollination is considered to be one of the main causative factors for the evolutionary success of flowering plants (see Pellmyr 2002, Regal 1977, Willmer 2011). Recruitment of other animals and abiotic agents (wind and water) for pollination occurred later (Faegri and van der Pijl 1971). Although biotic pollination evolved in two groups of gymnosperms(Cycadales and Gnetales) as early as Permian period (about 250 million years ago) it is present only in a limited number of extant gymnosperms. Beetles, flies, thrips and wasps are the most common pollinators in gymnosperms (see Pellmyr 2002). However, biotic pollination does not seem to be very successful in gymnosperms as the diversification rate in groups with biotic pollination is not more than that in those using wind as the pollinator. This may probably be due to the lack of effective attractants for pollinators and thus lack of improvement in pollinators' diversity and pollination efficiency in species with biotic pollination over wind-pollinated species.

An essential requirement for effective biotic pollination is the need to maintain sustainable visits of the pollinators to conspecific flowers. For this, flowers have to offer some rewards to attract pollinators and advertise the presence of the rewards. Biotic pollination is largely mutualistic and benefits both the partners. This has been referred to as 'biological barter' (Ollerton 2006). Plants exchange their rewards for pollination services rendered by floral visitors. Pollination is incidental, not intentional, as far as the floral visitors are concerned. Plants make use of their visits to achieve pollination by positioning their anthers and the stigma of the flower in such a way that pollen grains from dehisced anthers are deposited on the visitors' body during their visit and the stigma comes in contact with the part of the visitor on which the pollen grains are deposited, thus enabling pollen transfer. Figure2.1 presents some pollinators with reference to the arrangement of the anthers and the stigma. The details of pollinator behaviour and their foraging activities are elaborated in Chapter 4 (Sinu, this volume).

The pollination may be autogamous or geitonogamous or xenogamous (see Box 1.1). For effective autogamy:

a. the anthers and the stigma should be positioned close to each other so that pollen grains come in contact with the stigma,
b. the pollen dispersal and the stigma receptivity should overlap so that autogamous pollen can function on the stigma and
c. the species should be self-compatible.

The autogamous pollination can take place before the flowers get a chance for out-cross pollination (prior autogamy) or along with out-cross pollination (competitive autogamy) or delayed until all the opportunities for out-cross pollination are over (delayed autogamy). Autogamy may occur on its own without the need for floral visitor or during the visit of pollinator to the flower (facilitated autogamy).

The geitonogamy may occur when more than one flower opens on a plant at a given time. This situation permits the movement of pollen along with the movement of the pollinator from one flower to the other flower of the same plant. Geitonogamy is common in tree species as large number of fresh flowers open on each day. Xenogamy requires movement of pollinators from one plant to the other. The extent of xenogamy is reduced under scarcity of conspecific plants and/or their pollinators. Fragmentation of the habitat also reduces cross-pollination as the pollinators tend to move within the fragment rather than across the fragments. The type of pollination that can occur depends on the sexuality of the flowers, plants and populations, structure of the flower and the breeding system of the species (see Box 1.1 for definitions).

Pollination limitation/failure is common in natural habitats due to anthropogenic disturbances (Burd 1994, Knight et al. 2005, Shivanna 2014b, Wilcock and Neiland 2002). If pollination limitation continues for longer period, the species is pushed to vulnerable category (rare, endangered and threatened) and eventual extinction. Thus, pollination plays an important role in the stability and management of our plant diversity.

2.3 Floral Advertisement

As pointed out earlier, flowers have to advertise the presence of the rewards so that the visitors would recognize and visit the flowers. Flowers use two methods for advertisement, visual and olfactory. Flowers and inflorescences act as visual cues for pollinators. Flowering plants show amazing variations in the size, shape and colour to make them visible to the pollinators in the background of green leaves. Floral colours are essentially due to the presence of pigment such as anthocyanins, carotenoids, flavonols and betaxanthins (see Pellmyr 2002). Most of the pollinating insects can see in UV light also (Chittka and Kevan 2005, Kevan 1983). Therefore, their perception of flower colour is different from that of the humans.

Olfactory signals are in the form of emission of fragrance which is specific to each species (see Dobson et al. 2005, Dudareva and Pichersky 2000, Knudsen et al. 2006, Shivanna 2014a). The fragrance is in the form of complex mixtures of a large number of volatile compounds. Flowers emit maximum fragrance in synchrony with the activities of their pollinators. Floral visitors are capable of recognizing specific fragrance and visit the flowers. The fragrance helps in attracting the visitor from long distances as the visitors can perceive the fragrance before seeing the flowers. When they come closer to the flower, both visual and fragrance cues help in guiding the pollinator to the flower.

2.4 Floral Rewards

Pollen grains and the nectar are the major rewards for floral visitors. For many insect pollinators such as honey bees, stingless bees and bumble bees, pollen and nectar serve as the exclusive food source both for adults and their larvae. Other floral visitors supplement floral resources with others such as insects and fruits. Pollen grains are highly nutritious and form a rich source of proteins, vitamins, amino acids and minerals (Roulston and Cane 2000, Schmidt and Buchmann 1992). The nectar is an aqueous solution made up of sugars (largely of sucrose, fructose and glucose) and a small amount of amino acids (Nicolson and Thornburg 2007); nectar of many species also contains some secondary metabolites such as alkaloids and phenolics (see Baker 1977, Heil 2011). The concentration of sugars in the nectar generally ranges from 15 to 45 per cent. Recently the nectar of several species has been shown to contain bacteria (Fridman et al. 2012) or yeast (de Vega et al. 2009, Herrera et al. 2009); their significance with reference to floral visitors is not yet clear. In some species nectar is coloured (Hansen et al. 2007). In several species pollen grains and/ or nectar also emit fragrance and help in attracting pollinators to the flower (de Verga et al. 2009, Dobson et al. 2005, Raguso 2004). Several species offer other rewards such as oils and resins, young seeds and fragrance compounds (see Armbruster 2012, Pellmyr 2002, Shivanna 2014a, Willmer 2011).

Table 2.1: Pollination syndromes (based on Faegri and van der Pijl 1971, Kevan 1997, Ghazoul and Sheil 2010, Willmer 2011, Shivanna 2014a).

Pollinator group (technical term)	Floral syndrome traits
Beetles (*Cantharophily*)	Flowers generally open, large, unspecialized, cylindrical or bowl-shaped, dull coloured, sex organs exposed, nectar guides absent with easily accessible rewards, strong odour, pollen deposited all over the body.
Flies (*Myophily*)	Flowers small, radial, dull coloured, nectar guides absent, decaying odour, sex organs generally hidden, pollen deposited all over the body.
Bees (*Melittophily*)	Flowers highly variable in morphology, radial, often bilateral with landing sites, open during day, brightly coloured, odours mild, nectar hidden but not deep, nectar guides present , sexual organs concealed, stamens few, precise pollen placement on body, offer nectar and pollen.
Butterflies (*Psychophily*)	Flowers tubular, open during day, mild odour, brightly coloured , flowers erect, nectar ample and deeply hidden in tubes/spurs, pollen deposited on proboscis and head.
Moths (*Phalaenophily*)	Flowers horizontal or pendent, nocturnal blooming, pale and heavily scented, nectar deeply hidden in long tubular corolla or spurs, pollen deposited on proboscis and head.
Birds (*Ornithophily*)	Flowers robust, rigid, open during day, vividly coloured often red, long tubular corolla, copious nectar with low sugar concentration, odours absent, pollen placed on beak, head or precisely on body, nectar sole reward.
Bats (*Chiropterophily*)	Flowers nocturnal blooming, generally placed at the tips of branches, pale and large, strong odour at night, large-mouthed and stiff, copious nectar, anthers often numerous, pollen placed on the head, nectar sole reward.

2.5 Biotic Pollinators

Insects are the principal pollinators in flowering plants amongst animals (Figure 2.1 and 2.2). Over 95 per cent of pollinating insects belong to just four orders of insects:

Hymenoptera (bees, wasps and ants), *Lepidoptera* (butterflies and moths), *Coleoptera* (beetles) and *Diptera* (flies). *Hymenoptera* includes nearly 50 per cent of the pollinating insect species. The remaining insect pollinators belong to the other three orders (see Willmer 2011). Thrips (*Thysanoptera*) have also been shown to pollinate a number of plant species (Ananthakrishnan 1993, Garcia-Fayos and Goldarazena 2008, Williams et al. 2001). Apart from insects, birds and bats form the major vertebrate pollinators. They are largely confined to tropical and semi-tropical areas. Both bird and bat pollinated flowers secrete larger amount of nectar when compared to insect-pollinated species (see Table 2.1). About 500 genera of plants contain bird-pollinated species (Willmer 2011). Sunbirds and sugarbirds are the major bird pollinators in the old world, hummingbirds are the principal bird pollinators in the New World; in Australia honeyeaters and in Hawaii honeycreepers are the main bird pollinators. UV vision is present in several birds also (see Willmer 2011). Bird pollinated flowers are usually larger, and red or orange in colour. As birds are not very sensitive to fragrance, bird-pollinated flowers generally lack any fragrance. They produce larger amount of nectar containing low sugar concentration.

Fig. 2.1. A. *Apis dorsata* foraging a flower of *Syzygium cumini*. This is an actinomorphic open flower and any forager can harvest the rewards. Pollen grains are deposited in a diffuse manner on the lower part of the body of the insect and the stigma comes in contact with pollen laden surface. B, C. A wasp foraging a flower of *Justicia betonica* (B) and *Apis dorsata* foraging on flower of *Lucas* sp. Both *Justicia* and *Lucas* are zygomorphic flowers and pollen grains are deposited at specific parts on the body of insect. Anthers touching the thorax of the insect and pollen grains deposited on its surface are clear in B. (Figs B and C, Courtesy Dr Giby Kuriokose).

About 250 genera contain bat-pollinated species (Willmer 2011). Only two families of bats, *Megachiroptera* (in tropical and subtropical regions of the Old World including Australia and Pacific islands) and *Microchiroptera* (in tropical and subtropical regions of the New World) contain bat species involved in pollination (see Fleming et al. 2009). Old World bats are larger with shorter tongues and do not hover but perch for nectar feeding whereas those of the New World are smaller with longer tongues and can forage nectar by hovering. Bat pollinated flowers are also larger with dull coloured or white flowers and generally well exposed at the margin of the canopy Some unusual pollinators such as cockroaches, mice, squirrels, snails and lizards have also been reported in some plant species (see Shivanna 2014a).

Fig. 2.2. A butterfly foraging on a flower of *Justicia betonica*. B. *Apis dorsata* on the flower of *Eletteria cardamomum* (cardamom) with pollen deposited on its head. C. *Xylocopa* sp. (carpenter bee) foraging the nectar of cardamom flower. D. *Amegilla* sp. foraging on the flower of *Rhynchoglossum notonianum*. E. Honeybees robbing the nectar of *Passiflora edulis* flower. *Passiflora* is also an open flower and any visitor can harvest the nectar. Honeybees are not the pollinators as they are too small and do not come in contact with the anthers or the stigma. (Figs A-D, Courtesy Dr Giby Kuriokose)

2.6 Nocturnal Pollination

A majority of plant species are pollinated during the day time (diurnal). But there are a number of species in which pollination takes place during the night (nocturnal) and the flowers of such species open during the night. Some species show both diurnal and nocturnal pollination and their flowers remain open during the day and night periods. As visual attraction is not reliable during the night, nocturnally pollinated species relay more on olfactory cues (see Table 2.1). Nocturnally-pollinated flowers are generally drab in colour but emit strong smell (Dafni et al. 2012, Dar et al. 2006, Muchhala et al. 2009, Young 2002). Moths and bats are the most important nocturnal pollinators.They have keen sense of smell that helps them in locating the flowers. New World Bats (*Microchiroptera*) can produce ultrasonic sound to locate flowers (von Helversen and von Helversen 1999). The ultrasound is reflected by the petals of bat-pollinated flowers and bats have the ability of recognize this reflected sound (echolocation).Echolocation ability is not well developed in most of the old world bats (*Megachiroptera*); they depend largely on olfactory and visual cues to locate the flowers.

2.7 Generalists and Specialists

Flowers of many species such as the members of *Myrtaceae* and *Rosaceae* are actinomorphic and 'open'. The rewards of such flowers are exposed. Any floral visitor can harvest the rewards (Figure 2.1A). Such flowers are visited by a number of floral visitors and many of them are effective in bringing about pollination. Such species are generally referred to as generalists as they are pollinated by a number of non-specialized floral visitors. Flowers of many other species show specialization to different degrees (see Endress 2011, Mitchell et al. 2009, Pellmyr 2002, Willmer 2011). Zygomorphy, one of the evolutionary innovations of the flower, is generally associated with floral specialization (Figure 2.1 B, C). In most of the specialized flowers, the nectar is not exposed but hidden in the corolla tubes or spurs of varying lengths. Flowers in which the nectar is hidden generally develop contrasting patterns, termed nectar guides, on their petals pointing towards the source of nectar. Several experimental studies have shown that nectar guides do help the visitor to locate the nectar (Leonard et al. 2011, Shivanna 2014a). All visitors cannot harvest the rewards from flowers with hidden nectar; only those animals that show functional fitness with the flower can harvest the nectar (see also More et al. 2007, Pellmyr 2002, Santos-Gally et al. 2013). For example if the nectar is located at the base of the corolla tube of about 1 cm, only those insects whose proboscis matches (i.e., its length is also about 1 cm) can harvest the reward. Those whose proboscis is less than 1 cm cannot harvest the reward. Such trends that restrict the visits to a limited number of pollinators represent specialized pollination systems. Those animals that cannot harvest the nectar, avoid visiting such flowers. As the flower gets more and more specialized, it restricts progressively the number of visitors that can harvest the rewards. The flower and the pollinator specialization may progress to such an extent that the flowers of each plant species can be effectively pollinated by just one species of pollinator. Such a condition is the result of co-evolution between the flower and the pollinator. Such specialization is found in several plant species. Species of figs pollinated by fig-wasps, yucca pollinated by yucca moth, and several orchids each pollinated by species-specific pollinator are some of the well known examples of extreme specialization (see Pellmyr 2002, Willmer 2011, Shivanna 2014a, see also Waser et al. 1996).

Species of *Ficus,* largely a tropical genus, exemplify one of the most highly specialized pollinating systems known (Cook and Rasplus 2003, Machando et al. 2005, Wang et al. 2013). Fig species are pollinated by fig-wasps; each *Ficus* species (>700) is pollinated by a specific species of fig-wasp. *Ficus* species produce flowers in special urn-like inflorescences called syconia. Each syconium has a small opening at the tip called ostiole and the flowers are arranged compactly on the inner surface of the syconium. Most of the *Ficus* species are monoecious (see Box 1.1). The female flowers are present toward the base and the male flowers are produced at the tip. When the female flowers are receptive, the syconium emits species-specific fragrance and attracts specific female fig-wasp. The wasp enters through the ostiole and moves around the syconium. During its movement it pollinates female flowers as it carries the pollen from its visit to the previous syconium. Simultaneously the wasp lays its eggs in the ovaries of some of the female flowers and dies. After the eggs are hatched the larvae consume the gall that develops in oviposited flowers and grow into adults (males and females). Non-oviposited but pollinated flowers develop fertile seeds. The males are wingless and their life is very short; as the ostiole is closed by this time, the males bore a hole in the wall of the syconium for the escape of females and die. By this time the syconium is in the male phase and the pollen grains get deposited on female wasps. They come out of the syconium through the hole (made by the males) and enter another receptive syconium that is in the female phase; it pollinates the stigmas of the new syconium and oviposit in some of the female flowers (Proffit et al. 2008). This cycle continues.

In several orchids the nectar is hidden in special extension called spur; the length of the spur may vary greatly from a few centimetre to several centimetres. In *Angraecum sesquipedale,* an orchid endemic to Madagaskar, the nectar is located at the base of a long spur of 25-30 cm. As early as 1862 Charles Darwin predicted that the pollinator of this orchid should be the one which has exceptionally long proboscis matching with the spur length. It was in 1903 (over 40 years after Darwin's prediction), a hawkmoth (*Xanthopan morgani* ssp. *praedicta)* with a proboscis length of 22 cm was shown to be the pollinator of *Angraecum* orchid (see Willmer 2011). The concept of generalist and specialist applies to pollinator species also. Those animals that visit flowers of a large number of plant species are generalists. Specialization leads to a reduction in the number of plant species the pollinators visit and in highly specialized systems one pollinator visits the flowers of just one species.

2.8 Pollination Syndromes

Pollination syndrome refers to a combination of floral traits associated with attraction and utilization of specific types of animals for pollination (Faegri and van der Pijl 1971, Ghazoul and Sheil 2010, Willmer 2011). Various pollination syndromes such as beetle-pollinated, bee-pollinated, butterfly-pollinated and bird-pollinated have been identified and described (see Table 2.1). Different pollination syndromes are often present in closely related species. Similarly, unrelated species also show similarities in floral traits and pollination syndromes. Thus, pollination syndromes represent neither closed compartments nor exclusiveness of the specific group of pollinators to the flowers of a particular syndrome. They also do not indicate the presence of all specific traits mentioned for each syndrome in all plant species of that syndrome. They simply represent traits that are over-represented in flowers that attract specific types of pollinators; they do not exclude other visitors (Faegri and van der

Pijl 1971, Pellmyr 2002). Thus, pollination syndromes do not represent a phylogenetic relationship of the species within each syndrome.

2.9 Deceptive Pollination

There are a number of plant species in which pollination takes place by deceit; they advertise the presence of rewards but receive pollination services without offering real rewards to the pollinators. There are various types of deceits such as food deception, sexual deception and brood-site deception (see Ayasse et al. 2003, Galazia et al. 2005, Renner 2006, Shivanna 2014a). Deceit pollination is very common in orchids and it is elaborated in Chapter 5 (Kuriokose, this volume). There are also some floral visitors that cheat the plants; they harvest the rewards without bringing about pollination (Figure 2E). They are generally referred to as nectar or pollen robbers (see Tandon, this vol).

2.10 Pollination Management in Crop Species

As pointed out earlier, effective pollination is the basic requirement for seed and fruit set. As seeds and/or fruits are the commercial products of a majority of our crop plants, pollination limitation, due to some reason or the other, reduces seed and fruit set and thus the yield of crop species. Additional fertilizers or improved agronomic practices do not help in improving the yield unless pollination is adequate. Except cereals which are pollinated by wind, most of our agri-horticultural crops are pollinated by animals particularly a variety of insects, largely bees (Free 1993, James and Pitt-Singer 2008, McGregor 1976, Roubik 1995). Historically pollination requirements of crop species were met by wild pollinators living on and around the farm landscape. However, in recent years wild pollinators are insufficient to take care of pollination requirements of crop plants because of major changes in agricultural practices and also anthropogenic impacts on the habitat. The following are the major causes for pollination constraints of crop species:

- Reduction in the density and diversity of natural pollinators as a result of habitat loss and extensive use of pesticides (see Basu and Chakraborty, this vol), herbicides and a marked increase in the level of pollutants for which insects are very sensitive.
- Monoculture cropping system leading to an enormous increase in the area covered by the same crop, often extending to hundreds of hectares.
- Climate change resulting in disruption of long-term adaptations between plants and pollinators.

In multi-seeded fruit crops such as tomato and cucurbits, pollination limitation not only reduces the yield but also the quality of fruits and thus their market value. Farmers, particularly in Western countries, have been following a number of effective approaches to sustain pollination in crop plants. Some of these approaches are elaborated in the following pages.

2.11 Use of Managed Pollinators

2.11.1 Honey bees

Honey bees are the principal pollinators in a majority of crop plants (Delaplane and Mayer 2000, James and Pitt-Singer 2008). One of the effective approaches to sustain pollination services is to use managed pollinators. Bee keeping for obtaining honey and bee wax has been in practice by humans since long. Keeping honey bee colonies in the crop field during its flowering season is one of the effective and convenient methods to sustain pollination (Currie 1997, Delaney and Tarpy 2008, James and Pitt-Singer 2008, Shivanna 2003). This approach has been followed in Western countries, particularly in USA since 1940s and has grown into a huge, well-organized industry. *Apis mellifera* (European honey bee), *A. cerana* and *A. florea* are the most important managed honey bees and have been introduced to many countries outside their natural ranges for pollination services. Honey bee colonies are used routinely for over 15 crops such as almond, apple, pears, plums, tomatoes, watermelons, canola, clover, alfalfa and sunflower. The farmers sign contract with the bee keepers specifying the number of colonies needed and the period for which they have to be maintained in the farm for pollination services. The bee keepers bring the colonies in huge trucks and keep them in the field during the flowering period of the crop. In USA alone, about 2.0-2.5 million colonies are rented every year for pollination services (Bennett 2013). The bee colonies are moved from one crop to the other coinciding with the flowering season.

The number of bee colonies available in the US for pollination services has come down by 50 per cent in recent years (from five million in 1950s to 2.5 million at present) because of the spread of various diseases and other causes (James and Pitt-Singer 2008, Vergera 2008). The demand for the colonies has increased due to the increase in the acreage of the crops that require managed pollinators. For example almond is one of the major crops in California and over 60 per cent of the available bee colonies in the country are used to pollinate this crop (Anonymous 2012). Reduced number of colonies available and increased demand has increased the rental value of honey bee colonies. For this crop, average rental value for each hive has jumped from about 54 USD in 2004 to about 140 USD in 2006 (Burgett et al. 2010, James and Pitt-Singer 2008). For bee keepers the main income comes from renting the hives for pollination services; the income from honey and wax has become secondary.

Often many other species particularly weeds growing in and around the crop fields may also flower along with the crop species. When the co-flowering species offer better rewards to managed pollinators, they may desert crop plants and start visiting coflowering species. To prevent such competition many investigators have studied the potential of spraying crop plants with bee attractants in the form of sugar solution (Beeline, Bee-Q and Bee-Lure), pheromones (Bee-Here, Bee Scent and QMP) or plant volatiles isolated from nectar or pollen on crop plants to attract and/or retain pollinators when co-flowering species compete with target crop species (*see* Currie 1997, Delaney and Tarpy 2008, Deshpande and Naik, this vol, Viraktamath and Naik, this vol). The results have not been consistent and there are still several problems for their regular use; these approaches have not yet become a commercial success.

2.11.2 Bumble Bees

Although honey bees are the most efficient pollinators for some crops such as almond, apple and canola, they are not the efficient pollinators for several other crops such as tomato, alfalfa and clover (Currie 1997, Holm 1966); for these crops bumble bees are the better pollinators because of their long tongue, large size, and their ability to vibrate flowers (required for some of these crops), and to fly at relatively low temperatures (Currie 1997, James and Pitt-Singer 2008). Honey bees were being used for these crops also in the absence of any other better managed bees. As they were less efficient, the density of colonies kept in the farm had to be increased considerably to obtain satisfactory pollination. Since late 1980s it has become possible to domesticate several bumble bee species and supply them commercially (just like honey bee colonies) for pollination services (Velthuis and Doorn 2006). Since then bumble bees are being used for pollination services of some of these crops (in which honey bees were not efficient) in most of the countries in Europe, America and some Asian countries such as China and Japan.

In temperate countries many of the fruits and vegetables cannot be grown under field conditions around the year because of the low temperature in winter months; they have to be grown in heated green houses. Tomato (*Lycopersicon esculentum*), pepper (*Capsicum annuum*), watermelon (*Citrullus lanatus*), melon (*Cucumis melo*), zucchini (*Cucurbita pepo*), cucumber (*Cucumis sativus*), eggplant (*Solanum melongina*) and *Frageria* sp (Guerra-Sanz 2008) are the major crops grown in green houses. Pollination is a problem in green house grown crops as they are not exposed to natural pollinators. They had to resort to manual pollination to harvest fruits and it used to be expensive. For tomato, one of the major greenhouse-grown crops that needs buzz pollination (the pollinator has to vibrate flowers to release pollen from the anthers), wind and insects, particularly bumble bees bring about pollination under field conditions. Until 1990s tomato growers in green houses were carrying out manual pollination by using battery-operated vibrators ('magic bees'). Using the vibrator, flowering twigs/flowers used to be vibrated manually to bring about pollination. Since 1990s bumble bees have become the most efficient and popular pollinators for green house-grown tomatoes. Over 95 per cent of the sales of bumble bee colonies are used to pollinate greenhouse tomatoes. *Bombus terrestris* is the most common species used in several countries (Banda and Paxton 1991, Kearns and Inouye 1993, Straver and Plowright 1991).

2.11.3 Native Solitary Bees

Apart from bumble bees, a few other native solitary bees such as mason bee (*Osmia* spp), alkali bee (*Nomia melanderi)* and alfalfa leaf cutting bee (*Megachile rotundata*), which are more efficient for some crops, have become available commercially in recent years for pollination services (Currie 1997, James and Pitt-Singer 2008). Farmers have started using these solitary bees for pollination services. In fact, alfalfa leaf cutting bees and alkali bees have now become more popular than honey bees for alfalfa pollination (James and Pitt-Singer 2008).

2.11.4 Integrated Use of Native and Managed Pollinators

Several studies in recent years have highlighted the role of wild pollinators even in the presence of managed pollinators in sustaining the yield of crop species (Brittain et al. 2013, Garibaldi et al. 2013, Greenleaf and Kremen 2006). Limited number of bee species that have so far been domesticated for commercial pollination services, a steady reduction in the number of available colonies of managed pollinators (because of the number of prevailing diseases and other causes) over the years and marked increase in their rental value has created interest to make use of native wild species for pollination services. At the same time great reduction in the density and diversity of wild pollinators has raised global concerns in safeguarding pollination services (Biesmeijer et al. 2006, Burkle et al. 2013, Potts et al. 2010, *see also* Tylianakis 2013, Winfree et al. 2011). Attempts are being made to domesticate many more wild pollinators. Some studies have shown the potential of such an approach for domestication of many stingless bees such as species of *Trigona* and *Melipona* (Slaa et al. 2006). Attempts are also being made (Kremen 2008) to conserve and enhance wild pollinator populations on and around agricultural landscapes by following pollinator-friendly farming practices (James and Pitts-Singer 2008, Garibaldi et al. 2013). These approaches include increasing the availability of nest sites by maintaining uncultivated strips along field margins, providing artificial nesting sites, and reducing the use of pesticides and other unfriendly agrochemicals. The future trend seems to be the integrated use of wild and managed pollinators to safeguard pollination services of crop plants.

2.12 Pollination Management in Self-incompatible Species

Management of pollination services in self-incompatible (de Nettancourt 2001) species (in which self-pollen grains from the same plant or the same clone cannot induce seed set, examples: almond and apple) need additional precautions. In such species, pollinators have to move from plant to plant to bring out cross-pollination with pollen collected from compatible plants. Many orchards may not have suitable proportion of cross-compatible plants or their density and distribution in the orchards may not be sufficient for effective cross-pollination. Also, the flowering period of the pollen donors and pollen receivers may not be synchronous. These conditions create pollination constraints even in the presence of managed pollinators; most of the pollen deposited on the stigma may belong to the same plant/line.

One of the standard methods used to overcome such problems is to supply pollen of compatible line/variety in pollen dispensers fitted to the opening of the bee hives. The honey bees are forced to move through pollen dispensers during their exit from the hive so that compatible pollen grains from the dispensers are deposited on their body parts (Ferrari 1990, Legge 1976). Some of this pollen eventually gets deposited on to the stigmas of the compatible plants during the bees visit (Griggs and Iwakiri 1960, Williams et al. 1979). In Western countries there are a number of commercial pollen supply companies which sell the pollen of the known variety along with pollen dispensers. The orchard owners buy pollen of the desired variety and use them in pollen dispensers.

2.13 Hybrid Seed Production

Hybrid seed technology is one of the most important and highly organized agricultural industries in the world. Increased yield as a result of hybrid vigour is being exploited in a number of crop species. One of the major limitations of this technology is that the seeds collected from hybrid plants cannot be reused for sowing the next crop as half of the resulting plants will segregate into male and female lines; therefore, hybrid seeds have to be produced every year and distributed to the farmers. The success of hybrid seed technology depends on the cost of producing hybrid seeds. Hybrid seed production involves prevention of self- and intraline-pollination and effective cross-pollination of the female line with the pollen of the male line in hybrid seed production plots. In most of the seed crops, development of male sterile lines has facilitated effective prevention of self- and intraline-pollinations (see Frankel and Galun 1977, Shivanna and Sawhney 1997, Shivanna 2003). Wild and/or managed bees are used to achieve cross-pollination between male sterile female line and male fertile male line in crops needing biotic pollination.

In many of the vegetable crops such as tomatoes, eggplant, sweet pepper and cucurbits, hybrids are preferred because of their vigour, high yield and longer shelf-life. However, suitable male sterile lines are not available in many of them and even when available they lack effective pollinators to transfer pollen from the male to the female lines. In such crops manual emasculation and/or pollination is carried out to raise hybrids. As this method is labour intensive, hybrid seed production becomes expensive. In these crops hybrid seeds are produced mostly in developing countries such as India and China where the labour cost is comparatively cheap.

2.14 Pollination Management in Developing Countries

Use of managed pollinators is not being exploited fully in the developing countries including India. The farmers still depend on natural pollinators for pollination services of crop plants. Many studies in India have shown that pollination is a constraint in a range of field and orchard crops (Deodikar and Suryanarayana 1977, Keven 1995, Roubik 1995, Savoor 1998). Limited studies available so far have also indicated that maintaining honey bee colonies in the field increases the yield in several crops such as sunflower, mustard, cucumber, melon, grape and onion (Abrol 2011, Anonymous 2013, Mohapatra et al. 2010, Prasad et al. 1989, Partap et al. 2001, Partap 2010, Sihag 1995). However no systematic efforts have so been made to utilize this low-tech, eco-friendly approach of using managed pollinators for pollination services.

One of the crops in which considerable progress has been made in using managed honey bees for pollination services is apple (Partap 1998, 2010). Orchard owners in Himachal Pradesh have started renting honey bee colonies (*Apis cerana* or *A. mellifera)* for pollinating their apple orchards. In Maoxian County of the Hengduan Mountains in China also, apple and pear are the important crops. Extensive use of pesticides over the years has reduced natural pollinators and bee keepers are reluctant to rent their bee colonies for pollination services. They follow hand pollination regularly to sustain apple productivity (Partap and Partap 2000). Almost every member of the family gets engaged in hand pollination during the flowering season.

Vanilla planifolia orchid is another crop in which hand pollination has been in regular practice (Roubik 1995, Willmer 2011). This species is a native of Southern Mexico and Central America where it is pollinated by a euglossine bee. *Vanilla* is being grown extensively in many parts of tropical Asia particularly in some parts of Kerala and Karnataka where the pollinator is absent. Hand pollination is routinely carried out to sustain the yield.

2.15 Concluding Remarks

As apparent from the above discussion, pollination ecology is of fundamental importance in sustenance of our wild plant species in their natural habitat and of productivity of our crop species. Although our country is rich in biodiversity with a high proportion of endemic species, there is very little information on pollination ecology of a majority of our plant species, particularly endemic and endangered species. This is unlike the situation in developed countries where the biodiversity is rather limited but baseline data on pollination ecology is available on most of those species. Thus, they are in a better position to conserve and manage their biodiversity more effectively.

A number of approaches are being exploited in developed countries to sustain pollination services of crop plants. Several recent studies highlighting the decline of native pollinators and their role in pollination services even in the presence of managed pollinators. Further, the limited number of managed pollinator species available and their dwindling numbers have raised serious concern on the sustainability of pollination services of crop plants in the coming decades. The research in this area is being focused on two lines.

1. Domestication of additional number of pollinator species for commercial utilization.
2. Increasing the density and diversity of native pollinators by making agri-horticultural farms and orchards eco-friendly by facilitating nesting places of pollinators and reducing the use of pesticides and herbicides. More and more dependence on natural pollinators together with judicious use of managed pollinators is going to be the future trend to sustain pollination services.

Unfortunately, use of managed pollinators, which is eco-friendly, simple and sustainable is not being exploited fully in developing countries. They still depend on natural pollinators which are likely to be inadequate under several ecological conditions. There are many reasons for this.

1. Lack of awareness among farmers and bee keepers about the potential use of managed bees in pollination services.
2. Lack of research back up for effective management and maintenance of bee colonies for pollination services.
3. Lack of even baseline data on pollination ecology of a number of crop species. We do not have the authentic data on even the pollinators of our crop species, and their temporal and spatial variability.

Unlike Western countries, where they have been able to convincingly document the decline of native pollinators and initiated measures to restore their density and diversity, we have no knowledge on the density and diversity of our native pollinators and what is happening to them in recent years. As habitat degradation in developing countries is more severe than that in developed countries, we should expect more drastic decline of our native pollinators when compared to Western countries. It is important to initiate relevant research needed to make use of available technologies to sustain pollination services of crop species. The focus of bee keepers from honey-centric mind set has to be shifted to exploitation of the colonies for pollination services as a profitable venture. Deceleration of productivity in agri-horticultural crops since the mid 1990s has been the focus of agricultural scientists. However, there is hardly any concern about the possibility of inadequate pollination acting as a possible constraint in non-cereal crop production. It is important to rule out this possibility before making other rational approaches to increase productivity.

One other area of pollination ecology not covered in this review is its relevance to crop introduction. When a crop has to be introduced to new areas within the country or to other country, it is important to make sure that its natural pollinators are available in the area to be introduced for its sustenance. Otherwise, it cannot survive and sustain in the new location. Equally important is the role of pollination ecology in the release of genetically modified plants for commercial cultivation. One of the main concerns for the release of genetically modified crops has been the escape of engineered genes to the wild. Before any genetically modified plants are released, it is imperative to provide data on all relevant details on pollination ecology such as its effective pollinators, extent of pollen flow under natural conditions, the possibility of pollen of transgenic plants pollinating related crop and wild species, and the possibility of hybrid production between wild and transgenic crop to the authorities concerned to rule out the possibilities of its escape to the wild (see Armstrong et al. 2005).

In spite of its importance in fundamental and applied aspects of plant biology, pollination ecology is one of the neglected areas of research and teaching in developing countries in general and India in particular. For the size of the country and the vastness of plant diversity, the number of laboratories working on pollination ecology is very few. Baseline data on pollination ecology can be collected with very little infrastructure or equipment so that teachers and students of colleges located in biodiversity-rich areas can undertake these studies (see Shivanna and Tandon 2014). It is high time that our young researchers get interested and initiate studies on pollination ecology of our wild species as well as crop species to sustain diversity of wild species and the productivity of crop species.

Acknowledgements

I thank the Indian National Science Academy for the award of a position of INSA Honorary Scientist.

References

Armbruster WS 2012. Evolution and ecological implications of "specialized" pollinator rewards. pp. 44–57. In Patiny S (Ed) Evolution of plant-pollinator relationships. The Systematic Association Special Vol. 81, Cambridge Univ Press, New York.

Armstrong TT, Fitzjohn RG, Newstrom LE et al. 2005. Transgenic escape: What potential for crop-wild hybridization. Mol Ecol 14: 2111–2132.

Ayasse M, Schiestl FP, Paulus HF, Ibarra F, Francke W 2003. Pollinator attraction in a sexually deceptive orchid by means of unconventional chemicals. Proc R Soc B 270: 517–522.

Baker HG 1977. Non-sugar chemical constituents of nectar. Apidologie 8: 349–356.

Banda HJ, Paxton RJ 1991. Pollination of greenhouse tomatoes by bees. Acta Hort 288: 194–198.

Bennett C 2013. Honey bees to rent? Demand will only grow. http://westernfarmpress.com/blog/honey-bees-rent-demand-will-only-grow (accessed on 20 December 2013).

Biesmeijer JC, Roberts SPM, Reemer M, Ohlemuller R 2006. Parallel declines in pollinators and insect-pollinated plants in Britain and the Netherlands. Science 313: 351–354.

Brittain C, Williams N, Kremen C, Klein A-M 2013. Synergistic effects of non-Apis bees and honey bees for pollination services. Proc R Soc B 280: 2012–2767.

Bronstein JL, Ruben Alarcon R, Geber M 2006. Tansley review: The evolution of plant–insect mutualisms. New Phytol 172: 412–428.

Burd M 1994. Bateman's principle and plant reproduction: the role of pollen limitation in fruit and seed set. Bot Rev 60: 83–139.

Burgett M, Daberkow S, Randal RR, Thurman W 2010. U.S.Pollination markets: Recent changes and historical perspective. Am Bee J 150: 35–41.

Burkle LA, Marlin JC, Knight TM 2013. Plant-pollinator interactions over 120 years: Loss of species, co-occurance and function. Science 339: 1611-1615.

Chittka L, Kevan PG 2005. Flower colours as advertisements. In Dafni A, Kevan PG, Husband BC (Eds) Practical pollination biology. Enviroquest, Ltd, Cambridge, Canada, pp 157–196.

Cook JM, Rasplus JY 2003. Mutualists with attitude: coevolving fig wasps and figs. Trends Ecol Evol 18: 241–248.

Currie RW 1997. Pollination constraints and management of pollinating insects for crop production. In Shivanna KR, Sawhney VK (Eds) Pollen biotechnology for crop production and improvement, Cambridge Univ Press, New York, pp 121–151.

Dafni A, Marom-Levy T, Jurgens A, Dotterl S, Shimrat Y et al. 2012. Ambophily and "super generalization" in *Ceratonia siliqua* (Fabaceae) pollination. In Patiny S. (Ed). Evolution of plant-pollinator relationships. The Systematics Association Special Vol 81. Cambridge Univ Press, Cambridge, pp 345–373.

Dar S, Ma del C, Arizmendi MD, Valiente-Banuet A 2006. Diurnal and nocturnal pollination of *Marginatocereus marginatus* (Pachycereeae: Cactaceae) in Central Mexico. Ann Bot 97: 423–427.

Delaney DA, Tarpy DR 2008. The role of honey bees in apple pollination. Apiculture program, North-Carolina State Univ, USA Bee Keeping Notes 3.03. 9/2008: 1–6. http://www.cals.ncsu.edu/entomology/apiculture/PDF%20files/3.03.pdf (accessed 20 January 2014).

Delaplane KS, Mayer DF 2000. Crop pollination by bees. CABI Publishing, Cambridge.

de Nettancourt D 2001. Incompatibility and Incongruity in Wild and Cultivated Plants. Springer, Heidelberg.

Deodikar GB, Suryanarayana MC 1977. Pollination in the services of increasing farm production in India. In Nair PKK (Ed) Advances in pollen spore research, Today and Tomorrow Printers and Publishers, New Delhi, pp 60–82.

de Vega C, Herrera CM, Johnson SD 2009. Yeasts in floral nectar of some South African plants: Quantification and associations with pollinator type and sugar concentration. South African J Bot 75: 798–806.

Dobson HEM, Raguso RA, Knudsen JT, Ayasse M 2005. Scent as an attractant. In Dafni E, Kevan PG, Husband BC (Eds), Practical pollination biology. Enviroquest, Ltd, Ontario, Canada, pp 197–230.

Dudareva N, Pichersky E 2000. Biochemical and molecular aspects of floral scents. Plant Physiol. 122: 627–634.

Endress PK 2011. Evolutionary diversification of flowers in angiosperms. Am J Bot 98: 370–396.

Faegri K, van der Pijl L 1971. The Principles of pollination ecology. 2nd edn, Pergamon Press, Oxford.

Ferrari TE 1990. 'Enpollination' of honey bees with pre-collected pollen improves pollination of almond flowers. Am Bee J 130: 801.

Fleming TH, Geiselman C, Kress WJ 2009. The evolution of bat pollination: a phylogenetic perspective. Ann Bot 104: 1017–1043.

Frankel R, Galun E 1977. Pollination mechanisms, reproduction and breeding. Springer, Berlin, Heidelberg, New York.

Free JB 1993. Insect pollination of crops, 2nd edn, Academic Press, Orlando, FL.

Friedman J, Barrett SCH 2009. Wind of change: new insights on the ecology and evolution of pollination and mating in wind-pollinated plants. Ann Bot 103: 1515–1527.

Fridman S, Izhaki I, Gerchman Y, Halpern M 2012. Bacterial communities in floral nectar. Environ Microbiol Rep 4: 97–104.

Galizia CG, Kunze J, Gumbert A, Borg-Karlson A-K et al. 2005. Relationship of visual and olfactory signal parameters in a food-deceptive flower mimicry system. Behav Ecol 16: 159–168.

Garcia-Fayos P, Goldarazena A 2008. The role of thrips in pollination of *Arctstaphyllos uva-ursi*. Int J Plant Sci. 169: 776–781.

Garibaldi LA, Steffan-Dewenter I, Winfree R, Aizen MA et al. 2013. Wild pollinators enhance fruit set of crops regardless of honey bee abundance. Science 339: 1608–1611.

Ghazoul J, Sheil D 2010. Tropical rain forest ecology, diversity, and conservation. Oxford Univ Press, New York.

Greenleaf SS, Kremen C 2006. Wild bees enhance honey bees' pollination of hybrid sunflower. Proc Natl Acad Sci USA 103: 13890–13895.

Griggs WH, Iwakiri BT 1960. Orchard test of beehive pollen dispensers for cross-pollination of almonds, sweet cherries and apples. Proc Am Soc Hort Sci 75: 114–128.

Guerra-Sanz JM 2008. Crop pollination in greenhouses. In James, RR and Pitts-Singer TL (Eds) Bee pollination in agricultural ecosystems. Oxford Univ Press, New York, pp 27–47

Hansen DM, Olesen JM, Mione T, Johnson SD, Muller CB 2007. Coloured nectar: distribution, ecology, and evolution of an enigmatic floral trait. Biol Rev 82: 83–111.

Heil M 2011. Nectar: generation, regulation and ecological function. Trends Plant Sci 16: 191–200.

Herrera CM, de Vega C, Canto A, Pozo MI 2009.Yeasts in floral nectar: a quantitative survey. Ann Bot 103: 1415–1423.

Holm SN 1966 The utilization and management of bumble bees for red clover and alfalfa seed production. Ann Rev Entomol 11: 155–182.

James RR, Pitts-Singer TL 2008. Bee Pollination in agricultural ecosystems. Oxford Univ Press, New York.

Kearns A, Inouye DW 1993. Techniques for pollination biologists. Univ Press Colorado, Colarado.

Kevan PG 1983. Floral colours through the insect eye: what they are and what they mean. In Jones CE, Little RJ (Eds), Handbook of experimental pollination biology. Scientific and Academic Editions, van Nostrand Reinhold Company Inc, New York, pp 3–29.

Kevan PG (Ed) 1995. The Asiatic hive bee: Apiculture, biology, and role in sustainable development in tropical and subtropical Asia. Enviroquest Ltd, Cambridge, Obtario, Canada.

Knight TM, Steets JA, Vamosi JC, Mazer SJ et al.2005. Pollen limitation of plant reproduction: pattern and process.Ann Rev Ecol Evol Syst 36: 467–497.

Knudsen J, Eriksson R, Gershenzon J, Ståhl B 2006. Diversity and distribution of floral scent. Bot Rev 72: 1–120.

Kremen C 2008. Crop pollination services from wild bees. In James RR, Pitts-Singer TL (Eds), Bee pollination in agricultural ecosystems. Oxford Univ Press, New York, pp 10–26.

Legge AP 1976. Hive inserts and pollen dispensers for tree fruits. Bee World 57: 159–167.

Leonard AS, Papaj DR 2011. 'X' marks the spot: The possible benefits of nectar guides to bees and plants. Funct Ecol 25: 1–9.

Machado CA, Robbins N, Gilbert MT P, Herre EA 2005. Critical review of host specificity and its coevolutionary implications in the fig-fig-wasp mutualism. Proc Natl Acad Sci USA 102: 6558–6565.

McGregor SE 1976. Insect pollination of cultivated crop plants. Agricultural handbook No. 495, USDA-ARS, US Gov Print Office, Washington, DC.

Mitchell RJ, Irwin RF, Flanagan RJ, Karron JD 2009. Ecology and evolution of plant-pollinator interactions. Ann Bot 103: 1355–1363.

Mohapatra LN, Sontakke BK, Ransingh N 2010. Enhancement of crop production through bee pollination. Orissa Rev Sept 2010, 44–47.

More M, Sersic AN, Cocucci AA 2007. Restriction of pollinator assemblage through flower length and width in three long-tongued hawkmoth-pollinated species of *Mandevilla* (Apocynaceae, Apocynoideae). Ann Missouri Bot Gard 94: 455–504.

Muchhala N, Caiza A, Vizeute JC, Thomson JD 2009. A generalized pollination system in the tropics: bats, birds and *Aphelandra acanthus*. Ann Bot 103: 1481–1487.

Nicolson SW, Thornburg RW 2007. Nectar chemistry. In Nicolson S et al. (Eds), Nectaries and nectar. Springer, Heidelberg, pp. 215–263.

Ollerton J 2006. "Biological barter": Patterns of specialization compared across different mutualisms. In: Waser NW, Ollerton J (Eds), Plant-pollinator interactions; From specialization to generalization. Univ Chcago Press, Chicago, pp 411–435.

Ollerton J, Winfree R, Tarrant S 2011. How many flowering plants are pollinated by animals? Oikos 120: 321–326.

Partap U 1998. Successful pollination of apples in Himachal Pradesh.Beekeeping and Development 48: 6–7.

Partap U 2010. Honey bees as providers of pollination services. Mountain Form Bull, Jan 2010, pp 42–44.

Partap U, Partap T 2000. Pollination of apples in China.Beekeeping and Development 54: 6–7.

Partap U, Partap T, Yonghua H 2001. Pollination failure in apple crop and farmers' management strategies.Acta Hort 561: 225–230.

Patiny S (Ed) 2012. Evolution of plant-pollinator relationships. The Systematics Association Special Vol. 81, Cambridge Univ Press, Cambridge.

Pellmyr O 2002. Pollination by animals. In Herrera CM, Pellmyr O (Eds), Plant-animal interactions: An evolutionary approach. Lackwell Publishers, Malden, MA; Oxford, UK, pp 157–184.

Potts SG, Biesmeijer JC, Kremen C, Neumann P et al. 2009. Global pollinator declines: trends, impacts and drivers. Trends Ecol Evol 25: 345–353.

Prasad D, Hameed SF, Singh R et al. 1989. Effect of bee pollination on the quantity and quality of rai crop (*Brassica juncea* Coss.). Indian Bee J 51: 45–47.

Proffit M, Chen C, Soler C, Bessiere J-M, Schatz B, Hossaert-McKey M 2009. Can chemical signals responsible for mutualistic partner encounter promote the specific exploitation of nursery pollination mutualisms? – The case of figs and fig wasps. Ent Exp et Appl 131: 46–57.

Raguso RA 2004. Why are some floral nectars scented? Ecology 85: 1486–1494.

Regal PJ 1977. Ecology and evolution of flowering plant dominance. Science 196: 622–629.

Renner SS 2006.Rewardless flowers in the angiosperms and the role of insect cognition in their evolution. In Waser NW, Ollerton J (Eds), Plant-Pollinator Interactions: From Specialization to Generalization. Univ Chcago Press, Chicago, pp 123–144.

Roubik DW (Ed) 1995. Pollination of cultivated plants in the tropics, FAO Agriculture Service Bull 118.

Roulston TH, Cane JH 2000. Pollen nutritional content and digestibility for animals. Plant Syst Evol 222: 187–209.

Santos-Gally R, Perez-Barrales R, Simon VI, Arroyo J 2013. The role of short-tongued insects in floral variation across the range of a style-dimorphic plant. Ann Bot 111: 317–328.

Savoor RR 1998. Pollination management: an ecofriendly green revolution eludes India. Curr Sci 74: 121–125.

Schmidt JO Buchmann SL 1992. Other products of the hive. In Graham JM (Ed), The hive and the honey bee. Dadant & Sons, Hamilton, IL, pp 927–982.

Shivanna KR 2003. Pollen biology and biotechnology. Science Publishers, Inc Enfield, USA; Plymouth, UK.

Shivanna KR 2014a. Biotic pollination: How plants achieve conflicting demands of attraction and restriction of potential pollinators. In Ramawat KG. Merillon J-M Shivanna, KR (Eds), Reproductive biology of plants, CRC Press, pp 218–267.

Shivanna KR 2014b. Recruitment limitation: a major driver for species vulnerability. In: Kapoor R, Kaur I, Koul M (Eds), Plant reproductive biology and conservation, IK International, Delhi, pp 69–89.

Shivanna KR 2015. Management of pollination services to enhance crop productivity. In Bahadur, B et al. (Eds), Plant biology and biotechnology. Springer, New York pp 697-711..

Shivanna KR, Sawhney VK (Eds) 1997. Pollen biotechnology for crop production and improvement. Cambridge Univ Press, New York, NY.

Shivanna KR, Tandon R 2014. Reproductive ecology of flowering plants: A Manual. Springer India, New Delhi.

Sihag RC (Ed) 1995. Pollination biology: Pollination, plant reproduction and crop seed production. Rajendra Scientific Publishers, Hissar, India.

Slaa EJ, Sanchez-Chaves LA, Malagodi-Braga KS, Hofstede FE 2006. Stingless bees in applied pollination: practice and perspectives.Apidologie 37: 293–315.

Straver WA, Plowright RC 1991. Pollination of greenhouse tomatoes by bumble bees. Greenhouse Canada 11: 10–12.

Tylianakis JM 2013. The global plight of pollinators. Science 339: 1532–1533.

Velthuis HHW, van Doorn A 2006. A century of advances in bumble bee domestication and the economic and environmental aspects of its commercialization for pollination. Apidologie 37: 421–451.

Vergera CH 2008. Environmental impact of exotic bees introduced for crop pollination. In James RR, Pitts-Singer TL (Eds) Bee pollination in agricultural ecosystems. Oxford Univ Press, New York, pp 145–165.

von Helversen D, von Helversen O 1999. Acoustic guide in bat pollinated *flcous*. Nature 398: 759–760.

Wang G, Compton SG, Chen J 2013. The mechanism of pollinator specificity between two sympatric fig varieties: a combination of olfactory signals and contact cues. Ann Bot 111: 173–181.

Waser NM, Ollerton J (Eds) 2006. Plant-Pollinator interactions: From specialization to generalization. Univ. Chcago Press, Chicago.

Waser NM, Chittka L, Price MV 1996. Generalization in pollination systems, and why it matters. Ecology 77: 1043–1060.

Wilcock C, Neiland R 2002. Pollination failure in plants: why it happens and when it matters. Trends Plant Sci. 7: 270–277.

Williams GA, Adam P, Mound LA 2001. Thrips (Thysanoptera) pollination in Australian subtropical rainforests, with particular reference to pollination of *Wilkiea huegeliana* (Monimiaceae). J Nat Hist 35:1–21.

Williams RR, Church RM, Wood DES, Flook VA 1979. Use of an anthocyanin progeny marker to determine the value of hive pollen dispensers in apple orchards. J Hort Sci 54: 75–78.

Willmer P 2011. Pollination and floral ecology. Princeton Univ Press, New Jersey.

Winfree R, Bartomeus I, Cariveau DP 2011. Native pollinators in anthropogenic habitats. Ann Rev Ecol Evol Syst 42:1–22.

Young HJ 2002. Diurnal and nocturnal pollination of *Silene alba* (Caryophyllaceae). Am J Bot 89: 433–440.

Chapter 3

Plant-Pollinator-Robber Interaction

*Vineet Kumar Singh, Divya Mohanty, Chandan Barman and Rajesh Tandon**
tandon.raj@gmail.com

Abstract

In nature, mutualistic relationship between plants and pollinators is often intruded by the so-called 'nectar robbers' who illicitly acquire nectar from the flowers and thereby making it unavailable to the pollinators. Generally, the nectar robbers exploit this interaction and have been presumed to adversely influence the fecundity of the robbed plants and the foraging behaviour of their pollinators. Nectar robber may, either directly or indirectly, impact the plant's reproductive fitness. Plants, in response, employ several strategies to minimize the negative effects of robbery. However, remarkably, under certain situations nectar robbers have also proved to be beneficial for the existence of the host plant. It appears that in such a scenario the breeding system of plants may strongly contribute to the net outcome of robbing. Plant, pollinator and robber have been successful in demarcating a thin line of co-existence and have co-evolved together. In this chapter, we provide an overview of the present understanding of the dynamics of plant-pollinator-robber interactions, the consequences on the three components and evolutionary significance of nectar robbing.

3.1 Introduction

Plant-pollinator interaction is one of the crucial mutualistic interactions, which has co-evolved during the course of evolution and range expansion of the flowering plants (Mitchell 2009). Nearly 90 per cent of the angiosperms are pollinated by a variety of animal vectors (Ollerton 2011), and in return the vectors are benefitted with floral nectar and/or pollen as food reward (Faegri and van der Pijl 1971). Besides the pollinators, many a times, the sugary nectar also attracts non-pollinating animals. The ability of some floral foragers to take away floral nectar and/or pollen without facilitating pollination is termed floral larceny (Inouye, 1980); robbing of nectar and pollen are its subsets. The larcenists/robbers illegitimately

DOI: 10.1201/9781003505471-3

collect the rewards and often damage the floral parts in the process. On the basis of the type of reward stolen and the mode of stealing, the specific terms used are listed in Table 3.1.

Table 3.1: Terms associated with illegitimate collection of the type of floral rewards by the foragers (Inouye 1980).

Type/ Terminology	Description
Nectar robbing	Nectar is obtained illegitimately by piercing a hole in the corolla (includes floral damage)
Nectar thieving	Nectar is obtained legitimately by the floral opening but this theft does not lead to pollination due to morphological mismatch between the two.
Base working	Limited to plants with polypetalous corolla; in this case neither a hole is made nor is the floral opening used. Reward is obtained without carrying out pollination.
Pollen robbing	Collection of pollen by damaging the anthers; pollination does not happen.
Pollen theft	Pollen is collected without damaging any floral part and there is no pollination.

As nectar constitutes the major floral reward to the pollinators, in this chapter, we confine the details of plant-pollinator-robber interaction to nectar robbing and its consequences on plant reproduction. We also discuss the ecological and evolutionary consequences of nectar robbing and the mechanism by which plants circumvent and minimize the adverse effects of nectar robbing on their fitness.

The earliest record of nectar robbing can be traced to the seminal work entitled *The Secret of Nature in the Form and Fertilization of Flowers Discovered* by Christian Konard Sprengel (1793), who mentioned that "...bumblebees pierce the nectar spurs illegitimately to obtain reward". Charles Darwin (1876) was bewildered by the robbing bees that caused detrimental effects in red clover and stated that "all plants must suffer in some degree when bees obtain their nectar in a felonious manner by biting holes through the corolla". The robbers usually obtain nectar either by piercing through the corolla tube (nectar robbing) or entering through the opening of flower without bringing out pollination. The phenomenon of nectar robbing is reported from many, taxonomically unrelated species of angiosperms that have adapted to stock the nectar in a tubular or spurred corolla (Irwin and Maloof 2002, Irwin et al. 2010, Maloof and Inouye 2000).

Such an exploitation of floral rewards has both ecological and evolutionary consequences on the reproductive fitness of plants. However, the strength and development of plant-pollinator-robber interaction is generally context-dependent (Bronstein 2001, Irwin et al. 2010) and may vary within a species over space and time (Morris et al. 2010). Also, the range of flowers robbed in a plant may vary from none to 100 per cent and the degree of robbing rates is dependent on the existing alternative nectar resources in the vicinity (Burkle et al. 2007, Irwin et al. 2010).

One of the most expected consequences of presenting empty flowers in rewarding plants is the possible decline in the floral visits by the pollinators. Robbing on a large scale may adversely affect the reproductive success of the robbed plants (Bronstein 2001, Inouye 1980, Maloof and Inouye 2000). The net consequence of this interaction depends upon many factors. It may be influenced by the morphology of the flower and their visitors (phenotypic traits), mating system of the host plant, availability and quality of rewards and replenishment of nectar, and variation in the abundance of flowers, robbers and pollinators (Burkle et al. 2007, Castro et al. 2008, Irwin et al. 2010).

Nectar robbers are generally considered the exploiters of pollination mutualism. It is an economically profitable strategy for obtaining floral rewards, therefore there is a possibility that pollinator may switch to nectar robbing and destabilize the pollination system of a species (Wang et al. 2012). However the plant-pollinator mutualism appears to be thriving well even in the presence of nectar robbers and the interaction appears to have stabilized during the course of evolution. The strong persistence and frequent occurrence of plant-pollinator-robber interaction suggests that the functional consequence of nectar robbing is not always deleterious to the fitness of plants.

3.2 The Nectar Robbers

For the reason that a robber of one plant species may behave as pollinator of the other and that the robbers may consume rewards without performing any pollination service in return, they are often described as "cheaters" (Bronstein 1994, Darwin 1841, Richardson 1995, Thompson 1982). Gerner (1972) differentiated nectar robbing into two categories: primary and secondary nectar robbing. Primary robbers are the first to arrive and make fresh holes in flowers to steal nectar, whereas the secondary robbers use holes made previously by the primary robbers. Depending upon the opportunity and ease of accessing rewards, the robbers may shift their roles between primary and secondary (Irwin and Maloof 2002, Irwin et al. 2010, Maloof and Inouye 2000). This varied foraging behaviour may also lead to robbing and pollination in the same or different floral visits (Higashi 1988, Navarro 2000). Primary nectar robbing is most common and is usually observed in flowers with tubular corolla, while secondary nectar robbing is generally an infrequent behaviour. Although the robbers mostly prefer biting holes into the fresh flower, there are minimal chances that a robber will become specialized as a primary or secondary robber (Irwin et al. 2010).

Nectar robber could be an insect, bird or even a mammal. Among insects, hymenopterans (bees, wasps and ants) are the most common primary robbers. They use their specialized mouth parts to make a hole into the corolla to obtain nectar. Bees are quick learner and they show a social transfer of the robbing behaviour i.e., they can emulate other bees involved in robbing (Leadbeater and Chitka 2008, Sherry 2008). They generally use their maxillae and toothed mandible to make a slit into corolla to get nectar (Inouye 1983, Roubik 1982). Ants are also known to obtain nectar by chewing the floral parts and then getting entry into the tubular corolla. Birds make a hole in the corolla by using their pointed beak which may sometimes damage the reproductive organs of the flower (Arizmendi et al. 1996, Navarro 2001). Birds are also frequently involved in nectar thieving, which is less commonly observed among the bees. Hummingbirds (New World), sunbirds (Old world) and orioles (Australia) represent some of the common primary nectar robbers. Secondary nectar robbing is most frequent and evident in those plants that are usually robbed by birds. Rarely, squirrels have been also reported to steal nectar (Deng et al. 2004).

Nectar robbers may directly influence the plant reproductive success either by damaging the floral parts (Galen 1983, Utelli and Roy 2001) or by acting as pollinator in the same or different floral visit (Morris 1996, Zhu et al. 2010). Indirectly, the robbers may benefit the fitness of plants either by increasing the pollen flow or influencing the pollinators to visit additional plants in a foraging bout (Heinrich and Raven 1972, Miller and Travis 1996). In some instances, robbers can prevent pollinators to forage on a common resource in their territorial encounters (Roubik 1982).

3.3 Plant-pollinator-robber Dynamics and Their Coexistence

It is well known that mutualistic interactions are established through various attributes of selection process in nature. In this context, it is possible to assume that selection on the main components of pollination mutualism i.e., pollinators and plants, was stronger than robbing and the mechanism has somehow sustained robbing. In order to do so, some plants accommodate nectar robbing, either by resisting or avoiding them in such a way that they do not deter the legitimate floral visitors (Adler 2000, Irwin et al. 2004, 2010). As the flower should remain an intact unit of pollination and fertilization success, any damage to the flower or its parts has a direct impact on the fitness of plant. Therefore, it is believed that plants induce strong selection to protect the floral organs either by manipulating the quantity of nectar components or by reinforcing the protective tissues of flower (Irwin et al. 2008, Irwin 2009).

For sustaining plant-pollinator-robber interaction, plants have to maintain a balance between resistance to robbers and attraction to pollinators at the same time for their survival. Darwin proposed that the reproductive success of a species is categorized by its population density and the stable equilibrium in plant-pollinator systems (Hofbauer and Sigmund 1998). Persistence of plant-pollinator mutualism in the presence of a nectar robber may strongly depend upon the density and efficiency of the legitimate visitors and also on whether the plant applies resistance or deters the robbers (Wang et al. 2012). Wang (2012) has suggested that the plant-pollinator-robber interaction coexists under certain conditions such as – (i) plant alone can sustain robbers, when there is no pollinator limitation and pollinator visits on the flowers are relatively sufficient; (ii) if the plant is pollinator limited, then even if it can sustain robbing, plant-pollinator mutualism will be directed towards extinction; (iii) when plant cannot sustain the robbers alone then the plant-pollinator-robbers interaction could co-exist only when the robber contributes to the total fitness of the plant; and (iv) pollinator does not become extinct, if the plant is able to sustain the robber and the outcome of plant–pollinator interaction exceeds the negative effect of plant-robber interactions.

3.3.1 Consequences of Nectar Robbing on Plant Reproductive Success

Interaction of plant with pollinators and robbers can have significant outcome on the reproductive fitness of plants (Bronstein 2001, Irwin 2003, Irwin et al. 2010, Maloof and Inoye 2000, Richardson 2004). The net impact of robbing may range from negative to neutral and even positive (Irwin et al. 2001, 2010, Maloof and Inouye 2000, Morris 1996, Singh et al. 2014). There are several factors that may have both direct and indirect implications on the fitness of male and female components of the plant (Irwin et al. 2010, Maloof and Inouye 2000, Morris 1996, Zhu et al. 2010). Many studies involving effects of nectar robbing on plant fitness show that these complex ecological interactions are spatially and temporally dynamic and occasionally difficult to untangle (Bronstein 2001, Irwin et al. 2010). The following factors are known to determine the net effect of robbing on the reproductive fitness of plants.

Table 3.2: Role of the type of (identity) robber/pollinator partners in different combinations on the reproductive fitness of plants.

Robber	Pollinator	Expected impact	Reason
Insects/ Bird	Birds	Strongly negative	Some birds, especially hummingbirds, can detect the presence of nectar and they tend to avoid the robbed flowers and even leave the population if energy requirements are not met. Insects such as bees are more in number and may add to the negative effects (Irwin 2000, Irwin et al. 2010, Irwin and Maloof 2002)
Insect/ Bird	Insect/ Bees	Weak Positive	Bees, unlike birds are not able to discriminate between robbed and unrobbed flowers; as a result they are unable to avoid robbed flowers. As bees are more in number they can visit more flowers as pollinators (Richardson 2004, Irwin et al. 2010)

3.3.1.1 Identity of Plant Foragers

Many studies demonstrate that the identity or the type of functional group of floral visitors has a substantial effect on plant reproduction. For example, piercing a hole to obtain nectar from the flower, especially by birds, sometime may damage the nectaries and even the ovaries, which directly and negatively impacts the female fitness of the plant. In some cases, the robber may prevent the pollinator by their aggressive territorial behaviour. In doing so, the common resource is protected but imparts a negative effect on the overall fitness of the plant (Roubik 1982). In several other situations, the robber may also act as pollinator which mitigates the negative effect of robbing (Morris 1996, Zhu et al. 2010). Table 3.2 presents the possible impact of robbers on plant reproduction under different scenarios of identity of pollinators and robbers (Arizmendi et al. 1996, Burkle et al. 2007, Irwin 2000, Irwin et al. 2001, Irwin et al. 2010).

3.3.1.2 Breeding System of Plants

Breeding system of a plant is a crucial factor and determines the direct impact of robbing on the reproductive success (Burkle et al. 2007, Irwin and Maloof 2002, Irwin et al. 2010). In autonomously self-pollinated plants, the effect of robbing is negligible because fruit-set is ensured even in the absence of legitimate pollinators. In mixed-mating plants, autonomous selfing can ensure fruit-set but there can be a strong influence of inbreeding depression on the fitness of the offspring (Burkle et al. 2007). Pollinator mediated self-pollination and obligate outcrossing also have a direct impact on their fitness, which could vary between the two extremes depending on the foraging behaviour of pollinators and their abundance (Burkle et al. 2007, Maloof and Inouye 2000, Zimmerman 1985). In preferential or obligate outbreeding plants, a fewer visits to the flowers of the same plant may reduce stigma clogging through geitonogamy (Burkle et al. 2007, Castro et al. 2008, Maloof and Inouye 2000, Singh et al. 2014).

3.3.1.3 Pollinator Limitation

Pollinator limitation can certainly influence the net outcome of robbing on the fitness of plants. A limited number of visits by the pollinators would result in poor pollen deposition on the stigma, thereby imparting a direct negative effect on the fitness of pollinator-dependent robbed plants (Burkle et al. 2007, Castro et al. 2008, Irwin et al. 2010). In the absence of pollinator limitation, the robbed plants may receive sufficient pollen to fertilize all the ovules and thus, the negative effects of robbing are mitigated. Such circumstances of the absence of pollinator limitation may establish when (i) the competitive resources in the nearby vicinity are scarce (Burkle et al. 2007, Irwin et al. 2001, Singh et al. 2014), (ii) the pollinators are unable to discriminate between the robbed and unrobbed flowers (Goulson et al. 1998, Irwin et al. 2010, Singh et al. 2014) and (iii) the robbers can pollinate the flowers in the same or different visits (Higashi et al. 1998, Irwin et al. 2010). The net positive effect may also result if the production of seed is resource-limited rather than pollinator-limited (Campbell and Halama 1993, Singh et al. 2014).

3.3.1.4 Dynamics of Nectar Production and Replenishment

Certain functional floral traits may alleviate the adverse effects of robbing on plant reproduction (Irwin et al. 2010, Maloof and Inouye 2000). Nectar replenishment in the robbed flowers is one such trait. Several studies show that if a plant produces high amount of nectar or if the nectar is replenished at regular intervals during the receptive phase of stigma, then the overall negative effect of robbing is negated (Navarro 2001, Newman and Thomson 2005, Singh et al. 2014). Although plant has to invest more energy in replenishing the nectar, it is essential for ensuring the fitness of plant by continuously engaging the pollinators and making them visit even the robbed flowers. If a plant produces low amount of nectar or the nectar produced is not replenished, then the negative effect on female fitness is quite high. There is also a significant negative impact on plants when the entire amount of nectar is consumed in a single visit by the robber.

3.3.1.5 Longevity of Flowers

In plants with long-lived flowers, there are greater chances of repeated legitimate visits by the pollinators (Navarro 2001, Newman and Thomson 2005). However, in short-lived flowers nectar robbing may adversely affect fruit-set in plants if a legitimate visit is not met in time.

3.3.1.6 Fruit and Seed-set

The consequence of robbing on the natural fecundity of plants is directly measured by the amount of fruits and seeds set after robbing. Lower the flower to fruit ratio, greater would be the reproductive success of the plant. Most of the studies to establish the effect of robbing employ nectar removal experiments (Burkle et al. 2007, Maloof. 2001, Traveset et al. 1998, Zhang et al. 2011). In these experiments, nectar is either removed with the help of micro-capillary or syringe to mimic the natural mechanism of robbing. The outcome of robbed individuals is correlated with that of unrobbed conditions. Unrobbed conditions are produced by caging the plants and allowing only the pollinators to visits the flowers (Maloof 2000, Zhang 2011) or by applying a tape, such as cello-tape, to the basal region of corolla tube where robbing usually occurs (Singh et al. 2014, Traveset et al. 1998).

Table 3.3: The possible influence of behavioural changes observed among the pollinators in response to nectar robbing.

Behavioral changes in pollinators upon robbing	Expected result in robbed plant
Increased inter tree flight distance	Pollen flow distance and quality of pollen deposited on stigma increases
Fewer flowers visited in a tree	Reduces stigma clogging in self-incompatible plants i.e., reduces the chances of geitonogamy
Visits additional flowers across the population in unit time.	Increases pollen flow rate
Reduced flower handling time	Poor pollen transfer or load on stigma
Avoidance of robbed flowers.	Decreases female fitness but can increase pollen flow
Avoiding population or switching to different species.	No pollen transfer has severe negative effect on the fitness of plant.

3.4 Foraging Behaviour of the Pollinators upon Nectar Robbing

The male component of plant reproduction is more susceptible to changes in pollinator behaviour and their visitation pattern. The effect on male component determines the vigour and quality of offspring sired and hence it is an indication of the reproductive fitness of the plant. By reducing the nectar standing crop,i.e., the total amount of nectar available to the foragers), the robbers indirectly influence the pollinator flight distance (Irwin et al. 2010, Maloof and Inouye 2000, Maloof 2001). An increase in flight distance reduces the chances of geitonogamy thereby increasing the pollen flow among the individuals within a population, quality of pollen received and vigour of the offspring (Burkle et al. 2007, Maloof and Inouye 2000, Richardson 2004). However, robbing may reduce the total number of fruits produced in the self-compatible and pollinator-dependent species. A few studies that validate the increased pollen flow distance in response to robbing emphasize the use of fluorescent dyes and radio-transmitter to track the movement of the pollinator and the extent of pollen flow in both robbed and unrobbed patches of the population (Castro et al. 2008, Irwin and Brody 1999, Mayer et al. 2014). It is important to note that an increase in pollinator flight distance alone is not a measure of the positive effect on plant reproduction. It has to be correlated with other factors like pollen presentation or pollen removal by pollinators, which can be assessed by determining the stigmatic pollen load (Irwin 2003, Irwin et al. 2010). The amount of pollen deposited on a conspecific stigma reflects the wellness of male component of the plant; it is usually directly proportional to the pollinator abundance and availability. Correlating the pollen load with fruit set can determine whether or not the plant male fitness is getting affected due to robbing. Table 3.3 shows the possible behavioural changes in the pollinators in response to nectar robbing (Maloof and Inouye 2000).

3.5 Evolutionary Significance of Nectar Robbing

Nectar robbing is believed to be an inherited behaviour while the position on flower to make a hole is learnt through experience (Irwin et al. 2010, Oleson 1996, Zhang et al. 2007). Owing to its behavioural context, it is tempting to believe that evolution of robbing involved greater selection on animals for the process, which is believed to be through two ways. First, as a consequence of the inability of the forager, such as short-tongued bees, to access nectar in tubular flowers and second, to cut the energy cost involved in legitimate foraging (recued flower-handling time) (Inouye 1980, Maloof and Inouye 2000, Zhang et al. 2007). Recently, evidences have been collected to understand the selection of floral traits that have contributed to the evolution of nectar robbing.

Flowers with long spur and tubular corolla is usually a selected trait for nectar robbing but it may also depend on the overall effect on plant fitness. For example, in *Quassia amara*, flowers have significantly smaller corolla in populations where nectar robber was absent than at sites where nectar robbers (bird or bees) were present (Maloof and Inouye 2000, Roubik et al. 1985). Many flowers that are subjected to nectar robbing generally produce either copious amount of nectar or replenish the nectar (Maloof and Inouye 2000, Navarro 2001, Pyke 1982, Singh et al. 2014, Zhang et al. 2007). Hence, the relationship between high nectar-robbing and high nectar production is the outcome of co-evolution amongst plant and nectar-robber (Maloof and Inouye 2000, Morris 1996). However, when nectar-robbing constantly imparts a negative bearing on plant fitness, defensive mechanisms might evolve in the plant (Inouye 1983, Irwin et al. 2004, 2010, Maloof and Inouye 2000).

3.6 Plant Defence against Robbing

Floral characters and secondary metabolites play an important role in minimizing the effect of nectar robbing on fitness. According to the optimal defence theory, quantitative provisioning of resources to defences within plant is associated with those tissues that contribute significantly to plant fitness (Moriera et al. 2012, Zangerl and Bazzaz 1992). The theory is also applicable in case of nectar foraging. Plants tend to invest more in those floral tissues that minimize the adverse effects on its reproductive fitness (McCall and Irwin 2006).

3.6.1 Structural Defences against Nectar Robbers

Structurally, plants may provide defence to the standing nectar crop by clumping the flowers together in a tight cluster that makes the calyx or spur less penetrable to the robbers (Proctor et al. 1996). Thicker and bigger calyx in plants also prevents the robbers to pierce a hole in it (Roubik 1982, Watts et al. 2012). Liquid secretion from bracts prevents robbing by deterring the ants (Wootton and Sun 1989). Nectar guide patterns also minimize the effect of robbing on plant fitness. For example, bumblebees are shown to legitimately visit the flower with the aide of nectar guide but rob nectar from the flowers that lack the guide marks (Leonard et al. 2013).

3.6.2 Escape in Time and Space

Plants can avoid nectar robbing temporally by producing flowers or nectar at times of least or no activity of robbers. This escape is generally observed in species where pollinators and robbers forage the plant at different times. In *Castilleja linariaefolia*, nectar production and pollination activity is confined to the cooler periods (morning and evening). During these times, the robber bumble bee remains inactive (Carpenter 1979). Avoidance in space also helps in minimizing nectar robbing, such as by associating with other plant which is more suitable to robbers (Irwin et al. 2004).

3.6.3 Defence by Chemical Resistance

Repellence through certain chemicals is believed to be the most common type of resistance produced by plants against robbing. Plants vary in quality of nectar sugars i.e., sucrose rich or glucose rich (Baker and Baker 1982, 1986, 1990) and their concentration i.e., low or high sugar (Baker and baker 1982, Bolten and Feinsinger 1978), which decides, from an array of available visitors that may develop constancy with the flower. Similar to the defences against herbivory, plant produces some secondary compounds which deter the unwanted floral visitors to obtain nectar; the most common being the alkaloids and phenolics. Alkaloids may deter Lepidopterans but favour honey bees (Baker and Baker 1978). Phenolic compounds strongly deter the honey bees and bumblebees whereas they favour the birds. In *Aloe vryheidensis,* presence of phenolics in nectar minimizes robbing by bumblebees (Johnson et al. 2006). In this case, presence of secondary metabolites in nectar makes it unpalatable to the bees and sunbirds but allows certain insectivorous birds like bulbul to visit the flower. Such secondary compounds are reported from the nectars of nearly 21 families (Adler 2000), suggesting the evolution of plant defence against illegitimate foraging. Many volatile organic compounds play an important role in minimizing the negative effect of robbing by emitting fragrance attractive to pollinators but repelling to robbers. For example, in *Nicotiana attenuata*, benzyl acetone and nicotine repel the ants, which act as robbers and favour the hummingbird and hawkmoth pollinators (Kessler and Baldwin 2007).

3.6.4 Indirect Defence

Many plants are known to attract predators of nectar robbers, which reduce robbing. Extrafloral nectaries present on plants attract ants but prevent some bees and birds to forage those flower (Rudgers and Gardener 2004). For example, Keeler (1977) observed that ants feeding on the extrafloral nectaries of *Ipomea carnea* deter the robber carpenter bees and affect the female fitness of plant. In some plants, secondary compounds may also deter the legitimate visitors thereby causing damage to the fitness of plants (Adler and Irwin 2012, Irwin et al. 2004). Also, the overall investment in resisting the robber may increase the cost of reproduction. The cost of reproduction incurred on the production of the secondary metabolites and other structures (e.g., extrafloral nectaries) balances with the cost of attraction essentially needed to lure the legitimate pollinators.

3.7 Costs and Benefits of Nectar Robbing

To understand the co-evolutionary consequences of nectar robbing, knowledge of animal behavioural (pollinator and robbers) and their fitness implications is crucial. In nature, many floral visitors may become opportunistic and opt for secondary robbing (Dedej and Delaplane 2004). The cost of legitimate floral visits depends upon average flower-handling time, optimal amount of nectar intake and the flight distance. For most of the floral visitors, chewing the corolla tube is the only strategy to obtain nectar because of their inability to have access through the tubular corolla. For example, the short-tongue bumblebees often pierce at the base of long tubular corolla to obtain nectar. Thus, these bees always prefer robbing rather than spending energy through legitimate visits (Irwin and Brody 1998, 1999). Also, it has been seen that the frequency of secondary nectar robbing is strikingly increased when more than 70 per cent of flowers have already been robbed once (Olesen 1996). The robber honey bees also spend much lesser time probing the flower than foraging them legitimately as pollinators (Dodej and Delaplane 2004).

According to Darwin (1876), "The motive which impels bees to gnaw nectar by making holes though the corolla seems to be saving of time, for they lost much time in hovering and climbing into and out of the large tubular flowers". This implies that nectar robbing can have incredible benefits to the foragers. In plants where nectar gets replenished, even the effective pollinators can act as secondary robbers thereby minimizing their cost incurred on prolonged handling of flowers. Shift in foraging strategy may also emerge from competition for the available resources (Maloof 2001). For example, the length of tongue of bumblebees may influence the strategy to be employed for taking nectar. In cases where the tongue lengths overlap, bees prefer separate niches to avoid species exclusion (Irwin and Maloof 2001, Pyke 1982).

In terms of nectar energetics, it is difficult to understand shifts between primary and secondary robbing, as it has been argued that why a forager must spend energy on flowers from which the resources have already been extracted (Irwin et al. 2010). Primary nectar robbing is considered a beneficial strategy than the secondary; the latter is more prevalent in the species where the plant can replenish nectar (Irwin et al. 2010, Navarro 2001). The benefits gained by foraging the flowers illegitimately impart direct fitness to plant reproductive success and their interaction with the legitimate pollinators.

According to optimal foraging theory (Pyke 1984), the pollinator will decide to leave a patch if its net energy gained from a flower falls below the average of the energy gained from that patch or population. Avoidance of robbed flowers can reduce the cost incurred in foraging by the pollinator (Irwin and Brody 1998, Irwin et al. 2010). For example, hummingbirds can easily discriminate between robbed and unrobbed flowers which minimize their flower handling time (Irwin and Brody 1998). While bees as pollinators are not able to avoid the robbed flowers and thus, they have to spend more energy by visiting additional flowers to furnish their nutritional requirement.

Ecologically, nectar robbing is a complex phenomenon and has a significant and diverse effect on plant-pollinator mutualism. It is notable that there is paucity of information on nectar robbing from the Old World, even though there is enormous richness and diversity of tubular and spurred flowers. Studies are needed to understand not only the prevalence but also the dynamics of mutual interaction at higher level of organizations and its contribution to maintenance of diversity.

References

Adler LS 2000. The ecological significance of toxic nectar. Oikos 91: 409–420.

Adler LS, Irwin RE 2012. Nectar alkaloids decrease pollination and female reproduction in a native plant. Oecologia 168: 1033–1041.

Arizmendi MC, Dominguez CA, Dirzo R 1996. The role of an avian nectar robber and of hummingbird pollinators in the reproduction of two plant species. Funct Ecol10: 119–127.

Baker HG, Baker I 1978. Ants and flowers. Biotropica 10: 80

Baker HG, Baker I 1982. Chemical constituents of nectar in relation to pollination mechanisms and phylogeny. In: Nitecki MH (Ed) Biochemical Aspects of Evolutionary Biology. Chicago: The University of Chicago Press, pp 131–171.

Baker HG, Baker I 1986. The occurrence and significance of amino acids in floral nectar. Plant Syst Evol 151: 175–186.

Baker HG, Baker I 1990. The predictive value of nectar chemistry to the recognition of pollinator types. Israel J Bot 39: 157–166.

Bolten AB, Feinsinger P 1978. Why do hummingbird flowers secrete dilute nectar? Biotropica 10: 307–309.

Bronstein JL 1994. Our current understanding of mutualism. Quart Rev Biol 69: 31–51.

Bronstein JL 2001. The exploitation of mutualisms. Ecol Lett 4: 277–287.

Burkle LA, Irwin RE, Newman DA 2007. Predicting the effects of nectar robbing on plant reproduction: implications of pollen limitation and plant mating system. Am J Bot 94: 1935–1943.

Campbell DR, Halama KJ 1993. Resource and pollen limitations to lifetime seed production in a natural plant population. Ecology 74: 1043–1051.

Carpenter EL 1979. Competition between hummingbirds and insects for nectar. Am Zool 19: 1105–1114.

Castro S, Silveira P, Navarro L 2008. Consequences of nectar robbing for the fitness of a threatened plant species. Plant Ecol 199: 201–208.

Darwin C 1841. Humble-Bees. In: Barrett PH (Ed). The Collected Papers of Charles Darwin, Vol 1. University of Chicago Press, pp 142–145.

Darwin C 1876. The Effects of Cross and Self-Fertilisation on the Vegetable Kingdom. New York: Appelton.

Dedej S, Delaplane KS 2004. Nectar-robbing carpenter bees reduce seed-setting capability of honey bees (Hymenoptera: Apidae) in rabbit eye blueberry, *Vaccinium ashei*, 'Climax'. Environ Entomol 33: 100–106.

Deng XB, Ren PY, Gao JY, Li QJ 2004. The striped squirrel (*Tamiops swinhoel hainanus*) as a nectar robber of ginger (*Alpinia kwangsiensis*). Biotropia 36: 633–636.

Faegri K, van der Pijl L 1971. The Principles of Pollination Ecology, 2nd ed New York: Pergamon Press.

Galen C 1983. The effect of nectar-thieving ants on seed set in floral scent morphs ofPolemonium viscosum. Oikos 41: 245–249.

Gerner W 1972. Blüteneinbruch durch Apiden.Zool Anz 189: 34–44.

Goulson D, Hawson SA, Stout JC 1998. Foraging bumblebees avoid flowers already visited by conspecifics or by other bumblebee species. Anim Behav 55: 199–206.

Heinrich B, Raven PH 1972. Energetics and pollination ecology. Science 176: 597–602.

Higashi S, Ohara M, Arai H, Matsuo K 1988. Robber-like pollinators: overwintered queen bumblebees foraging on *Corydalis ambigua*. Ecol Entomol 13: 411–418.

Hofbauer J, Sigmund K 1998. Evolutionary Games and Population Dynamics. Cambridge University Press.

Inouye DW 1980. The terminology of floral larceny. Ecology 61: 1251–1252.

Inouye DW 1983. The ecology of nectar robbing. In: Beattey B, Elias T (Eds) The Biology of Nectaries. New York: Columbia University Press, pp 153–173.

Irwin RE 2000. Hummingbird avoidance of nectar-robbed plants: spatial location or visual cues? Oikos 91: 499–506.

Irwin RE 2003. Impact of nectar robbing on estimates of pollen flow: conceptual predictions and empirical outcomes. Ecology 84: 485–495.

Irwin RE 2009. Realized tolerance to nectar robbing: compensation to floral enemies in *Ipomopsis aggregata*. Ann Bot 103: 1425–1433.

Irwin RE, Brody AK 1998. Nectar robbing inIpomopsis aggregata: effects on pollinator behavior and plant fitness. Oecologia116: 519–527.

Irwin RE, Brody AK 1999. Robbing Bumble Bees Reduce the Fitness of *Ipomopsis aggregata* (Polemoniaceae). Ecology 80: 1703–1712.

Irwin RE, Maloof JE 2002. Variation in nectar robbing over time, space, and species. Oecologia 133: 525–533.

Irwin RE, Adler LS, Brody AK 2004. The dual role of floral traits: pollinator attraction and plant defense. Ecology 85: 1503–1511.

Irwin RE, Brody AK, Waser NM 2001. The impact of floral larceny on individuals, populations, and communities. Oecologia 129: 161–168.

Irwin RE, Bronstein JL, Manson JS, Richardson L 2010. Nectar robbing: Ecological and evolutionary perspectives. Annu Rev Ecol Evol S 41: 271–292.

Irwin RE, Galen C, Rabenold JJ, Kaczorowski R, McCutcheon ML 2008. Mechanisms of tolerance to floral larceny in two animal-pollinated wildflowers, *Polemonium viscosum* and *Ipomopsis aggregata*. Ecology 89: 3093–3104.

Johnson SD, Hargreaves AL, Brown M 2006. Dark, bitter-tasting nectar functions as a filter of flower visitors in a bird-pollinated plant. Ecology 87: 2709–2716.

Keeler KH 1977. The extrafloral nectaries of *Ipomoea carnea* (Convolvulaceae). Am J Bot 64: 1182–1188.

Kessler D, Baldwin IT. 2007.Making sense of nectar scents: the effects of nectar secondary metabolites on floral visitors of *Nicotiana attenuata*. The Plant J 49: 840–854.

Leadbeater E, Chittka L. 2007. Social learning in insects—from miniature brains to consensus building. Curr Biol 17: R703–R713.

Leonard AS, Brent J, Papaj DR, Dornhaus A 2013. Floral nectar guide patterns discourage nectar robbing by bumble bees. Plos One8: e55914.

Maloof JE 2001. The effects of a bumble bee nectar robber on plant reproductive success and pollinator behaviour. Am J Bot 88: 1960–1965.

Maloof JE, Inouye DW 2000. Are nectar robbers cheaters or mutualists? Ecology 81: 2651–2661.

Mayer C, Dehon C, Gauthier A, Naveau O, Rigo C, Jacquemart A 2014. Nectar robbing improves male reproductive success of the endangered *Aconitum napellus* ssp. *lusitanicum*. Evol Ecol 28: 669–685.

McCall AC, Irwin RE 2006. Florivory: the intersection of pollination and herbivory. Ecol Lett 9: 1351–1365.

Miller TE, Travis J 1996. The evolutionary role of indirect effects in communities. Ecology 77: 1329–1335.

Mitchell RJ, Irwin RE, Flanagan RJ, Karron JD. 2009. Ecology and evolution of plant–pollinator interactions. Ann Bot 103: 1355–1363.

Moreira X, Zas R, Sampedro L 2012. Differential allocation of constitutive and induced chemical defenses in pine tree juveniles: a test of the optimal defense theory.Plos One 7: e34006.

Morris WF 1996. Mutualism denied? Nectar-robbing bumble bees do not reduce female or male success of bluebells. Ecology 77: 1451–1462.

Morris WF, Vázquez DP, Chacoff NP 2010. Benefit and cost curves for typical pollination mutualisms. Ecology 91: 1276–1285.

Navarro L 2000. Pollination ecology of *Anthyllis vulneraria* subsp. *Vulgaris* (Fabaceae): nectarrobbers as pollinators. Am J Bot 87: 980–985.

Navarro L 2001. Reproductive biology and effect of nectar robbing on fruit production in *Macleania bullata* (Ericaceae). Plant Ecol 152: 59–65.

Newman DA, Thomson JD 2005. Effects nectar robbing on nectar dynamics and bumblebee foraging strategies in *Linaria vulgaris* (Scrophulariaceae). Oikos 110: 309–320.

Olesen JM 1996. From naivete to experience: bumblebee queens (*Bombus terrestris*) foraging on *Corydalis cava* (Fumariaceae). J Kan Entomol Soc 274–286.

Ollerton J, Winfree R, Tarrant S 2011. How many flowering plants are pollinated by animals? Oikos 120: 321–326.

Proctor M, Yeo P, Lack A 1996. The Natural History of Pollination. Portland: Timber Press.

Pyke GH 1982. Foraging in bumblebees: rule of departure from an inflorescence. Can J Zool 60: 417–428.

Pyke GH 1984. Optimal foraging theory: a critical review. Annu Rev Ecol Evol S 15: 523–575.

Richardson SC 1995. Nectar-robbing bees as cheaters on a plant–pollinator mutualism. Bull Ecol Soc Amer 76: 380–381.

Richardson SC 2004. Are nectar-robbers mutualists or antagonists? Oecologia 139: 246–254.

Roubik DW 1982. The ecological impact of nectar-robbing bees and pollinating hummingbirds on a tropical shrub. Ecology 63: 354–360.

Roubik DW, Holbrook NM, Parra GV 1985. Roles of nectar robbers in reproduction of the tropical treeletQuassia amara (Simaroubaceae). Oecologia 66: 161–167.

Rudgers JA, Gardener MC 2004. Extrafloral nectar as a resource mediating multispecies interactions. Ecology 85: 1495–1502.

Sherry DF 2008. Social learning: nectar robbing spreads socially in bumble bees. Curr Biol 18: R608–R610.

Singh V, Barman C, Tandon R 2014.Nectar RobbingPositively Influences the Reproductive Success of *Tecomella undulata* (Bignoniaceae). Plos One 9: e102607.

Sprengel CK 1793. The Secret of Nature in the Form and Fertilization of Flowers Discovered. Berlin: Friedrich Vieweg Altere.

Thompson JN 1982. Interaction and Coevolution. New York: Wiley and Sons.

Traveset A, Willson MF, Sabag C 1998. Effect of nectar-robbing birds on fruit set of *Fuchsia magellanica* in Tierra Del Fuego: a disrupted mutualism. Funct Ecol 12: 459–464.

Utelli AB, Roy BA 2001. Causes and consequences of floral damage in *Aconitum lycoctonum* at high and low elevations in Switzerland. Oecologia 127: 266–273.

Wang Y, Wua H, Sun S 2012. Persistence of pollination mutualisms in plant–pollinator–robber systems. Theor Popul Biol 81: 243–250.

Watts S, Ovalle DH, Herrera MM, Ollerton J 2012. Pollinator effectiveness of native and non-native flower visitors to an apparently generalist Andean shrub, *Duranta mandonii* (Verbenaceae). Plant Sp Biol 27: 147–158.

Wootton T J, Sun IF 1989. Bract liquid as a herbivore defense mechanism for *Heliconia wagneriana* inflorescences. Biotropica 22: 155–159.

Zangerl AR, Bazzaz FA 1992. Theory and pattern in plant defense allocation. In: Fritz RS, Simms EL (Eds). Plant Resistance to Herbivores and Pathogens: Ecology, Evolution and Genetics.Chicago: University of Chicago Press, pp 363–391.

Zhang C, Irwin RE, Wang Y1, Ping He YP, Yang YP, Duan YW2011. Selective seed abortion induced by nectar robbing in the selfing plant *Comastoma pulmonarium*. New Phytol 192: 249–255.

Zhang YW, Wang Y, Guo YH 2007. Effects of nectar robbing on plant reproduction and evolution. Front Biol China 2: 443–449.

Zhu XF, Wan JP, Li QJ. 2010. Nectar robbers pollinate flowers with sexual organs hidden within corollas in distylous *Primula secundiflora* (Primulaceae). Biol Lett 6: 785–787.

Zimmerman M, Cook S. 1985. Pollinator foraging, experimental nectarrobbing and plant fitness inImpatiens capensis. Am Mid Natural 113: 84–91.

Chapter 4

Natural History and Behavioural Ecology of Bees

Palatty Allesh Sinu, Sangeetha Varma,*
U Prashanth Ballullaya, G Asha, Prabitha Mohan
sinupa@gmail.com

Abstract

Pollination is a fine example of mutualism, where the flowering plants and the visiting animals are equally benefited. The costs involved in the interaction are the floral rewards such as pollen, nectar and oil for the plants, and the time and energy spent to harvest the rewards for the visitors. Flowers are in sexual isolation till they are pollinated.The sexual isolation is due to the floral morphological reasons or due to the ethological reasons of the visitors. The morphological reasons are with respect to the position of the sexual parts (androecium and gynoecium) in a flower and compatibility of the self-pollen. The ethological reasons of the visitors are with respect to the foraging behaviour of the visitors, such as the mode of floral access of the visitors for the rewards, degree of floral constancy maintained by the visitors and so on. Bees constitute the major group of pollinators in any given type of ecosystem. Earlier, we considered that managed honey bees can support the pollination demand of our crops; but, our current understanding is that maintaining the functional diversity of the pollinators is essential for a desirable output. Every wild bee, including the solitary and social bee contributes to the functional diversity of the pollinators. Therefore an understanding on their diversity, foraging plants, nesting ecology, and their foraging behaviour is essential to manage pollinators in a given system. This chapter gives a detailed account on the diversity of bees, the adaptations of the bees to collect and transport pollen grains, the floral constancy of the bees and important behavioural modification of the bees that determine the pollination success in plants to certain extent.

4.1 Introduction

Over 90 per cent of nearly 250,000 described flowering plants (Soltis et al. 2005) require an animal vector to carry out pollination (Bawa 1990, Ollerton et al. 2011). A recent study

DOI: 10.1201/9781003505471-4

finds that 62–73% of the wild flowering plants are pollen-limited, meaning that the number of compatible pollen landing on the stigma is less than that required for adequate seed set (Ashman et al. 2004). About 35 per cent of our global plant-based food production is reliant on pollinator fauna (Aizen et al. 2008, Klein et al. 2007). Among the animals, insects are the major group of pollinators because of their abundance, morphological, functional and behavioural diversity (Losey and Vaughan 2006). Among the insects, bees are the most successful pollinators, because their diet is exclusively based on floral rewards, such as pollen, nectar and oil. There are about 20,000 species of bees described all over the world (Ascher and Pickering 2012), and the bee – flowering plant interaction is considered as one of the most successful coevolved mutualism (Danforth et al. 2013). Bees therefore have evolved morphological and behavioural adaptations to transport pollen on their bodies. Fidelity to partner flowering plants enhances the role of bees as efficient pollinators. Bees also can be trained and managed easily in agroecosystems (James and Pitt-Singer 2008, Klein et al. 2007, Roubik 1989, Shivanna, this volume). In plant-pollinator mutualism, pollen, nectar and oil are the key commodities the plants offer. Thus they form the key attractants for insect pollinators. However, the other floral traits that are important to improve pollination efficiency are scent and colour to attract pollinators, and the morphology of the flower, particularly the position of the anthers and the stigma to achieve pollen transfer during insect foraging (Harder and Barrett 1996, Harder and Routley 2006, Nicholls and de Ibarra 2014).

Since, pollination is a mutualistic interaction between flowering plants and animals (Cauich et al. 2004, Mitchell et al. 2009),the legitimate foraging behaviour of the animals in flowers is of particular relevance for successful pollination (Bronstein 2001, Cauich et al. 2004). Since bees constitute major pollinators, an understanding of their origin, evolution, classification, natural history and foraging adaptations is very important in pollination studies. This chapter elucidates some of these important aspects of bees.

4.2 Evolution and Diversity of Bees

The current understanding of the evolution of bees is that they have evolved from a wasp-like ancestor belonging to apoid (Melo 1999) or sphecoform groups (Michener 2007) that comprise four families: Heterogynaidae, Ampulicidae, Crabronidae, and Sphecidae. According to the phylogeny studies based on the morphology and molecular data, apiformes (bees) evolved from the family Crabronidae in the early to mid-Cretaceous period approximately 140 to 110 million years ago, roughly coincident with the origins and early diversification of flowering plants (Brothers 1975, Danforth et al. 2013). Today, Sphecid wasps are the closest relative of apiformes (Figure 4.1).The apiformes (bees) and spheciformes (sphecoid wasps) were previously treated as two paraphyletic groups under the holophyletic group called as Apoidea (Brothers 1975). The recent morphological (Michener 2007) and molecular data (Danforth et al. 2006, Ohl and Bleidorn 2006) analyses suggest that bees are clearly a monophylectic group. Currently bees are placed in the Anthophila clade under the superfamily Apoidea (Michener 2007). Thus the extant bees are represented as Hymenoptera: Apoidea: Anthophila. The clade Anthophila has seven families Apidae, Andrenidae, Colletidae, Halictidae, Megachilidae, Melittidae and Stenotritidae (Michener 2000) and 25 sub-families (Michener 2007).

Fig. 4.1. *Ammophila sp* (Sphecidae), a wasp species carries a paralyzed host larvae to the nest hole. The long and slender legs adapted for predatory wasps and less pubescence may be noticed.

The seven families of bees are first classified based on the tongue length: Apidae and Megachilidae are long-tongued bees, and the remaining five are short-tongued bees (see Chapter 17 for more details). The other broad classification is based on the sociality of bees: Social bees and Solitary bees. Solitary bees, as the name implies, live alone and the bee nest is managed by a single reproductive female bee, which provisions brood cells, lays eggs and cleans the nests. This behaviour is the most primitive form of bee behaviour, and is considered to be derived from their ancestors. Nevertheless, 90 per cent of the bees are solitary in their behaviour. Sociality is seen in different forms in bees. The most advanced social behaviour, eusociality, is reported in the corbiculate (explained under structural adaptations) members of Apidae. Advanced eusocial colony is characterized by the occupancy of the nest by a group of females of multiple generations, division of labour based on the reproductive capacity of the female occupants and cooperative brood care. The colony is primarily constituted by a queen (reproductively fertile female) and workers with some males. Sometimes considerable variation in morphological characteristics occurs among different castes of a given bee species (Figure 4.2). Workers are reproductively sterile females. Queen and workers are morphologically different; queens lack structures needed to collect and manipulate pollen which the workers possess in honey bees. Advanced eusociality is typically seen in Apid bees of the Meliponini and Apini tribes. In a primitive eusocial colony as seen in the Apid bees of Allodapini and Bombini, the queens and workers are distinguished only by the size of the bees; they are structurally identical. Another form of sociality is the temporary association between reproductive females of two generations and temporary division of labour during active breeding season, in which some females serve as guards and others forage and lay eggs. It is reported in Xylocopini, Ceratinini, Allodapini and Euglossini tribes of Apidae (Cardinal and Danforth 2011) and in Hylaeinae subfamily of Colletidae (Spessa et al. 2000). Communal nesting is considered as a different form of sociality reported in many bee families, such as Halictidae, Andrenidae, Apidae, Megachilidae and Melittidae. Here, multiple reproductively-active females of the same generation finds and share a common site for nesting. But, there is neither reproductive division of labour nor cooperative brood care observed. The origin and evolution of sociality in bees received much attention since the studies of Cameron (1993) (Brady et al. 2006,Cardinal and Danforth 2011, Danforth 2002, Danforth et al. 2003, Schwarz et al. 2007, Schwarz et al. 2011, Spessa et al. 2000, Thompson and Oldroyd 2004). The third class of bees based on the sociality is parasites. Some bees use nests of other bees during their breeding period and lay their eggs in a partially-provisioned brood cell of the host bee. This sort of brood parasitism is known as cleptoparasitism. Since there are evidences that cleptoparasites attack the nests of closely-related species for egg-laying, it is believed that the cleptoparasitc bees arose from their host bee species. Cleptoparasitism is reported in 12 per cent of all bees and 30 per cent of bees of Apidae (Danforth et al. 2013). It is reported in members of Halictidae (Halictini, Augochorini), Megachilidae (Megachilini, Osmiini and Anthidiini), Colletidae (Hylaeinae) and Apidae (Euglossini, Ctenoplectini) (Cardinal et al. 2010, Gibbs et al. 2012). A minor form of bee parasitism is social parasitism. In this type, both the parasitic and host bee belongs to the social class only. The female social parasitic bee after excluding or killing the host queen manages the nest and the worker groups.It is found that both the parasitic and host bees involved in the social parasitism are phylogenetically closely related. Social parasitism is reported in bees belonging to eusocial class of Apidae (Bombini) and Halictidae (Halictini, Augochorini and Sphecodini).

Fig. 4.2. Pictures show the morphological and phenotypic variations in queen (a) and worker (b) castes of a social bumblebee species, *Bombus haemorrhoidalis* (Apidae).

Fig. 4.3. a) The long tongue of a long-tongued bumblebee, *Bombus haemorrhoidalis* (Apidae); b) an open bee hive of *Apis dorsata*; c) the nest of Apis cerana in a tree branch hole; d) the teeth marks of a leaf-cutting *Megachile* bee on a leaf; e) a solitary resin bee (*Anthidium* sp: Megachilidae) cleaning the ground nest hole; f) the nest hole of a carpenter bee (*Xylocopa* sp); g) pollen pellets collected from the nest of *Apis dorsata*.

4.3 Bee families

4.3.1 Apidae (honey bees, bumblebees and stingless bees)

It is one of the species-rich eusocial bee family with primitive and advanced species as its members. Over 5,700 described species belonging to 209 genera are reported from the world (Ascher and Pickering 2012). The greatest diversification has occurred in Oriental and Neotropical regions. This family includes the most successfully managed European honey bee species, *Apis mellifera*. This family includes three subfamilies, Apinae, Xylocopinae and Nomadinae (Danforth et al. 2013). They have ground nesting (e.g., Emphorini and Eucerini), open-nesting (e.g., Apini), wood nesting (e.g., Xylocopini) and stem nesting (e.g., Ceratinini) bees. Some of them also construct nests using the plant resins (e.g. Meliponini and Euglossini) (Figure 4.3).

4.3.2 Megachilidae (leaf-cutting, resin- and mason-bees)

It is the second largest family of long-tongued bees represented by over 4,000 species placed in four subfamilies: Fideliinae, Lithurginae, Pararhophitinae and Megachilinae. All of them are solitary or communal (live together in individually made nest constructed close to each other). Leaf-cutting bees use leaf discs to construct nests (e.g., *Megachile* spp). Resin bees belong to the tribe Anthidiini (e.g., *Anthidium* spp) construct nests using plant resin (Figure 4.3). Mason bees belong to the tribe, Osmiini (e.g., *Osmia* spp) construct nests using mud.

4.3.3 Andrenidae

This family includes nearly 2,900 species of bees in three subfamilies (Andreninae, Oxaeniae and Punerginae). Although it has wide distribution, greatest diversity is reported in semi-arid zone of southwest North America, South America and Palearctic region. It is a family comprising of only solitary bees, and ground dwellers. Most of the Andrenid bees are oligolectic in behaviour (i.e., with narrow host plant range) (Danforth et al. 2013). Most of the genera are species poor and confined to small geographical regions, particularly in North America; but, *Andrena* is a cosmopolitan genus.

4.3.4 Halictidae

With over 4,300 described species and four subfamilies, Halictidae is the largest cosmopolitan family in the world. Its distribution is more or less even across biogeographic zones. In contrast with the andrenids, the halictids are largely polylectic at species level. The members of the family show diversity in social behaviours at the adult female level that includes, eusocial, semi social, communal and solitary (Schwarz et al. 2007). *Halictus* and *Lasioglossum* are the two predominant genera of Halictidae.

4.3.5 Colletidae

It has over 2,500 reported species in seven sub-families. It is widely distributed across continents. Many species in Colletidae are hairless, therefore carry pollen internally in crop and are cleptoparasitic in nature. Some bees have scopae developed on the hind legs to transport the pollen, but in most bees scopae is highly degenerated. The bifid glossa (tongue) is unique to this family. *Colletes* and *Hylaeus* are the cosmopolitan genera of Colletidae.

4.3.6 Bee Families not Reported from India

4.3.6.1 Melittidae

It is a small family comprising of nearly 200 species and 15 genera nested within three sub-families. Its distribution is confined to the Mediterranean and temperate regions of the old and new world. The greatest diversity occurs in the semi-xeric climates of Southern Africa.

Oil-foraging is a general character of Melittid bees, which is less-reported in other families. Most species are oligolectic in nature (Michez et al. 2008).

4.3.6.2 Stenotritidae

It is an endemic bee family to Australia, which is represented by 21 species and two genera. Molecular data analyses (Danforth et al. 2013) has shown that it is a sister group of Colletidae. The bees are large, strictly solitary and ground-dwellers. Most of them have primary host-relationship to the flowers of Myrtaceae.

4.4 Reproductive Isolation in Flowers

Until and unless a stigma receives its conspecific pollen, the flower is considered reproductively isolated. The reproductive isolation can occur because of two different sets of reasons:

1) Morphological reasons: Even though the pollen grains and ovules are present in the same flower (bisexual flowers), the floral architecture may not allow its pollen reaching its own stigmas (self pollination). The constraints include the distant positioning of stigma from the anthers, distally opening stigma cup from the anthers, encircling protective hairs around the stigma cup, and so on. The adaptations such as protandry (male phase followed by female phase) and protagyny (female phase followed by male phase) in bisexual flowers also limit the extent of self-pollination considerably (Shivanna 2012).
2) Ethological isolation: This situation arises when the behaviour of a visitor is not legitimate enough to carry out effective pollination (Bronstein 2001). Flowering plants tend to evolve mechanisms and processes to facilitate cross-pollination between conspecific plants. The best strategy that a flower has developed is luring animals by offering pollen and the nectar as rewards. Plants in a community compete or sometimes facilitate each other to share the available riches and resources, which, according to evolutionary biologists is mediated by the pollinators (Ackerly 2003, Armbruster et al. 1994). Recent researches support the facilitative principle rather than the competitive principle to explain the coexistence of functionally-similar flowering plants (De Jager et al. 2011, Ghazoul 2006, Moeller 2004). Floral density, floral diversity, inflorescence size, floral display size and co-flowering species shape the pollinator fauna in a given habitat (De Jager et al. 2011,Ghazoul 2006, Makino et al. 2007, Ohara and Higashi 1994).

In the process of pollination, the pollen grains get transported by various visitors eventually to the conspecific flower stigmas. To optimize this transfer, flowering plants have evolved mechanisms to ensure multiple visits of legitimate visitors (Harder and Thomson 1989). Simultaneously, they have also developed counter measures to reduce the pollen loss to illegitimate visitors and to hetero-specific stigmas. Floral phenology, timing of daily floral display, anthesis time, pollen-release to individual visits, maturity period of male and female reproductive parts of the flower, nectar production etc., evolved within the floral community and individual species level to attain the best outcome from the interaction between flowers and pollinators (Grindeland et al. 2005, Robertson and Macnair 1995).

4.5 Pollen Resource: An Essential Commodity for Plants and Pollinators

Pollen grains are the haplotypic (n) male gametophytes that require a vector or medium to reach stigmas of conspecific flowers. The size of pollen varies between 5μm and 215μm with an average of 34μm (Roberts and Vallespir 1978, Roulston et al. 2000). Protein is one of the major ingredients of the pollen that ranges widely between 2.3 per cent and 61 per cent among plant species (Buchmann 1986) and between 12 per cent and 61 per cent in angiosperms species. Protein content of pollen is highly conserved within plant genera, families and even divisions (Roulston et al. 2000). Solanaceous plants are the richest in terms of pollen protein (34.1%-53.4%) and conifers are the poorest (2.3%-9.8%). Pollen grains of wind-pollinated plants are generally poor in nutritional quality when compared to the animal-pollinated plants (Roulston et al. 2000). Small pollen grains are generally rich in protein content. Most pollen proteins are enzymes facilitating pollen germination and pollen tube growth. All available evidences indicate that pollen protein content is largely correlated with the distance between the stigma and ovule (Roulston et al. 2000).

Pollen is a rich source of protein for many bees, which is either consumed directly by the flower-visiting animals or harvested and transported to the nests to feed their young ones (Hargreaves et al. 2009) (Figure 4.3). Pollen is directly consumed by most of the non-bee visitors, such as beetles (Coleoptera), wasps (Hymenoptera), Syrphid flies (hover-flies, Diptera), bats (Chiroptera), thrips (Thysanoptera), birds and mammals. Bees, with very few exceptions collect pollen to feed their larvae. Several studies have shown that consumption of pollen with high protein content improves the survival, size and longevity of bees (Génissel et al. 2002, Greenberg 1982, Pech-May et al. 2012, Schmidt et al. 1987). Great variability in the quantity and quality of the pollen produced has direct implications in shaping plant-animal interactions in animal-pollinated plants (Roulston et al. 2000). Although most of the wind-pollinated (anemophilous) plants produce abundant pollen; a few such plants attract the animals. On the other hand some zoophilous (animal-pollinated) plants that reward only pollen, attract many visitors. For instance, many solanaceous plants are nectar-less and buzz-pollinated(see below), but superior in terms of pollen quality and quantity, and are pollinated primarily by the bees (Burkart et al. 2013, Roulston et al. 2000, Smith and Knapp 2002).

4.6 Foraging Adaptations of Bees

4.6.1 Structural adaptations

The apiformes when diverged from the ancestral habit of provisioning nests with animal food, began acquiring the habit of provisioning plant-based food, such as pollen, nectar and oil in the nests. In other words, the spheciformes retained "non-vegetarian" habit of the ancestor species and the apiformes became "vegetarians". However, both the forms shared the common ancestral habit of provisioning nests with the foraging materials for the nestlings (Thorp 1979). The sphecids' (digger wasps) provision includes the arthropods that primarily include caterpillars of lepidopteran insects, mole crickets, spiders etc. The sphecid wasps paralyze the prey insect and drag them to the pre-built nest holes using the mandibles or the legs. So, they have excellent mandibulate mouth parts and spiny or bristled long slender legs to help them carry the prey to the nest (Figure 4.1). Once the apiformes become pollen and nectar acquisitionists, they develop marked behavioural and structural modifications in their body; this happens largely in the female bees. One important difference between the apiformes and spheciformes is the plumosity of the body hairs. Apiformes have dense pubescent and plumose branched hairs on all parts of their body including face, thorax, abdomen and legs (Figure 4.2) in contrast to the sparsely distributed simple hairs in the spheciformes. Some of their body hairs are specialized to enable them lifting and carrying pollen, which are called scopae (Figure 4.4). In contrast to the hairy pollen collecting apparatus, some bees have glabrous (smooth) plate like structure to transport pollen, termed corbicula (Figure 4.4), and the bees possessing them are called corbiculates (see Box 4.1). Some bees carry pollen internally in crop (e.g., Hyaelinae and Euryglossinae bees of Family Colletidae).

Box 4.1

The corbiculae is a bare concave plate delimited by curved specialized hairs on some or all sides of the plate. It is believed that the corbiculate clade (*Centris* and *Epicharis* group) arose in the New World and diverged about 84 million years ago (Martins et al. 2014). Molecular evidences and fossil records show that the ancestral state prior to the evolution of corbiculae was a hairy hind leg adapted for oil-collection as seen in *Centris* and *Epicharis*. Hairy hind leg is seen on the propodeum or femora in Andrenidae, Halictidae and Colletidae. But, a typical corbicula is seen on the hind tibia of bees belonging to Apidae (stingless bees, honey bees and bumblebees), which are commonly known as pollen basket bees. In pollen-collecting female *Apis* spp, the hind tibia is apically (at the tip) widened and the outer side of it forms a slightly concave plate. It is surrounded by long curved simple (unbranched) hairs; near the apico-posterior corner on the outer side of tibia a single long hair is present; between the apico-posterior hair and the apical margin of the tibia a row of minute apically flattened hairs are present. The basal (=base) part of the basitarsus (first tarsal segment from the body) also forms a plate covered with small tubercles, which is angled toward the outer face of the tibia (corbicula). The inner side of the hind tibia has stout setae, which form the comb of the pollen basket. Structural variation in corbiculae is seen in non-Apis bee species belonging to the tribe Euglossini and the subfamily Meliponinae. But, the corbicular structure of *Bombus* is similar to that of the *Apis* (Michener et al. 1978). In some bees belonging to the family Colletidae the scopae are highly reduced as they carry pollen internally inside the crop or honey stomach (it is a part of the foregut in bees that generally stores nectar and regurgitates the matter in the nest).

Fig. 4.4. Scopal apparatus on different body parts of bees; a) scopae on the ventral side of the gaster of Megachilid bee; b) scopae on the hind femur and tibiae in Xylocopa sp; c) scopae on the hind coxa and femur on an Amegilla sp., d) scopae on the entire hind leg of a halictid bee; e) scopae on the hind leg of Ceratina sp., f) pollen transporting corbicula (outer smooth concave plate) on hind tibia and basitarsus of *Apis cerana*; on the opposite side of the leg, the pollen comb is also seen in the inner side of basitarsus.

Scopa is primarily seen on different segments of the legs of all but the megachilid bees. Generally it is seen on femora, trochanter, tibia, basitarus and coxa segments of hind legs and on propodeum part of pollen-collecting female bees (Figure 4.4). In megachilidae, the scopa is seen on the ventral side of abdomen (Figure 4.4). The scopal hairs generally transport dry pollen (pollen not moistened with nectar or honey). The specialized scopal hairs are capable of collecting pollen irrespective of great variation in their number, size and surface features (Roberts and Vallespir 1978). The host-plant association and pollen specialization have shaped the structural variation of the scopae (Danforth et al. 2013). Structural variations include the position, spacing, length and plumosity of scopal hairs in bees. The scopal hairs become dense and multi-branched in bees specialized to collect extremely small pollen grains. The scopal hairs are unbranched and stouter when the bee is a specialist of large pollen grains. Thorp (1979) gave an excellent account on the adaptive structural variations in the scopae of North American bees with respect to the host plants. The corbiculae on the other hand transports pollen in sticky form, where the pollen is usually moistened with the nectar or honey (nectar blended with the salivary secretions of bees). Since corbiculae transport the pollen grains as a moist lump or pellet packed on its smooth outer surface, the bees possessing corbiculae can collect pollen from a range of unrelated plants whose pollen size and architecture can be extremely varied (Thorp 1979). This is considered as a fitness character in the social bees as their large colonies will have to survive on a range of unrelated host plants in different seasons and regions. If the corbiculae had to be replaced by the specialized scopal hairs, the bees' host plant range would have been narrowed.Although the pollen-transporting structures in bees are developed mostly on the rear part of the bee body, modifications are also found on the face, mouth parts and forelegs to pick and groom the pollen with the oil and transport that to the scopae or corbiculae at the rear portion of the bee. This includes setaceous mouth parts, hairy facial foveae, and combed fore- and mid-legs and pectinate (serrate or hairy) tibial spurs. We lack information on the structural modifications in the pollen-transporting structures (scopae, crop (honey stomach), corbiculae) in Indian bees at the species level, and their relationship to the host plant specificity.

4.6.2 Behavioural Adaptations

The ability of foraging floral rewards by floral visitors is considered as a feature developed through adaptation. There are three progressive stages of adaptation seen in all flower-visiting insects on the basis of morphological and behavioural characteristics:

- Allotropous insects – These are the flower visitors that are least adapted, either morphologically or behaviourally to effect pollination in plants. They treat floral resources as one among a range of their preferred food sources. They do not have a regular movement pattern on the flowers or structures to carry pollen grains externally.

Examples: Flower-visiting beetles (Coleoptera); flies of Muscidae and Empididae families (Diptera); flower-visiting wasps of Vespoidea, Ichneumonoidea and Chalcidoidea superfamilies (Hymenoptera)

Fig. 4.5. Some potential non-bee pollinators; a) a Sarcophagid fly; b) a Syrphid fly; c) a Calliphorid fly; d) a Muscid fly; e) a Scoliid wasp; f) a Vespid wasp.

- Hemitropous insects – These flower-visiting insects are partially adapted for transporting pollen grains on their body. They have both partly developed structural features and behavioural adaptation to maintain floral constancy. They have a regular and skilful pattern of movement on the flowers adapted for pollen collection and transport.
  Examples: some Cetonid beetles (Coleoptera); flies of Syrphidae and Bombyliidae families (Diptera); wasps of Vespidae, Pompilidae and Sphecidae and some short-tounged bees (Hymenoptera); most flower-visiting butterflies and moths (Lepidoptera) (Figure 4.5)
- Eutropous insects – These are the insects adapted completely structurally and behaviourally for transporting pollen grains on their body. They possess well-specialized structures and habits to collect and transport pollen on their bodies and adapted well for carrying out all types (auto, geitono and xenogamous) of pollination. One such behavioural adaptation of these insects is floral constancy.
  Examples: Most flower-visiting bees (Hymenoptera) and hawk-moths (Sphingidae: Lepidoptera).
  With respect to bees, the three stages of adaptations in collecting the floral resources (both pollen and nectar) are as follows:
- Monotropic bees – Individuals of agiven monotropic bee species are behaviourally adapted to visit the flowers of only a single species of plant in a community of flowers
  Example: bees of *Andrena* (Andrenidae).
- Oligotropic bees –An oligotropic bee visits a few related species of plants during their respective visitation bouts and the members maintain a broad floral constancy in a floral community.
  Examples: bees of many families.
- Polytropic bees –A polytropic bee visits many unrelated species of plants during each of their visitation bouts. They lack the property of maintaining floral constancy. This is explained well in the following classification.
  Examples: bees of many families.

4.6.3 Oligolecty vs Polylecty

To make a finer level of classification based on the behaviour of only the pollen-collecting female bees Robertson (1925) has referred the former three categories as monolectic, oligolectic and polylectic bees. *Apis cerana* at species level is considered polylectic as the workers of a given colony handle flowers of unrelated plant species for the common benefit of the colony. But, at an individual level, a given worker bee is monolectic, because it rarely switches from a single floral source to another in its visitation bouts (Free 1963).

Oligolecty is a case where a bee at species and population level is a pollen (=host) specialist, and consistently and regularly collect pollen from a single or few related plant species in the presence of other pollen sources (Linsley and MacSwain 1958, Robertson 1925). Polylecty implicates that a pollen-foraging bee at species and population level is a generalist, therefore, forages pollen from flowers of even distantly related plants. Recent studies on the phylogenetic relationships based on the morphological and molecular data and host-plant evolution clearly suggests that oligolecty was the ancestral trait and polylecty was a derived condition (Litman et al. 2011, Melo 1999, Michez et al. 2008, Müller 1996, Patiny et al. 2008). Although this can strengthen our belief that all plant-pollinator interactions were coevolved, evidences suggest that extreme pollinator specialization by the plants and extreme floral specialization for resources by the pollinators was unusual and scarce (Waser et al. 1996). We have only very few plant species that have single or a very small range of pollinator species. Yucca, figs, some orchid species, and a guild of red-flowered plants confined to the Fynbos of South Africa are species showing extreme pollinator specialization today (Bogler et al. 1995, Johnson and Bond 1992,Nilsson 1992, Wiebes 1979).

Oligolecty is predominantly seen in Melittidae (Dasypodainae, Melittinae), Andrenidae (Andreninae, Panurginae), Colletidae (Scrapterinae), Halictidae (Rophitinae), Megachilidae (Fideliinae) and Apidae (Emphoroni, Eucerini); but completely absent in Xylocopinae (Cane and Sipes 2006, Danforth et al. 2013, Dötterl and Vereecken 2010, Wcislo and Cane 1996). Since most of the extant bees in Melittidae are highly oligolectic and the family stands very close to the root of the phylogenetic tree of bee families (Danforth et al. 2013), the argument that oligolecty was the ancestral state of bees is widely accepted.

4.7 Pollen Collection by Bees

As discussed before, bees (excluding some clepto-(brood) parasitic bees) are covered with profuse branched hairs almost all over the body. These bodily hairs are used to trap and groom pollen from anthers. Bees having facial hairs to trap pollen often insert their face into the anthers, and the entangled pollen grains on the face are groomed to the scopae or corbiculae using their appendages. Some bees use flight musculature and vibration of wings to release pollen from “poricidal” flowers, the mechanism of which is explained under “buzz pollination” in this chapter.

Bees are not necessarily the pollen collectors in all the flower types they visit. Since plants require the visitors essentially to distribute their pollen grains to the stigmas of the conspecific flowers, position their anthers strategically in the flower so that the pollen will be trapped on the body of visitors even when they are specialized for nectar collection. Bees occasionally spend time to remove the “passively-trapped” pollen from their body using their mid legs (Neff and Simpson 1997, Sinu and Shivanna 2007a, Thorp 2000). However, the pollen grains trapped on the mid line of the bee's thorax are often spared from the grooming process, and will be available for pollination. This sort of pollen collection on the bee's body is known as passive pollen collection. In *Amomum subulatum* flowers, the bumblebee species, *Bombus haemorrhoidalis* is a nectar collector, but the strategic placement of anther-stigma column above in the upper lip of this narrow flower, makes the bee trap large amount of pollen grains on the head and thorax when it accesses the flowers for nectar collection (Figure 4.2). A pollen-covered bumblebee during its subsequent visits to other flowers of the same species inserts some pollen grains into the stigma cup, which is also positioned in the upper anther-stigma column (Sinu and Shivanna 2007).

An actively pollen-collecting bee often digs, probes, bites, milks, strokes, drums or strikes the anthers to release the pollen grains using their mouth parts and appendages (Thorp 2000). Honey bees using their forelegs either drum the anther tips of flowers to release the pollen or scrape the pollen grains from the dehisced anthers. A Megachilid bee during the pollen collection grasps the staminal column with its forelegs and mandibles and sits and presses its mid venter on the anthers below to collect the pollen directly to the scopae on the bottom surface of the venter. In some brush-type flowers a pollen specialist bee usually walks over a plane of anthers and gets its body covered by pollen, which is then groomed to the scopae, a behaviour known as scrabbling. During mandibular scraping, a bee uses its mandible to scrape the pollen from the dehisced anthers, which is often encountered in flowers of Passifloraceae (Neff and Rozen 1995).

4.7.1 Buzz pollination

4.7.1.1 Structural adaptation of flowers

In majority of the flowering plants, the anthers open by splitting along the entire length longitudinally, which helps the floral visitors to access and collect the pollen easily. Against this, flowers of a number of unrelated plant species across the plant kingdom release pollen grains through a terminal pore formed at the tip of the anther (Figure 4.4 c and e). This type of anther is called as poricidal anthers. In dicot subclass, 19,000 species, 473 genera and 64 families have poricidal anthers, and in monocot subclass, 1,150 species, 71 genera and eight families have poricidal anthers (Buchmann 1983). Buzz-pollinated flowers are generally pendant so that the pollen readily fall on the pollinator during buzzing.

The buzz pollinating flowers have abundant pollen grains, which are non-sticky in nature, and have high pollen-ovule ratio. Since the pore is terminally-located in the anthers, access to the pollen is possible only through the tip of the anthers. Flowers of a wide range of morphology are adapted for buzz-pollination; however, a typical buzz-pollinated flower is a Solanum-type flower (Figure 4.4 c and e) (Solanoid flower) (Luca et al. 2013). A Solanoid flower has centrally-located short-filamented anther. Some species of *Solanum* produce heterantherous flowers, i.e., each flower has two types of anthers: feeding anther and pollinating anther (Vallejo-Marin et al. 2010). The feeding anthers are located centrally as a group of several anthers, which are used by the buzz-pollinators to grasp and collect pollen grains. The pollinating anther is a single large anther deflected from the feeder group, and opens opposite to the style (Vallejo-Marin et al. 2009), and contributes to the fertilization.

Although the *Solanum* is the single largest genus adapted well for buzz-pollination among poricidal type of flowers (Bowers 1975, Buchmann 1978, Buchmann and Hurley 1978, King et al. 1996, Michener 1962) it is also reported in the legume genus *Cassia* (Buchmann 1974, Gottsberger and Silberbauer-Gottsberger 1988, Thorp and Estes 1975), and some members of Melastomataceae (Larson and Barrett 1999, Renner 1989). Buzz-pollination is also reported in some plant species having non-poricidal flowers (in which anthers split longitudinally to release the pollen), such as some species of Rosaceae, Scrophularaceae, Asteraceae, Papavaraceae, Myrtaceae and Cucurbitaceae.

4.7.1.2 Diversity and adaptations of bees

A large number of insects are reported as the visitors of a typical buzz-pollinated flower; this includes beetles, flies and bees. They collect pollen through the apical pore of the anthers either by chewing or milking the conical anthers. This abnormal pollen-releasing behaviour of the visitors sometimes causes widespread damage to the anthers (Renner 1983). An important, legitimate pollinator group of buzz-pollinated flowers is the sonicating bees (King and Buchmann 2003). According to Buchmann (1985), all the bees (excluding the leafcutter bees (Megachilidae)) that forage on poricidal flowers have a stereotyped pollen-harvesting mechanism, which is known as buzzing or sonication. Upon alighting on a typical buzz-pollinated flower, the bee curls its ventral side of the body around the anther cone (feeding anthers in Solanoid flower), or grasp clusters of stamens of long filaments (in non-poricidal flowers), shiver the large flight muscles housed in the thorax that trigger intense vibration of the wings, which produces a characteristic buzzing sound of high pitch (generally refers to as sonication) that transmits strong vibration to the floral androecium (Buchmann and Hurley 1978; King et al. 1996). In response to this vibration, the pollen is expelled through the miniature pore at the tip of the anther within 0.1-1.0 Seconds that strikes and sticks to the venter of the bee, which is later groomed by the bee intermittently and packed into the scopae (Cane and Buchmann 1989, Macior 1964).

Only active pollen foraging female bees perform sonication. Globally, bee species belonging to the families, Colletidae, Melittidae, Andrenidae, Halictidae, and Apidae are structurally and behaviorally adapted to release pollen from a typical poricidal flower adapted for buzz-pollination. The sonication behaviour in the members of the family

Apidae is restricted only to non-Apis species (e.g.,*Bombus, Euglossa, and Melipona*). Bees belonging to the genera, *Bombus, Xylocopa* and *Xenoglossa* are the typical buzz-pollinated bees (see Buchmann 1985 for the review). In India, I have frequently seen the bees of *Xylocopa, Anthophora, Amegilla, Bombus* and *Ceratina* performing buzz-pollination in both Solanacean and non-solanaceous flowers (pers observ). Body size, duration and amplitude of sonication affect the number of pollen grains released in a given buzzing attempt of a bee (Luca et al. 2013). Buzzing is exclusively performed by the members of bees; but, occasionally it is also reported in hover flies too (Buchmann 1983).

4.8 Floral Constancy

Behavioural adaptations for pollen and nectar foraging involve learning, memory and modification of behaviour to respond to floral signals, floral diversity, floral competition, floral shortage, seasonal variations and pollinators' breeding requirement. Floral signals such as colour and scent, pollen odour, and nectar quality and quantity are the primary factors that elicit the attention of pollinators, particularly the bees. One important adaptive behaviour with respect to bees is the floral constancy or flower constancy.

In simple words, floral constancy is the behaviour of a foraging bee that sequentially visits one flower type bypassing all other equally or more rewarding flower types (Chittka et al. 1999, Free 1963, Free 1970, Grant 1950, Heinrich 1976, 1979, Ishii and Masuda 2014, Menzel et al. 1993, Waser 1986). This is well demonstrated in nectar-collecting bees. As pointed out earlier, floral constancy in bees is spontaneously subjective and adaptive, and is maintained at the individual level, not at the species level *per se* (Hill 1997). For example, members of a honey bee colony foraging at different sites will have their own specialized (constant) flowers. According to Grant (1950), this behaviour in honey bees was first described by Aristotle in 350 BCE and was later drawn the attention of Darwin in 1876. Floral constancy is different from the other classification based on the pollen foraging behaviour of bees, i.e., oligolectic and polylectic bees. First, floral constancy is a quality possessed through a learning process; so it is maintained at individual level, not at the species level. Therefore, floral constancy is a behavioural state that is plastic in nature, and not an evolved state. Second, the terminology, floral constancy is used for both the pollen- and nectar- foraging bees; however, the oligolectic-polylectic classification is based on pollen specialization.

4.8.1 Implications of Floral Constancy

Floral constancy of bees is considered a favourable state from the perspective of plant reproductive fitness as long as their visits result in pollen transfer to the stigma (Levin 1978). Sinu and Shivanna (2007a), in their study on *Amomum subulatum* (Zingiberaceae) found that *Apis cerana*, despite, sequentially visiting the flowers for pollen collection, do not transfer pollen to the stigma. Thus the floral constancy maintained by the individual honey bee is not favourable for the plant, but rather becomes deleterious for the plant as the bees behave like pollen robbers. The scopae or corbiculae, specifically, and the body in general of a floral-constant bee are likely to be covered only by the conspecific pollen (Waser 1978).

One consequence of floral inconsistency by the bees is that it increases pollen wastage and stigma clogging by the foreign pollen (Feinsinger 1987). The other consequence of floral inconsistency of the bees is the possibility of hybrid production, which are sometimes less viable than the parental species (Free 1966). Thus a floral constant bee is facilitating the plant reproductive fitness in the following ways:

- It transports the conspecific pollen to stigmas of multiple flowers of the same species and therefore, increases its pollination efficiency (Waser 1986).
- It prevents pollen wastage and avoids possible hybridization between related species (Free 1966, Grant 1952, Goulson and Jerrim 1997).
- It helps to maintain the integrity of plant species (Levin 1978).
- It is one of the contributory factors in sympatric speciation (Free 1963, Grant 1949, 1992, 1993, 1994, Manning 1956).

Studies have shown that managed and wild honey bees (*Apis* spp: Apidae) have a remarkable capacity to maintain intraspecific floral constancy (De Jager et al. 2011, Grant 1950, Hills 1997). According to Grant (1950) *Apis* spp. at individual level maintains 83.3%-100% floral constancy in their visits. Floral constancy is relatively low in non-*Apis* Apidae members; for instance bumblebees (*Bombus* spp) maintains only between 51 per cent and 69.7 per cent constancy during an individual bee's foraging trip. Floral constancy is low in solitary bees. In Anthophorini (Apidae), only 20 per cent of the bees have pure pollen load in the pollen pellet retrieved from the scopae of the bees. In some Andrenid bees 43%-64% of the bees had homogeneous pollen load on their bodies. Among solitary bees the proportion of floral constancy is relatively high in some Halictidae and Megachilidae – 75-84% (Halictidae) and 54.5-75% (Megachilidae) (Grant 1950).

4.8.2 Factors Affecting Floral Constancy

How the floral constancy is being maintained by the bees has received much attention from bee physiologists and behavioural specialists. Food retrieval system in bees is compared to predator-prey interaction, which is always density-dependent unless the predator is a specialist, which is rare. Therefore, a generalist predator prefers a most common prey over a rarer prey. In plant-pollinator interaction too, a foraging bee gives preference to most common flower, and that the bee is not a specialist in food demand. Hence, a foraging bee will always acquire a search image of a site during foraging (Chittka et al. 1999, Ishii and Masuda 2014).

Bees learn several floral cues and keep them in the memory when they are in the foraging environment. The major floral cues that elicit the sensory response of the pollinators are visual, chemical and tactile. The first cue that elicits the sensory response of a naïve bee in a foraging ground is always visual, which is primarily the colour (Giurfa et al. 1995) and the characteristics of the flowers (shape and form) and its surroundings(Dukas 2008). The innate colour preferences of bees range between the wavelengths 400-420 nm and 510-520 nm. Insect vision is trichromatic just like human beings, but their primary colours are a) a yellow group, which includes vermillion, orange, yellow and yellow-green; b) blue-green group; which includes blue, purple and violet; and c) ultraviolet.The bees also have a fine sense of form. Rather than colour, the bees quickly recognize and have a long-term memory of geometrical shapes in nature (Dukas 2008). At a finer scale, bees recognize and remember the lines on the flower petals which are often referred to as nectar guides, than the colour shades. Since they have a very good memory of forms and architecture of the flowers, a flower-constant bee works more efficiently than an inconstant one, because the former bee has already learnt how to handle the flowers that includes the use of landing platform and proboscis insertion in the flower (Arenas et al. 2007, Free 1969,Takeda 1961). The flower inconsistent bee, since it visits flowers of different shapes, colour and size has to learn each type of flower and thus finds foraging of new flower, a bit cumbersome, and therefore, becomes relatively less efficient in foraging (Gumbert 2000, Raine and Chittka 2007). Subsequently, the chemical (olfactory) cues play a major role in eliciting further response of the bees to floral types.

Bees use both colour and quality of the rewards to either specialize or generalize their host/s. An experienced bee's memory is therefore tuned to such floral types (Chittka et al. 1999). Since, a nectar-collecting bee tastes and metabolizes the nectar in the field individually, it assesses the quality of the reward (primarily the sugar level) quickly and learns and updates its memory with the associated visual cues of the flower and its surroundings for its further perception (Dukas 2008, Scheiner et al. 1999, 2004). A floral constant nectar-collecting bee can quickly recall the quality of the nectar. These memories further improve the foraging efficiency of the bees (Arenas et al. 2007, Chaffiol et al. 2005). Whilst, this is an underlying mechanism that works primarily at an individual level in a nectar-collecting bee, it was largely unclear whether a pollen-forager can also maintain floral constancy, since a pollen forager does not ingest the pollen and taste it. However, a bee can sense the pollen using its antennae (Scheiner et al. 2004). Recent researches find that a pollen forager also learns the visual cues associated with the rewarding flower, the odour of pollen and the odour of the flower to maintain floral constancy (Arenas and Farina 2012, Cook et al. 2005, Grüter et al. 2008, Nicholls and de Ibarra 2014). Pollen-collecting bee can stimulate the response and orient the other members of a colony towards a specific pollen source after the bee flies back into the nest with the pollen pellet (Arenas and Farina 2012, Free 1969). The odour cues associated to the pollen pellet inside the colony reactivate the memory of the bee that triggers it for further foraging trip (Arenas and Farina 2012, Reinhard et al. 2004).

Since bees often encounter over a dozen of rewarding and unrewarding flowers sometimes of the same or contrasting floral signals (colour, shape etc.), it is necessary to compare the visual image with the stored memory every time to differentiate the rewarding and non-rewarding flowers. The ability of bees to recall a specific memory in a foraging ground is important to maintain floral constancy. Memory retrieval is quick when a bee is doing a single task, repeatedly (Chittka et al. 1999). This specific task is therefore stored in the short-term memory (Menzel 2001). Real (1991) believes that bees do not have long-term memory. However, Greggers and Menzel (1993) and Menzel (1999) believe that bees have an excellent capacity to store the memory for a longer term. Chittka et al. (1999) suggest that long-term memory is possible in bees, and the short-term memory of the bees is likely to be interfered by the novel floral stimuli and signals that the bee is acquiring in a learning process. Wherever a bee appears to be floral constant, it is believed that the short-term memory is programmed with cues to locate the flower using surrounding physical and floral signals (colour, size, structure, shape, pattern, odour and height from the ground) and retrieving some discrete behaviour, such as proboscis extension behaviour (Ishii 2005, Ishii 2006, Menzel et al. 1993). Sonicating bees practicing buzzing in a specific poricidal flower continue this buzzing behaviour in a non-poricidal flower for a while after it switches the flower type. However, this behaviour stops after it encounters a few non-poricidal flowers (Buchmann 1983). It is likely that a bee that had foraged only a monospecific environment can be more floral constant even if choices are offered (Wilson and Stine 1996). Since a non-constant bee visits multiple floral types in a visitation bout, it is more efficient to retrieve the necessary information from the remote long term memory when compared to a floral-constant bee.

In summary, floral constancy in bees is the culmination of its learned fidelity to the flowers, and is an adaptive state in their foraging behaviour to certain extent. The efficiency of a bee to maintain temporal specialization in one or multiple flower types is determined by how much the bee can store the information on the flower handling and manipulation in its short term memory, and how much a bee can pay attention to a focused flower type in its visual search image (Chittka et al. 1999, Goulson 2000, Guilford and Dawkins 1987, Ishii and Masuda 2014). Although it is believed that floral constancy is a behaviour maintained only by a few foraging bee species, some studies also report that floral constancy is maintained by hoverflies (Goulson and Wright 1998), butterflies (Lewis 1989, Goulson and Cory 1993), and pollen-feeding beetles (De Los Mozos Pascual and Domingo 1991).

4.9 Concluding Remarks

Pollination is critical for a flower to terminate its sexual isolation, which is imposed due to floral morphological reasons or due to the behavioral reasons of the visitors. Over 90 per cent of the flowering plants require an animal vector to bring out pollination indicating the exorbitant relevance of pollinators, their diversity, foraging behaviour and fidelity in nature. Bees constitute the major pollinator fauna of tropical and temperate plants including the crop plants. Although the managed honey bees belong to the *Apis* spp were considered our long-serving friends at least for many of our crop plants, recent researches suggest that wild bees belong to the non-*Apis* spp are the "trustworthy" group of pollinators of crop and wild plants (Breeze et al. 2011, Garibaldi et al. 2013). Recent researches that studied the efficiency of pollinators through monitoring the pollen distribution and fruit set brought enough evidences that pollination by wild bee results in better fruit set, which is assessed from the quality of the fruits (mass, seed quality and quantity) (Garibaldi et al. 2015, Garibaldi et al. 2013, Kuriakose et al. 2009, Sinu and Shivanna 2007, Sinu et al. 2012).

Floral fidelity and floral constancy are the behavioural modifications of the bees that may limit pollination efficiency. Reports suggest that honey bees qualify on these two aspects. But, it is now understood that it is not only the behavioural plasticity relevant in effecting pollination, but also the diversity of the functional morphology of the bees, such as the size of the bees, length of the mouth parts, buzzing ability, pollen-transporting mode (dry or sticky) and so on. Indeed, the functional morphology of the flowers also limits the efficiency of the visitors in bringing out the pollination. India, despite having a great diversity of flowering plants and native wild bees, our knowledge on the complexity of their interaction is meagre. The lack of the qualified bee taxonomists to collaborate in studies on pollination ecology may limit the scope and outcome of studies of this kind. Bee ecologist can contribute relevant information on the foraging plants of native wild bees, nesting ecology of the wild bees;the two critical factors affecting the population dynamics of pollinators in agroecosystems and wild habitats.

Acknowledgments

SV, UPB, GA thank Kerala State Council for Science Technology and Environment for PhD fellowship. PM thanks CSIR for the JRF award for PhD.

References

Ackerly DD 2003. Community assembly, niche conservatism, and adaptive evolution in changing environments. Int J Plant Sci 164: 165–184.

Aizen MA, Garibaldi LA, Cunningham SA, Klein AM 2008. Long term global trends in crop yield and production reveal no current pollination shortage but increasing pollinator dependency. Curr Biol 18: 1572–1575.

Arenas A, Fernández VM, Farina WM 2007. Floral odor learning within the hive affects honey bees' foraging decisions. Naturwissenschaften 94: 218–222.

Arenas A, Farina WM 2012. Learned olfactory cues affect pollen-foraging preferences in honey bees, *Apis mellifera*. Animal Behav 83: 1023–1033.

Armbruster WS et al. 1994. Floral character displacement generates assemblage structure of Western Australian trigger plants (Stylidium). Ecology 75: 315–329.

Ascher JS, Pickering J 2012. Discover life: bee species guide and world checklist (Hymenoptera:Apoidea:Anthophila) http://www.discoverlife.org/mp/20q?guide=Apoidea_species&flags=HAS

Ashman TL et al. 2004. Pollen limitation of plant reproduction: ecological and evolutionary causes and consequences. Ecology 85*:* 2408–2421.

Bawa KS 1990. Plant-pollinator interactions in tropical rain forests. Annu Rev Ecol Evol S 21: 399–422.

Bogler DJ, Neff JL, Simpson BB 1995. Multiple origins of the yucca-yucca moth association. Proc Natl Acad Sci USA 92: 6864–6867.

Bowers KAW 1975. The pollination ecology of *Solanum rostratum* (Solanaceae). Amer J Bot 62: 633–638.

Brady SG, Sipes S, Pearson A, Danforth BN 2006. Recent and simultaneous origins of eusociality in halictid bees. Proc R Soc B 273:1643–1649.

Breeze TD, Bailey AP, Balcombe KG, Potts SG 2011. Pollination services in the UK: How important are honey bees? Agr Ecosyst Env *142:* 137–143.

Bronstein JL 2001. The exploitation of mutualisms. Ecol Lett 4: 277–287

Brothers DJ 1999. Phylogeny and evolution of wasps, ants and bees (Hymenoptera, Chrysidoidea, Vespoidea and Apoidea). Zool Scripta 28: 233–249.

Buchmann SL 1974. Buzz pollination of *Cassia quiedondilla* (Leguminosae) by bees of the genera Centris and Mellipona. Bull So Calif Acad Sci 73: 171–173.

Buchmann SL, Hurley JP 1978. Biophysical model for buzz pollination in Angiosperms. J Theor Biol 72:639–657.

Buchmann SL 1983. Buzz pollination in angiosperms. In: Jones CE, Little RJ (Eds) Handbook of Experimental Pollination Biology. Scientific and Academic Editions, NY, pp 73–113.

Buchmann SL 1986. Vibratile pollination in Solanum and Lycopersicon: A look at pollen chemistry. In W G. D'Arcy, editor. Solanaceae: biology and systematics. Columbia University Press, New York, USA, pp 237-252.

Burkart A, Schlindwein C, Lunau K 2013. Assessment of pollen reward and pollen availability in Solanum stramoniifolium and Solanum paniculatum for buzz-pollinating carpenter bees. Plant Biol 16: 503–507.

Cameron SA 1993. Multiple origins of advanced eusociality in bees inferred from mitochondrial DNA sequences. Proc Natl Acad Sci USA 90: 8687–8691.

Cane JH, Sipes S 2006. Characterizing floral specialization by bees: analytical methods and a revised lexicon for oligolecty. In NM Waser, J Ollerton*(Eds) Plant-Pollinator Interactions: From Specialization to Generalization*. Chicago Univ Press, Chicago, London, pp 99–122.

Cane JH, Buchmann SL 1989. Novel pollen-harvesting behavior by the bee *Protandrena mexicanorum* (Hymenoptera, Andrenidae). J Insect Behav 2:431–436.

Cardinal S, Danforth BN 2011. The antiquity and evolutionary history of social behavior in bees. PLoS ONE 6: e21086

Cardinal S, Straka J, Danforth BN 2010. Comprehensive phylogeny of apid bees reveals the evolutionary origins and antiquity of cleptoparasitism. Proc Natl Acad Sci USA 107: 16207–16211.

Cauich O, Javier J, Quezada-Euan JJG, Macias-Macias JO, Reyes-Oregel V, Medina-Peralta S, Parra-Tabla V 2004. Behavior and Pollination Efficiency of *Nannotrigona perilampoides* (Hymenoptera: Meliponini) on Greenhouse Tomatoes (*Lycopersicon esculentum*) in Subtropical Mexico. J Econ Entomol 97: 475–481.

Chaffiol A, Laloi D, Pham-Delegue MH 2005. Prior classical olfactory conditioning improves odour-cued flight orientation of honey bees in a wind tunnel. J Exp Biol 208: 3731–3737.

Chittka L, Thomson JD, Waser NM 1999. Flower constancy, insect psychology, and plant evolution. Naturwissenschaften 86: 361–377.

Cook SM, Sandoz JC, Martin AP, Murray DA, Poppy GM, Williams IH 2005. Could learning of pollen odours by honey bees (*Apis mellifera*) play a role in their foraging behaviour? Physiol Entomol 30: 164–174.

Danforth BN, Sipes S, Fang J, Brady SG 2006. The history of early bee diversification based on five genes plus morphology. Proc Natl Acad Sci USA 103:15118–15123.

Danforth BN 2002. Evolution of sociality in a primitively eusocial lineage of bees. Proc Natl Acad Sci USA 99: 286–290.

Danforth BN, Conway L, Ji SQ 2003. Phylogeny of eusocial *Lasioglossum* reveals multiple losses of eusociality within a primitively eusocial clade of bees (Hymenoptera: Halictidae). Syst Biol 52: 23–36.

Danforth BN, Cardinal S, Praz C, Almeida EAB, Michez D 2013. The impact of molecular data on our understanding of bee phylogeny and evolution. Annu Rev Entomol 58: 57–78.

De Jager ML, Dreyer LL, Ellis AG 2011. Do pollinators influence the assembly of flower colours within plant communities? Oecologia 166: 543–553.

De Los Mozos Pascual M, Domingo LM 1991. Flower constancy in *Heliotaurus rufi collis* (Fabricius 1781, Coleoptera: Alleculidae). Elytron 5: 9–12.

Dötterl S, Vereecken NJ 2010. The chemical ecology and evolution of bee-flower interactions: a review and perspectives. Can J Zool 88: 668–697.

Dukas R 2008. Evolutionary biology of insect learning. Ann Rev Entomol 53: 145–160.

Feinsinger P, Beach JH, Linhart YB, Busby WH, Murray KG 1987. Disturbance, pollinator predictability, and pollination success among Costa Rican cloud forest plants. Ecology 68: 1294–1305.

Free JB 1963. Flower constancy of honey bees. J Anim Ecol 32: 119–131.

Free JB 1966. The foraging behavior of bees and its effect on the isolation and speciation of plants. In: Hawkes JG (Ed) Reproductive biology and taxonomy of vascular plants. Oxford, pp 76-92.

Free JB 1969. Influence of the odour of a honey bee colony's food stores on the behaviour of its foragers. Nature 222: 778.

Free JB 1970. The flower constancy of bumblebees. J Anim Ecol 39: 395–405.

Garibaldi LA et al. 2015. Trait matching of flower visitors and crops predicts fruit set better than trait diversity. J Appl Ecol 52: 1436-1444. doi: 10.1111/1365-2664.12530

Garibaldi LA et al. 2013. Wild pollinators enhance fruit set of crops regardless of honey bee abundance. Science 339: 1608–1611.

GénisselA, Aupinel P, Bressac C, Tasei J-N, Chevrier C 2002. Influence of pollen origin on performance of *Bombus terrestris* micro-colonies. Ent Exp et Appl 104: 329–336.

Ghazoul J 2006. Floral diversity and the facilitation of pollination. J Ecol 94: 295–304.

Gibbs J, Albert J, Packer L 2012. Dual origins of social parasitism in North American *Dialictus* (Hymenoptera: Halictidae) confirmed using a phylogenetic approach. Cladistics28:195–207.

Giurfa M, Núñez J, Chittka L, Menzel R 1995. Colour preferences of flower-naive honey bees. J Comp Physiol A 177: 247–259.

Gottsberger G, Silberbauer-Gottsberger I 1988. Evolution of flower structures and pollination in neotropical Cassiinae (Caesalpiniaceae) species. Phyton 28: 293–320.

Goulson D 2000. Are insects flower constant because they use search images to find flowers? Oikos 88: 547–552.

Goulson D, Cory JS 1993. Flower constancy and learning in the foraging behaviour of the

green-veined white butterfly, *Pieris napi*. Ecol Entomol 18: 315–320.

Goulson D, Jerrim K 1997. Maintenance of the species boundary between *Silene dioica* and *S. latifolia* (red and white campion). Oikos 78: 254–266.

Goulson D, Wright NP 1998. Flower constancy in the hoverflies *Episyrphus balteatus* (Degeer) and *Syrphus ribesii* (L.) (Syrphidae). Behav Ecol Sociobiol 9: 213–219.

Grant V 1949. Pollination systems as isolating mechanisms in angiosperms. Evolution3: 82–97.

Grant V 1950. The flower constancy of bees. Bot Rev 3: 82–97.

Grant V 1952. Isolation and hybridization between *Aquilegia formosa* and *A. pubescens*. Aliso 2: 341–360.

Grant V 1992. Floral isolation between ornithophilous and sphingophilous species of *Ipomopsis* and *Aquilegia*. Proc Natl Acad Sci USA 89: 11828–11831.

Grant V 1993. Origins of floral isolation between ornithophilous and sphingophilous plant species. Proc Natl Acad Sci USA 90: 7729–7733.

Grant V 1994. Modes and origins of mechanical and ethological isolation in angiosperms. Proc Natl Acad Sci USA 91: 3–10.

Greenberg L 1982. Year-round culturing and productivity of a sweat bee, *Lasioglossum zephyrum* (Hymenoptera: Halictidae). J Kan Entomol Soc 55: 13–22.

Greggers U, Menzel R 1993. Memory dynamics and foraging strategies of honey bees. Behav Ecol Sociobiol 3: 17–29.

Grindeland JM, Sletvold N, Ims RA 2005. Effects of floral display size and plant density on pollinator visitation rate in a natural population of *Digitalis purpurea*. Funct Ecol 19: 383–390.

Grüter C, Arenas A, Farina WM 2008. Does pollen function as a reward for honey bees in associative learning? Ins Sociaux 55: 425–427.

Guilford T, Dawkins MS 1987. Search images not proven: a reappraisal of recent evidence. Anim Behav 35: 1838–1845.

Gumbert A 2000. Color choices by bumble bees (*Bombus terrestris*):innate preferences and generalization after learning. Behav Ecol Sociobiol 48: 36–43.

Harder LD, Barrett SCH 1996. Pollen dispersal and mating patterns in animal-pollinated plants. In: Lloyd DG and Barrett SCH (Eds) Floral Biology. Chapman & Hall, New York, NY, pp 140–190.

Harder LD, Routley MB 2006. Pollen and ovule fates and reproductive performance by flowering plants. In: Harder LD and SCH Barrett (Eds) The ecology and evolution of flowers. Oxford University Press, pp 61– 80.

Harder LD, Thomson JD 1989. Evolutionary options for maximizing pollen dispersal of animal-pollinated plants. Am Natural 133: 323–344.

Hargreaves AL, Harder LD, Johnson SD 2009. Consumptive emasculation: the ecological and evolutionary consequences of pollen theft. Biol Rev 84: 259–276.

Heinrich B 1976. The foraging specializations of individual bumblebees. Ecol Monogr 46:105–128.

Heinrich B 1979. "Majoring" and "minoring" by foraging bumblebees, *Bombus vagans*: an experimental analysis. Ecology 60: 245–255.

Hill PSM et al. 1997. Spontaneous flower constancy and learning in honey bees as a function of color. Anim Behav 54: 615–627.

Ishii HS 2005. Analysis of bumblebee visitation sequence within single bouts: implication of overstrike effect on short-term memory. Behav Ecol Sociobiol 57:599–610.

Ishii HS 2006. Floral display size influences subsequent plant choice by bumble bees. Funct Ecol 20: 233–238.

Ishii HS, Masuda H 2014. Effect of flower visual angle on flower constancy: a test of the search image hypothesis. Behav Ecol 25: 933–944.

James R, Pitts-Singer TL 2008. Bee Pollination in Agricultural Ecosystems, Oxford University Press, New York

Johnson SD, Bond WJ 1992. Habitat dependent pollination success in a cape orchid. Oecologia 91: 455–456.

King MJ, Buchmann SL, Spangler HG 1996. Activity of asynchronous flight muscle from two bee families during sonication (buzzing). J Exp Biol 199: 2317–2321.

King MJ, Buchmann SL 2003. Floral sonication by bees: mesosomal vibration by Bombus and Xylocopa, but not Apis (Hymenoptera: Apidae), eject pollen from poricidal anthers. J Kan Entomol Soc 76: 295–305.

Klein A-M et al. 2007. Importance of pollinators in changing landscapes for world crops. Proc Roy Soc B 274: 303–313.

Kuriakose G, Sinu PA, Shivanna KR 2009. Domestication of cardamom (*Elettaria cardamomum*) in Western Ghats, India: divergence in productive traits and a shift in major pollinators. Ann Bot 103: 727–733.

Larson BM, Barrett SCH 1999. The pollination ecology of buzz-pollinated *Rhexia virginica* (Melastomataceae). Am J Bot 86: 502–511.

Levin DA 1972. The adaptedness of corolla-colour and outline in experimental and natural populations of *Phlox drummondii*. Am Nat 106:57–70.

Levin DA 1978. Pollinator behaviour and the breeding structure of plant populations. In: Richards AJ (Ed) The Pollination of Flowers by Insects, Academic Press, London, pp. 133–150.

Lewis AC 1989. Flower visit consistency in *Pieris rapae*, the cabbage butterfly. J Anim Ecol 58: 1–13.

Linsley EG, MacSwain JW 1958. The significance of floral constancy among bees of the genus Diadasia. Evolution 12: 219–223.

Litman JR, Danforth BN, Eardley CD, Praz CJ 2011. Why do leafcutter bees cut leaves? New insights into the early evolution of bees. Proc R Soc B 278: 3593–3600.

Losey JE, Vaughan M 2006. The economic value of ecological services by insects. Bioscience 56: 311–323.

Luca PAD, Bussiére LF, Souto-Vilaros D, Goulson D, Mason AC, Vallejo-Marín M 2013. Variability in bumblebee pollination buzzes affects the quality of pollen released from flowers. Oecologia 172: 805–816.

Macior LW 1974. Behavioral aspects of coadaptations between flowers and insect pollinators. Ann Mo Bot Gard 61:760–769.

Makino TT, Ohashi K, Sakai S 2007. How do floral display size and the density of surrounding flowers influence the likelihood of bumble bee revisitation to a plant? Funct Ecol 21: 87–95.

Manning A 1956. Some evolutionary aspects of flower constancy of bees. Proc R Phys Soc 25: 67–71.

Martins AC, Melo GAR, Renner SS 2014.The corbiculate bees arose from New World oil-collecting bees: Implications for the origin of pollen baskets. Mol Phyl Evol 80: 88–94.

Melo GAR 1999. Phylogenetic relationships and classification of the major lineages of Apoidea (Hymenoptera), with emphasis on the crabronid wasps. Univ Kans Mus Nat Hist Sci Pap 14:1–55.

Menzel R 2001. Behavioral and neural mechanisms of learning and memory as determinants of flower constancy. In: Chittka L and Thomson JD (Eds) Cognitive ecology of pollination. Cambridge (UK): Cambridge University Press. pp 106–126.

Menzel R, Greggers U, Hammer M 1993. Functional organization of appetitive learning and memory in a generalist pollinator, the honey bee. In: Papaj D, Lewis AC (Eds.) Insect learning: ecological and evolutionary perspectives. Chapman and Hall, New York, pp 79–125.

Michez D, Patiny S, Rasmont P, Timmermann K, Vereecken NJ 2008. Phylogeny and host-plant evolution in Melittidae s.l. (Hymenoptera: Apoidea). Apidologie 39:146–162.

Michener CD, Winston ML, Jander R 1978. Pollen manipulation and related activities and structures in bees of the family Apidae. Univ Kansas Sci Bull 51: 575–601.

Michener CD 2007. The Bees of the World. Baltimore, MD: Johns Hopkins Univ Press. 953 pp. 2nd Ed.

Michener CD 1962. An interesting method of pollen collecting by bees from flowers with tubular anthers. Rev Biol Trop 10: 167–175.

Mitchell RJ, Irwin RE, Flanagan RJ, Karron JD 2009. Ecology and evolution of plant pollinator interactions. Ann Bot 103: 1355–1363.

Moeller DA 2004. Facilitative interactions among plants via shared pollinators. Ecology 85: 3289–3301.

MüllerA 1996. Host-plant specialization in western Palearctic Anthidiine bees (Hymenoptera: Apoidea: Megachilidae). Ecol Monogr 66: 235–257.

Neff JL, Simpson BB 1997. Nesting and foraging behavior of Andrena (Callandrena) rudbeckiae Robertson (Hymenoptera: Apoidea: Andrenidae) in Texas. J Kan Entomol Soc 70:100–113.

Neff JL, Rozen JGJ 1995. Foraging and nesting biology of the bee *Anthemurgus passiflorae* (Hymenoptera: Apoidea), descriptions of its immature stages, and obser- vations on its floral host (Passifloraceae). American Museum Novitates 3138: 19 pp.

Nicholls E, de Ibarra NH 2014. Bees associate colour cues with differences in pollen rewards. J Exp Biol 217: 2783–2788.

Nilsson LA 1992. Orchid pollination biology. Trends Ecol Evol 7: 255–259.

Ohara M, Higashi S 1994. Effects of inflorescence size on visits from pollinators and seed set of *Corydalis ambigua* (Papaveraceae). Oecologia 98: 25–30.

Ohl M, Bleidorn C 2006. The phylogenetic position of the enigmatic wasp family Heterogynaidae based on molecular data, with description of a new, nocturnal species (Hymenoptera: Apoidea). Syst Entomol 31: 321–337.

Ollerton J, Winfree R, Tarrant S 2011. How many flowering plants are pollinated by animals? Oikos 120: 321–326.

Ollerton J et al. 2012. Overplaying the role of honey bees as pollinators: a comment on Aebi and Neumann (2011). Trends Ecol Evol 27: 141–142.

Patiny S, Michez D, Danforth BN 2008. Phylogenetic relationships and host-plant evolution within the basal clade of Halictidae (Hymenoptera, Apoidea). Cladistics 24: 255–269.

Pech-May FG, Medina-Medina L, de J. May-ItzáW, Paxton RJ, Quezada-Euán JJG 2012. Colony pollen reserves affect body size, sperm production and sexual development in males of the stingless bee *Melipona beecheii*. Insect Soc 59: 417–424.

Raine NE, Chittka L 2007. The adaptive significance of sensory bias in a foraging context: floral colour preferences in the bumblebee *Bombus terrestris*. PLoS One. 2: e556.

Real LA 1991. Animal choice behavior and the evolution of cognitive architecture. Science 253: 980–986.

Reinhard J, Srinivasan MV, Guez D, Zhang SW 2004. Floral scents induce recall of navigational and visual memories in honey bees. J Exp Biol 207: 4371–4381.

Renner SS 1989. A survey of reproductive biology in Neotropical Melastomataceae and Memecylaceae. Ann Missouri Bot Gard 76: 496–518.

Roberts RB, Vallespir SR 1978. Specialization of Hairs Bearing Pollen and Oil on the Legs of Bees (Apoidea: Hymenoptera). Ann Entomol Soc Am 71: 619–627.

Robertson AW, Macnair MR 1995. The effects of floral display size on pollinator service to individual flowers of *Myosotis* and *Mimulus*. Oikos72: 106–114.

Robertson C 1925. Heterotropic bees. Ecology 6: 412–436.

Roubik DW 1989. Ecology and natural history of tropical bees. Cambridge University Press, Cambridge, Massachusetts, USA

Roulston TH, Gane JH 2000. Pollen nutritional content and digestibility for animals. Plant Syst Evol 222: 187–209.

Roulston TH, Cane JH, Buchmann SL 2000. What Governs Protein Content of Pollen: Pollinator Preferences, Pollen-Pistil Interactions, or Phylogeny? Ecol Monogr 40: 617–643.

Scheiner R, Erber J, Page RE 1999. Tactile learning and the individual evaluation of the reward in honey bees (Apis mellifera L.). J Comp Physiol A 185: 1–10.

Scheiner R, Page RE, Erber J 2004. Sucrose responsiveness and behavioral plasticity in honey bees (*Apis mellifera*). Apidologie 35: 133–142.

Schmidt JO, Thoenes SC, Levin MD 1987. Survival of honey bees, *Apis mellifera* (Hymenoptera: Apidae), fed various pollen sources. Ann Entomol Soc Am 80:176–183.

Schwarz MP, Richards MH, Danforth BN 2007. Changing paradigms in insect social evolution: insights from halictine and allodapine bees. Annu Rev Entomol 52:127–150.

Schwarz MP, Tierney SM, Rehan SM, Chenoweth LB, Cooper SJB 2011. The evolution of eusociality in allodapine bees: Workers began by waiting. Biol Lett 7: 277–280.

Shivanna KR 2009. Pollination biology, breeding system and reproductive success of *Adhatoda vasica*, an important medicinal plant. Curr Sci 96: 408–412.

Sinu PA, Shivanna KR 2007a. Pollination biology of large cardamom (*Amomum subulatum*). Curr Sci 93: 548–552.

Sinu PA, Shivanna KR 2007b. Pollination ecology of cardamom (*Elettaria cardamomum*) in the Western Ghats, India. J Trop Ecol 23: 493–496.

Sinu PA, Kuriakose G, Shivanna KR 2011. Is the bumblebee (*Bombus haemorrhoidalis*) the only pollinator of large cardamom in central Himalayas, India? Apidologie 42: 690–695.

Smith SD, Knapp S 2002. The natural history of reproduction in Solanum and Lycianthes (Solanaceae) in a subtropical moist forest. Bull Nat Hist Mus B 32: 125–136.

Soltis DE, Soltis PS, Endress PK, Chase MW 2005. Phylogeny and Evolution of Angiosperms. Sunderland, MA: Sinauer Assoc. 370 pp.

Spessa A, Schwarz P, Adams M 2000. Sociality in *Amphylaeus morosus* (Hymenoptera: Colletidae: Hylaeinae). Ann Entomol Soc Am 93:684–692

Takeda K 1961. Classical conditioned response in the honey bee. J Ins Physiol 6: 168–179.

Thompson GJ, Oldroyd BP 2004. Evaluating alternative hypotheses for the origin of eusociality in corbiculate bees. Mol Phylogenet Evol 33:452–456.

Thorp RW 2000. The collection of pollen by bees. Plant Syst Evol 222:211–223.

Thorp RW 1979. Structural, behavioral, and Physiological adaptations of bees (Apoidea) for collecting pollen. Ann Missouri Bot Gard 66: 788–812.

Thorp RW, Estes JR 1975. Intrafloral behavior of bees on flowers of *Cassia fasciculata*. J. Kansas Entomol Soc 48: 175–184.

Vallejo-Marin M, Manson JS, Thomson JD, Barrett SCH 2009. Division of labour within flowers: heteranthery, a floral strategy to reconcile contrasting pollen fates. J Evol Biol 22: 828–839.

Vallejo-Marin M, Da Silva EM, Sargent RD, Barrett SCH 2010. Trait correlates and functional significance of heteranthery in flowering plants. New Phytol 188: 418–425.

Wcislo WT, Cane JH 1996. Floral resource utilization by solitary bees (Hymenoptera: Apoidea) and exploitation of their stored foods by natural enemies. Annu Rev Entomol 41: 257–286.

Wiebes JT 1979. Co-evolution of figs and their insect pollinators. Annu Rev Ecol Evol S 10:1–12.

Wilson P, Stine M 1996. Floral constancy in bumble bees: handling efficiency or perceptual conditioning? Oecologia 106: 493–499.

Waser NM 1978. Competition for hummingbird pollination and sequential flowering in two Colorado wildflowers. Ecology 59: 934–944.

Waser NM 1986. Flower constancy: definition, cause, and measurement. Am Nat. 127: 593–603.

Waser NM, Chittka L, Price MV, Williams NM, Ollerton J 1996. Generalization in pollination systems, and why it matters. Ecology 77: 1043–1060

Chapter 5

Pollination Diversity in Orchids: A Review

Giby Kuriakose
giby.kuriakose@gmail.com

Abstract

Orchids (family Orchidaceae) are fascinating angiosperm plants with high variation in life forms and flower morphology. They are considered to be one of the highly evolved groups of plants among angiosperms. Since Darwin, orchids draw special attention due to its unique mechanism of pollination. The variation in flower morphology, especially that of the petal, the lip, makes the pollination of orchids interesting. Almost all pollination syndromes so far reported from flowering plants, except a few (e.g., Hydrophily), are seen among orchids too. Orchid flowers offer wide array of resources for their floral visitors including nectar, wax, oil, starch, etc. Floral fragrance and pseudo-pollens are other two important unique resources offered by orchid flowers. Pollen is not a major reward for floral visitors. However, majority of orchid species do not offer any necessary floral reward for the visiting animals but the flowers get pollinated in different ways. This is said to be pollination by deception. Many deceiving orchid flowers possess cues for reward but don't really contain any reward. This is, according to several researchers, thought to be an energy saving process as production of rewards requires high energy. Although most of the orchids are self-compatible, they have diverse floral mechanisms and strategies to increase the chance for receiving cross-pollen rather than self-pollen. This increases chances of outbreeding among orchids. However, pollinator limitation always prevents high fruit set in orchids. The variation in floral morphology and the rewards make orchids a good system of plant-animal interaction researches. But, no serious attempts have been made so far to understand the pollination biology on Indian orchid species.

DOI: 10.1201/9781003505471-5

5.1 Introduction

Orchids are one of the most diverse and fascinating plant groups among flowering plants (angiosperms). They show a high degree of variation in their habit, morphology and life cycle strategies, which are considered as advance features amongst angiosperm families (Figures 5.1-5.3). They are distributed mostly in the tropical regions when compared to the temperate parts of the world. Orchid flowers possess diversity in their morphology and floral resources which enables them to develop specific strategies for survival and life supporting process.

Orchids are perennial (with a few exceptions of annuals such as *Epipogium roseum* (Figure 5.1 B), a saprophytic species) terrestrial or epiphytic, sometimes lithophytic, herbaceous plants. They are mainly autotrophic but some species are saprophytic. The stems are either sympodial or monopodial usually with foliage. However, in some species leaves are reduced to small scales (e.g. *Vanilla wightiana*). The number of leaves may range from one to many. Velamen roots are characteristics of orchids which help them to absorb water and nutrients. Leaves (sometimes stem too) of orchids are xerophytic in nature. Inflorescence may arise from the base, nodal region, or terminal part of individuals.

Fig. 5.1. Diversity in the general habit of some orchid species of the Western Ghats. Legend for the pictures: A) *Acampe praemorsa* (Roxb.) Blatt. & McCann, a common epiphytic orchid species; B) *Epipogium roseum* (D. Don) Lindl., a saprophytic orchid species; C) *Arides crispa*, Lindl. an epiphytic orchid; D) *Luisia zeylanica* Lindl., an epiphytic orchid species with terate leaf; E)_*Bulbophyllum tremulum* Wight, an epiphytic orchid species with bulbous leaf petiole; F) *Nervilia prainiana* (King & Pantl.) Seidenf., a terrestrial orchid species with single leaf.

Fig. 5.2. Diversity in the floral morphology, color, size and shape of selected orchid species of the Western Ghats. A) *Phalaenopsis mysorensis* C.J.Saldanha; B) *Aerides crispa*, Lindl.; C) *Bulbophyllum fischeri* Seidenf.; D) *Bulbophyllum sterile* (Lam.) Suresh; E) *Eulophia flava* (Lindl.) Hook.f.; F) *Luisia macrantha* Blatt. & McCann; G) *Flickingeria nodosa* (Dalzell) Seidenf.; H) *Bulbophyllum fimbriatum* (Lindl.) Rchb.f.; I) *Anoectochilus elatus* Lindl.; J) *Nervilia prainiana* (King & Pantl.) Seidenf.; K) *Cottonia peduncularis* (Lindl.) Rchb.f.; L) *Xenikophyton smeeanum* (Rchb.f.) Garay.

Reproductive events are crucial for the survival of the species. The major reproductive events of a plant species include flower production, pollination, fruit set and seed dispersal (Dafni 2000, Shivanna and Tandon 2014). Flower of angiosperm plants are the major structural component in sexual reproduction that helps in the succession of generations. Pollination is a pre-requisite for seed production and involves either biotic or abiotic pollen vector (Dafni et al. 2005, Shivanna and Tandon 2014).

Pollination in orchids is largely mediated through animals (biotic pollination). Abiotic pollination (anemophily and hydrophily) is not very common among orchids. This is because the pollen grains of orchid flowers are united together to form a pollinia which is not light enough to be carried by wind. None of the orchids so far described are aquatic and hence water-mediated pollination (hydroophily) does not arise. Therefore, it is important to possess or develop strategies to attract animal vectors to ensure pollination in general in flowering plants.

A majority of plants that depend on biotic pollination are generalists in that they are pollinated by a number of floral visitors. Some plant species are specialists;only one or a few specific vectors can forage resources from the flowers of such species and bring about pollination (Bawa 1990, Shivanna 2014). The pollinator-specific plants have evolved strategies, mostly by locating resources at specific sites that can be harvested only by specific vectors (Dafni 1984). This is a major driver of co-evolution of characters and interactions between the pollinator and the flower.The role of co-evolution between plants and pollinators in orchid pollinationhas been emphasised since the time of Darwin (1862). He predicted the pollinator of *Angraecum sesquipedale* based on the morphological features of its flower. Presence of an extremely long spur (about 35cm) in this species led him to predict the pollinator of this orchid as an insect with a proboscis long enough to forage nectar at the end of the spur. Darwin's prediction on the existence of such an insect was not reported until Rothschild and Jordan (1903) found a moth with a proboscis length of about 22cm that proved to be long enough to harvest nectar from the spur of *Angraecum sesquipedale*.This discovery confirmed the prediction of Darwin and paved way for theories of co-evolution.

This chapter attempts to understand the mode of pollination, pollinators, breeding systems in orchids and mechanisms of orchid flowers to attract effective pollinators in the absence of any reward.

Fig. 5.3. Diversity in the floral morphology, color, size and shape (A-H) and fruit structure (I&J) of selected orchid species of the Western Ghats: A) *Aerides ringens* (Lindl.) C.E.C.Fisch.; B) *Bulbophyllum tremulum* Wight; C) *Eulophia spectabilis* (Dennst.) Suresh; D) *Zeuxine longilabris* (Lindl.) Trimen; E) *Dendrobium wightii* A.D.Hawkes & A.H.Heller; F) *Dendrobium* sp.; G) *Seidenfadeniella filiformis* (Rchb.f.) Christenson & Ormerod; H) *Oberonia platycaulon* Wight; I) A typical orchid fruit (*Acampe praemorsa* (Roxb.) Blatt. & McCann); J) Dehisced orchid fruit with several micro seeds.

5.2 Floral Morphology

Raceme is the most common type of inflorescence. Each inflorescence may produce one or several flowers. The size, shape and colour of the flower is highly variable from species to species. The flowers are zygomorphic. They are hermaphrodite, sessile or pedicellate. Most of the orchid flowersshow resupination i.e., the flower twists by 180°at the time of anthesis so as to facilitate the lip to orient correctly to receive the incoming visitors. This is advantageous for appropriate pollinator to forage the reward and aid in pollination. Non-essential whorl, the perianth, may sometimes be differentiated into sepals and petals (three each). Sepals and petals (perianth) together act as the organs of advertisement of the flower. They vary in colour, shape and size.One of the petals is modified into a lip/labellum and it is the major attraction in orchid flowers. Androecium and gynoecium of an orchid flower adnate to form a gynostegium or column (Figures 5.4and 5.5). Pollen grains are aggregated in anthers to form a pollen mass called pollinium. The number of pollinia may vary from two to six.The stigma is tri-lobed with a modified mid-lobe called the rostellum. The stigma is receptive for a few to several days, depending upon the species, and the development of ovule sometimes is much delayed; in some species their differentiation is initiated only after pollination (Shivanna and Tandon 2014). Fruit is a capsule (Figure5.1 I) with a large number (up to a few million in some species) ofsmall, powdery seeds that lack endosperm (Figure5.1J).

Of these morphological features of the flower, three structural features, the column, pollinia along with zygomorphic nature of flowers distinguish orchid flower from the rest of the angiosperm flowers. Pollinia are seen in flowers of Asclepiadaceae as well, but, the flowers are actinomorphic. The androecium and gynoecium, of orchids are kept together in a structure called column and hence can be easily distinguished.

Fig. 5.4. Flower of *Diploprora championi* showing the floral morphology and parts of a typical orchid flower. S- Sepal, P- Petal, St- Stamen, L- Labellum/lip.

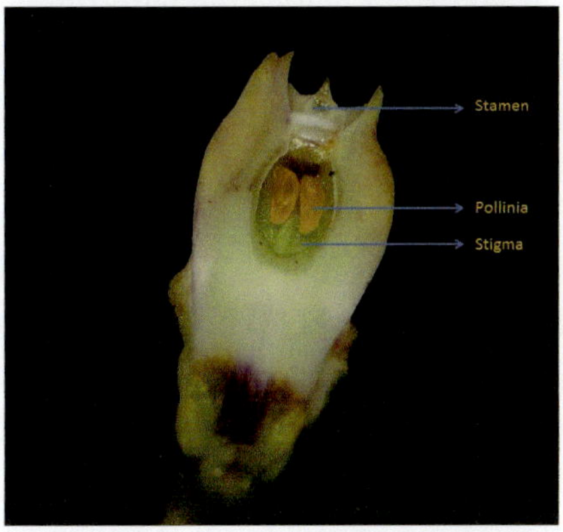

Fig. 5.5. A typical gynostegium of an orchid flower showing stamen, stigma and pollinia on the stigma after hand pollination.

5.3 Pollination in Orchids

Pollination mechanisms in orchids are generally considered to be highly specialised (Tremblay 1992, Johnson and Steiner 2000, Borba and Semir 2001, Dixon et al. 2003). For this, orchid flowers have evolved several strategies to achieve pollination. The major one is the structural organisation of the flower, i.e.,the floral symmetry. Orchid flowers are generally zygomorphic and are oriented in such a way as to accommodate specific and unique visitors (Goldblattet al. 2001, Freudenstein et al. 2002).The placement of androecium and gynoecium in the column of orchid flowers is peculiar that makes the flower a pollinator dependant for its pollen transfer. The column is the extention of ovary in the central axis of the orchid flower. Anther is placed at the top of the column protected by a rostellum and stigma is situated at the ventral side just below the anthers (Figure 5.4). Stigma is slightly sunken in the under surface of the column (Figure1). This unique placement of anthers and stigma might prevent the flower from autogamous pollination. Not only structural organisation of floral parts, but the floral attractants and rewards also play a major role in the evolution of pollination mechanism by specialized visitors (Mayer et al. 2011).

5.3.1 Floral Rewards

Orchids are well known for their diverse pollination mechanisms. This diversity is mainly due to the diversity of rewards and sometimes the absence of rewards in some species (Dressler 1981). Nectar and pollen are the major rewards offered to the visiting insects by flowering plants in general (Dafni 1992, Shivanna 2014). Whereas, in orchids, nectar is the major reward to the visitors (Dressler 1981, Pedron et al. 2012). But pollen has been reported to be a reward only in some species of *Cleistes* in North America (Gregg 1989, 1991). Nectar producing orchids show high fruit set or pollination success over nectarless species (Neiland and Wilcock 1998). Pollination efficiency is found to be high when the population size of orchids is large (Tremblay et al. 2005, Jacquemyn et al. 2007, Perez-Hernandez 2011). At times, it is further correlated with the flowering phenology of plants in a community (Jersakova et al. 2006). For example, there would be a collective effort in attracting the visitors when orchids produce flower alongwith other flowering plants that possess active flowers.However, increased availability of nectar (as a result of availability of a large number of flowers on the same plant) may also result in increased self-pollination as the flower visitor may limit its movements within a limited area and thus the distance of pollen transport gets reduced (Bawa and Webb 1983, Peakall 1989, Burd 1995, Hodges 1995).Also, when nectar is artificially provided in a non-rewarding orchid flower, insect mediated self-pollination and resulting fruit set is significantly increased (Jersakova and Johnson 2006)indicating that the lack of floral reward affects pollination in orchids. Other than nectar, some orchid flower may provide oil (Steiner 1989, Steiner and Whitehead 1991), wax (Dafni 1992, Mchatton 2011), starch or floral fragrance (Dodson et al. 1969,

Galizia et al. 2004, Kaiser 1993, Pemberton 2010, Salzmann et al. 2007, Williams 1982, Woodcock 2014) and pseudo-pollen (Dressler 1981).Some flowers possess small powdery projections or depositions of uni or multicellular, labellar trichomes on the surface of orchid lip known as pseudo-pollen (Davies et al. 2000, Davies and Turner 2004a). Pseudo-pollen is nutritious and is readily collected by bees (e.g., *Polystachya* spp) from orchid flowers (Reis et al. 2004). Orchid flowers that possess pseudo-pollen do not generally produce nectar (van der Pijl and Dodson 1969).

5.3.2 Pollination Syndromes

Almost 90per centof angiosperm flowers are pollinated by biotic vectors (Buchmann and Nabhan, 1996). Pollination in most of the orchids is mediated by highly specialised and unique pollinators (Dressler 1981); about 60per centof orchids are pollinated by single species specific pollinator (Tremblay 1992). This is possible mainly by elaboration of morphological features (Freudenstein et al. 2002) or through placement of floral rewards at specific sites or emission of species-specific fragrance (Mayer et al. 2011).Several other factors such as plant height (Lortieand Aarssen 1999), number of flowers per inflorescence/ plant (Mitchell et al. 2004) and flower size (Armbruster et al. 2005) are also pollination specific.

A large number of insects and birds are the major effective pollinators of orchids (Catling 1990). Chiropterophily (bat pollination) has not so far been recorded in orchid flowers (van der Pijl and Dodson 1966). As pointed out earlier, elaboration of morphological features reduces the guild size of pollinators.

5.3.2.1. Entomophily

Insects are the most diverse group amongst the animals (Danks and Smith 2009, Schultz 2001). Insect pollination is one of the major pollination syndromes in flowering plants (Dafni 1992). Bees, butterflies, moths, flies, wasps, beetles and ants are the major insect groups involved in pollination. Morphological peculiarities of flowers, floral advertisements and rewards and foraging behaviour are the characteristic adaptations of insects as effective pollinators. Among insects, bees are the most effective pollinators of orchids followed by dipteran flies (van der Pijl and Dodson 1966). Insects are attracted to orchid flowers mainly to forage nectar.

Flower morphology, colour, scent, flowering season, patterns and orientation of floral parts and time of floral events are the major features that attract visitors to the flowers. Bee pollinating flowers are generally zygomorphic, dark coloured and possess small tubular corolla, whereas, fly and moth pollinating flowers are light coloured (Dafni 1992). Bee pollinating flowers are active during day time, whereas, fly and moth pollinating flowers are generally noctornal that open during dusk, dawn and night. As insects visit flowers mainly for nectar, availability of these resources is significant for successful pollination. Many of the insects bring about pollination in species that do not provide any rewards but attract insects by deceit. This aspect is covered later.

The genus *Ophrys* is distributed in the Mediterranean regions up to Central and Northern Europe. The lip of all species of *Ophrys*mimics female bees, and hence they are commonly known as 'bee orchid'. In India, the lip of *Cottonia peduncularis* also resembles a bee and is known as 'Indian bee orchid'. Different species of *Ophrys* is pollinated by different bees belonging to Andrenidae, Anthophoridae, Colletidae, Megachilidae and Apidae (Stockl et al. 2005 and references therein). None of the species of *Ophrys,*so far studied, produce any floral rewards but the lip of the flowers are modified and mimic that of female bee and produce some odour that is similar to the pheromones produced by the female bees during copulation to attract its counterparts.

Generally insect pollinated orchid flowers do not produce nectar but they possess different deception mechanisms (see under deception). *Cyperipedium* sp, a'ladies slipper' orchid,is an example of insect pollinating orchids.It is pollinated by bees of Andrenidae, Halictidae, Megachilidae and Apidae families (Bänziger et al. 2005, 2008, Davis 1986, Li et al. 2006, 2008, Nilsson 1979, Singer and Sazima 1999, Stoutamire 1967, Sugiura et al. 2001, 2002).Nectar or any other reward for its visitors is absent in the flowers of this orchid (Li et al. 2012). *Changnienia amoena,* is a self-compatible orchid that depends on bumblebees (Apidae), for enhanced fruit set. Here also the flowers do not contain nectar but produce some odour which attracts the pollinator (Sun et al. 2003). However, a species of small carpenter bee (*Ceratina* sp) pollinates flowers of *Sirindhornia monophylla*, that offer nectar for the pollinator (Srimuang et al. 2013). Carpenter bees belonging to *Xylocopa* are other major insect pollinators of orchid flowers. So far, 24 species of orchids are reported to be pollinated by *Xylocopa* spp.(Sugiura 2013). Some orchid flowers offer nectar and wax for *Xylocopa* spp. However, a vast majority among these *Xylocopa*-pollinating orchid flowers are deceptive and possess different deception mechanisms (Sugiura 2013). Wasps are another major insect pollinator for some orchids (see Brys et al. 2008).Other than these,euglossine bees (Pedron et al. 2012) and moths were also reported as effective insect pollinators of *Habenaria* sp.

5.3.2.2 Butterflies and moths

Lepidopteran pollination mechanism in orchid flowers has been in discussion since Darwin (1862) predicted a long proboscid insect, probably a moth, as the pollinator for a long spurred orchid, *Angreicum sesquipedale*. Other than long spur, light colour with flat petal/ perianth lobes and less or no markings or nectar guides and the surface colour are also considered to be characteristics of lepidopteran flowers (Dafni 1992).

Several authors hypothesised that the length of the orchid spur determine the insect species (Boberg 2010, Darwin 1862, Johnson and Steiner 1997, Nilsson 1988, Pauw et al. 2009) that pollinates a specific orchid flower. Pollination would be less successful or would not happen if a long proboscid insect forages nectar frequently from a short spurred orchid (Boberg 2010).

Plantanthera sp., a hawk moth-pollinated orchid in the North America, has characteristic morphological features of nocturnal moths. It produces white coloured flowers with sweet fragrance which becomes stronger towards dusk. The flowers also possess a long spur filled with nectar.

While foraging floral rewards, moths and butterflies collect pollinia, either at the base of the proboscis or on the forehead near the base of the proboscis (Jordan et al. 2006). The placement of pollinia on insect body is correlated with the distance between the spur tip and column (Sheviak and Bowles 1986). If the spur is a little longer than the proboscis of the insect, the insect may need to struggle to forage nectar that is usually situated at the bottom of the spur (Dafni 1992).

Butterfly pollination is not a common syndrome among orchid flowers. *Habenaria foliosa*a terrestrial orchid that produces white flowers that are foetid or sweet scented with long (ca17cm) white spur containing nectaris reported to receive both butterfly (during clear day time) and moth (during cloudy or dusk time). Both these lepidopteran visitors collect and transfer pollinia from the orchid flowers (Dangat and Gaurav 2014).

5.3.2.3 Ornithophily

Bird pollination in orchid is not so common (van der Cingel 2001, van der Pijl and Dodson 1966). Only about three per cent of orchids, so far studied, are pollinated by birds (Siegel 2011). Mainly hummingbirds (Trochilidae)in the New World, Sunbirds (Nectariniidae) in the Old World, White-Eyes (Song birds, Zosteropidae) in Asia and Honey-Eaters in Australia are the frequent orchid flower visitors and pollinators (Fay 2009). All these are small sized birds and forage primarily nectar as their major source of energy. There are two different pollen transfer mechanisms in bird mediated pollen transfer in orchid flowers. The sticky pollinia of orchid flowers might stick to the smooth surfaced beak while inserting the beak inside the orchid flowers for foraging nectar(Johnson 1996, Singer and Sazima 2000). Sometimes birds, while foraging nectar, hang on the inflorescence by perching on the flowers with their claw. During this course, pollinia get attached to the claw and when the bird moves from one flower to the next the pollinia get transferred to the stigma (Johnson and Brown 2004). Generally, birds do not insert their beak completely in to the flower as this action depends upon the spur length. During this action, the chances are remote for the transfer of pollinia. Sometimes the pollinia thatis attached on the beak,might get removed while preening. Bird pollinated orchids are common in Neotropical and paleotropical regions (Stockl et al. 2005).

Bird pollinated flowers in general are brightly coloured (Dafni et al. 2005), and orchid flowers are no exception (Dressler 1981). Orchid flowers pollinated by hummingbirds and butterflies generally possess shades of red (Dressler1981) with tubular corolla (Dafni et al. 2005, Shivanna 2014). Bird pollinated flowers generally produce nectar as floral reward (Ackerman 2000). Also birds can locate red coloured flowers from a distance and forage nectar from the flowers.Although presence of nectar is a prerequisite for bird pollinating flowers, the amount of nectar produced in orchid flowers is less as compared to other bird pollinating flowers (Rodrígues-Robles et al. 1992). Several researchers have explained small amount of nectar available in orchid flowers is a strategy to reduce the resource allocation as nectar production is energetically expensive (Southwick 1984, Koopowitz and Marchant 1998). Deception is not yet recorded in bird pollinating orchids.

5.4 Deception in Orchids

About 30 per cent of all known orchid species are reported to be deceptive (Ackerman 1986, Dafni 1984, Little 1983). They attract their concerned pollinators by various devices without providing any rewards (Bell 1986, Dafni 1984). Resource allocation in plants for the production of floral rewards is a process that requires high energy. In order to save the cost of energy, orchids evolved strategies to reduce or stop producing rewards (Primack and Stacy 1998). Successful pollination would not happen in a pollinator dependent plant if there is no system, including floral rewards, to attract pollinators. Orchids successfully overcome this by producing flowers with no rewardsin offer but possess mechanisms that advertise rewards which is actually not present.Such non-rewarding orchids achieve pollination by deceit mechanisms include (Jersakova et al. 2006, Salzmannet al. 2007, Scopece 2010)generalized food deception, batesian mimicry, brood-site imitation, shelter imitation, pseudo antagonism, rendezvous attraction and sexual deception (Jersakova et al. 2006 and references therein) to attract its visitors.Deceptive orchids often produce chemicals that are responsible for odour. Odour produced by orchids in a small population is not intelligible, but if many such orchids grow together in a population and produce flowers at the same time the odour might be perceivable (Banziger et al. 2005).Sometimes, other floral morphological features such as fringes, hair, tassels and other appendages that move constantly may attract the attention of floral visitor (Mchatton 2011). *Bulbophyllum tremulum* (an endemic orchid of the Western Ghats, Figure 5.3B) may be considered as an example of such morphological features, though the importance of the hair present on the petals has not been studied yet.

5.4.1 Generalized Food Deception

Food deception is a mechanism developed in orchid flowers that helps in attracting food foraging insects. In generalized food deception orchid flowers may advertise signals indicating the presence of food. This includes, flower colour, odour, nectar guides on the flower petals, spurs (indicating presence of nectar) and pollen like papillae called the pseudo-pollen (Galizia et al. 2005).

Generalized food deception is considered to be the most common deceptive type of pollination occurring in about 90per centof deceptive orchids (Dafni 1984, Xiaohua 2012). When deceptive orchid flowers attract visitors that forage for food from flowers by producing mere signals (Jersaková et al. 2006) without producing rewards is known as generalised food deception. In generalized food deception orchids attract visitors by mimicking flowers of the rewarding species that grow sympatrically. The visitors, while visiting such flowers, would stray into the flower in search of food. During this process orchid pollinia may get loaded on the insect body and may get transferred to the stigma of flower of competing individuals. Few examples for generalised food deception are *Orchis*spp (Dafni 1987, Johnson and Nilsson 1999, Nilsson 1983), *Dactylorhiza*spp (Lammi and Kuitunen 1995), *Disa*spp (Johnson 1993, Johnson and Steiner 1995, Johnson et al. 1998), *Cypripedium* spp (Nilsson 1979, Stoutamire 1971).

Olfactory signals do not seem to have important role in food deceptive orchids (Galizia et al. 2005, Salzmann et al. 2007, Shivanna 2014). However, olfactory signals may play an indirect role in attracting insects from a distance and hence a combined effect of olfactory and visual signals is possible to attract insect pollinators to deceiving orchids. Pseudo-pollen is a common mode of food deception in orchids (see floral rewards). Here the insects that forage for pollen get attracted and move on the orchid lip. The insect would be moving vigorously on the lip and possibly dislodging the pollinia from the column. When the same insect visits flowers of the same species, the pollinia get transferred to the stigma. *Polystachya* (Davies et al. 2002), *Maxillaria* (Davies and Turner 2004a, van der Pijl and Dodson 1966), few species of *Dendrobium* (Davies and Turner 2004b, Kjellsson and Rasmussen 1987) and some species of *Eria* (Davies and Turner 2004c) are some well known examples of orchid flowers that produce pseudo-pollen.

A non-rewarding orchid species accomplish pollination by deceiving its visitors by producing flowers similar to that of a rewarding species; such devices are said to be food deceptive floral mimicry. Here, the orchids do not either possess the food or structures that look like food but mimic flowers of other plants that actually offer food. Such plants/ orchids are known as Batesian mimics (Brown and Kodric-Brown 1979, Bierzychudek 1981, Dafni 1984, Jersaková et al. 2006, Johnson 1994, Roy and Widmer 1999).

The rewarding plants in Batesian mimicry are often said to be the model plants. In Batesian mimicry colour is the major attraction rather than fragrance (Anderson et al. 2005, Jersaková et al. 2006, Johnson 2000, Johnson et al. 2003). An orchid would be successful with Batesian mimicry if the flower colour of mimic perfectly matches with that of the model (Jersaková et al. 2006). Therefore, it is believed that colour sense of pollinator is an important measure in the success of Batesian mimics (Anderson et al. 2005, Newman et al. 2012). Here, the model and mimic are competing with each other for a possible common pollinator (Brown and Kodric-Brown 1979, Dafni and Bernhardt 1990, Johnson et al. 2003, Jersaková et al. 2006).

5.4.2 Brood-site Deception

Several insects prefer specific sites to lay their eggs. Generally, dung, decaying faeces, fruiting body of fungi and carrion are some examples for such brood sites (Jersaková et al. 2006, Sakai 2002). Plants and plant parts are also preferred sites for many such insects and orchids are no exception. Deceiving orchids may possess some characteristic features of such brood sites by having characteristics of real brood sites. Insects, in search of brood site or food, may land on these plants and get deceived but the plants are benefited by getting pollinated. Apart from orchids, plants belonging to other families such as Annonaceae, Araceae, Aristolochiaceae, Asclepiadaceae and Rafflesiaceaeare also reported to have deceiving mechanisms with brood site characteristics (Jersaková et al. 2006, Kelly et al. 2013, Shivanna 2014, Urru et al. 2011). Generally the flowers of these plants mimic the odours of brood sites such as dung, decaying faeces and carrions (Shivanna 2014) to attract insects that seek site for egg laying. Insects belonging to Diptera and Coleoptera generally get attracted to such plants (Jersaková et al. 2006). Pale brown or red colour coupled with a foul smell is the characteristic feature of brood site mimicking orchids (e.g.,*Bulbophyllum* spp, *Paphilopedilum* spp). Pollination of an Australian orchid genus *Corybas,* that produce dark coloured flowers, is mediated by small flies that usually oviposit on fungus (Jones 1970). This is possible because the orchid mimics the fruiting body of some Basidiomycetes fungus (Jersaková et al. 2006). Besides, flowers of an American native orchid genus *Dracula*in which the lip mimicks a fungusand produces smell of rotten flesh (Vogel 1978). In such orchid species, the flowers are born almost close to the ground. In some species the flowers of brood site mimicry produces heat whichenhances emission of specific odour (Shivanna 2014).

5.4.3 Shelter Deception

Some orchid flowers may provide space to rest, hide or roost. The shelter may provide not only space but favourable environment to the insects/animals including temperature. A Mediterranean orchid, *Serapia spp,* where the flowers mimic nests of bees (Dafni et al.

1981) is an example for shelter mimicry. Since orchid flowers provide an analogous shelter with apt environment inside, its justificationto include under deception has been debated since long (Jersaková et al. 2006). An insect when enters such orchid flowers in search for shelter brings about the transfer of pollinia.

5.4.4 Sexual Deception

Sexual deception is another interesting type of deception found in orchids in which the orchid flower or its parts mimicthe female of a specific insect (Blanco and Barbozoa 2005). Also, the orchid flower emits chemicals, similar to that of the sex pheromones emitted by female bee (Kullenberg 1961), that attracts the male bee to 'mate'. Perhaps, sexual deception is the most studied deception mechanism among non-rewarding orchids (Jersaková et al. 2006). However, sexual deception is not as common as food deception in orchids (Gaskett 2011, Peakall et al. 2010) but the interaction between the orchids and the pollinator is important in evolutionary aspects.

Sexual deception is otherwise known as pseudo-copulation.This is because the male bee thinks that the lip of orchid flower is his counterpart ready for copulation. The orchid lip not only mimics the female bee but, more precisely, it also produces pheromones that the female insect produces during sexual copulation (Blanco and Barbozoa 2005, Kullenberg 1961, Shivanna 2014, Stokl 2010). On getting these cues, the male comes and tries to mate with the 'female insect' resulting in 'pseudo-copulation'. This is the most common sexual deceit mechanism among orchid flowers (Jersaková et al. 2006). In such flowers the lip of orchid flowersmimic exactly the female of a specific insect.

Sexual deception evolved to attract male bees is believed to be an exclusive mechanism in orchids (Ackermann 1986, Dafni 1984, Nilsson 1992). *Ophrys catalaunica*is is a classical example for sexual deceit by means of pseudo-copulation (Vereecken and Patiny 2005).This is achieved by attracting the pollinators with odour forming chemicals. Here the orchids use these olfactory cues to draw the attention of the potential pollinators (Bino et al. 1982). Volatile chemicals are essential attractant in *Ophrys fusca* (Stokl et al. 2005). *Cottonia peducularis* is a probable Indian species that possesses sexual deception mechanism. This epiphytic orchid flower has a peculiar lip that resembles a bee (Figure5.2 K).

In general, plants that are pollinated by deception need to have well developed deceit mechanisms or autogamy to achieve maximum fruit set (Tremblay et al. 2005) by attracting its pollinators. It needs to be studied further to understand the mechanisms of deception in orchids and to examine whether the mechanism functions at individual or community level for different deceit orchid species.

5.5 Breeding System

Most of the orchids are self-compatible and produce seeds when an orchid flower receives pollen from the same flower or flowers of the same individual. Autogamous pollination (self-pollination within a flower) is reported to be common among orchid flowers (Suetsugu 2013, van der Cingel 1995, 2001). So far, pollination system of about eight per cent orchid species has been studied, of which, 31 per cent were found to be autogamous (Peter and Johnson 2009, Zhou et al. 2012). *Pleurothallis ruscifolia*is an example for autogamous self-pollination in orchids (CaraDonna and Ackerman 2012). Autogamous pollination is reported mainly from the colder temperate regions (Catling 1990). However, autogamous pollination is dependent on pollinator abundance in the area. For example, autogamous pollination is less or does not occur if pollinator is abundant. When there is a scarcity or reduction in the abundance of pollinators, autogamous pollination prevails over other systems (Mooney and McGraw 2007). Orchids that show true self-incompatibility (SI) are very rare (see CaraDonna and Ackerman 2012, Micheneau et al. 2009, Neiland and Wilcock 1998) and when present SI is partial. In many other orchids self-pollination is restrictedmainly due to the complex morphological features of the flowers (Fay and Chase 2009).Rostellum, which separates androecium and gynoecium prevents selfing to a great extent in such orchids (Gamisch et al. 2014 and references therein).In autogamous species, the rostellum is poorly developed or absent (Catling 1990). Embryo formation as a result of self-pollination is reported to be significantly low (Darwin 1875, Tremblay et al. 2005).

Floral features and rewards in orchids enhances the possibilities of cross pollination (Darwin 1877), a pre-requisite for recombination. But due to pollinator specificity and the limited availability of these pollinators, successful fruit production is much less compared to other angiospermsso far studied. Most of the animal mediated transfer of orchid pollen results in cross-pollination rather than self-pollination (as the visitors move from one plant to the other after each visitation, especially when only one flower per orchid plant is active at a given point of time) and hence out crossing is the prevailing breeding system in orchids (Fay and Chase 2009) that ensures speciation and diversity. The breeding system and pollination in tropical orchids, especially from Asian countries, are poorly documented. A rare but interesting pollination system, rain pollination in *Acampe rigida* in China, has been revealed by Fan et al. (2012), in which the flowers that are produced in the rainy season get self-pollinated by rain water.Another interesting pollination system in orchids is Vermiculiphily (Kumar 2011), a recently reported pollination syndrome. It is reported only in *Epipactis veratrifolia* (Western Himalaya, India) in which the larvae of a hoverfly species *Ischiodon scutellar*isfed on the petals of *Epipactis veratrifolia*. After hatching, the larvae move vigorously inside the flower, especially on the column and pollinia get attached on the body of the larvae. The pollinia later gets transferred on to the stigma surface and results in pollination of the flowers of *Epipactis veratrifolia*.

5.6 Pollination Limitation

Although, orchids do have different pollination mechanisms and strategies, fruit set is reported to be low (Peakall and Handel 1993, Widmer et al. 2000) as a result of pollinator limitation (Ackerman1989,Calvo and Horvitz 1990, Nosrati et al. 2011, Zimmerman and Aide 1989) or breeding system limitation (see Tremblay, 2005 and references therein). Pollination limitation in orchids could be due to several reasons. According to Tremblay et al. (2005), effectiveness of pollinator, resource/reward constrains, removal and deposition of pollinia, abundance and diversity of pollinator, pollen quantity and quality and several ecological features are the major attributes to the pollination limitation in orchids.

5.7 Concluding Remarks

Orchid pollination includes diverse mechanisms thatare mediated through either morphological variation of the flower or variation in rewards. Mechanisms that have evolved to attract specific pollinators enable them to save their resource availability and allocation.A wide range of pollinators, from a range of insects to birds have been reported, but most of them are species specific. Self-compatibility in orchids is advantageous for reproductive assurance to some extent. Although, pollination of orchids has been studied extensively worldwide, no systematic attempts have been made so far to study the pollination biology and breeding systems of Asian orchids. Although India has a rich orchid diversity with about 1,300 species of wild orchids (Kumar and Manilal 1994, Gogoi et al. 2012, Jalal and Jayanthi 2012), no detailed studies on their reproductive ecology study have been carried out, other than few peripheral works that deal with floral visitors and pollen carriers. Understanding of breeding biology systems are essential to know the underlying facts about the reproductive constraints in plants, especially in orchids where the system is so complex. Evolution, geographical distribution pattern, ecological barriers, community dynamics and conservation of species are directly related to pollination and breeding systems. Without the baseline data on these aspects, managing diversity and understanding the functional aspects may not be easy. Pollination biology of Indian orchids is so poor that no details are available on deception in any Indian orchid. It is high time that Indian orchids are subjected to detailed and systematic studies on their pollination ecology and breeding system.

Acknowledgments

I acknowledge Professor KRShivanna (Ashoka Trust for Research in Ecology and the Environment, Bangalore) for his thorough review of the manuscript and suggestions at different stages of writing. I thankMr Aravind R Nair (Sacred Heart College (Autonomous), Thevara, Cochin) for reviewing the language of the manuscript. I acknowledge the Department of Science and Technology, India for SERB-DST Fast Track Young Scientist project on Orchid ecology.

References

Ackerman JD 1986.Mechanisms and evolution of food deceptive pollination systems in orchids.Lindleyana 1: 108-113.

Ackerman JD 1989.Limitations to sexual reproduction in *Encyclia krugii* (Orchidaceae). Syst Bot 14: 101-109.

Ackerman JD 2000.Abiotic pollen and pollination: ecological, functional, and evolutionary perspectives. Plant Syst Evol222:167-185.

Anderson B, Johnson SD, Carbutt C 2005.Exploitation of a specialized mutualism by a deceptive orchid.Am J Bot92: 1342-1349.

Armbruster WS, Antonsen L, Pelabon C 2005. Phenotypic selection on Dalechampia blossoms: honest signalling affects pollination success. Ecology 86:3323-3333.

BänzigerH, Sun HQ, Luo YB 2005. Pollination of a slippery lady slipper orchid in south-west China: *Cypripedium guttatum*(Orchidaceae). BotJLinnSoc148: 251-264.

Bänziger H, Sun HQ, Luo YB 2008. Pollination of wild lady slipper orchid *Cypripedium yunnanense* and *C. flavum* (Orchidaceae) in south-west China: why are there no hybrids? Bot J Linn Soc 156:51-64.

Bawa KS, Webb CJ 1983.Floral variation and sexual differentiation in *Muntingia calabura* (Elaeocarpaceae), a species with hermaphrodite flowers. Evolution37: 1271-1282.

Bawa KS 1990. Plant-pollinator interactions in tropical rain forests. Annu Rev Ecol Evol S 21: 399-422.

Bell G 1986.The evolution of empty flowers. J Theor Biol118:253-258.

Bierzychudek P 1981. Asclepias, Lantana and Epidendrum: a floral mimicry complex? Biotropica (Suppl) 1: 54-58.

Bino RJ, Dafni A, Meeuse ADJ 1982. The pollinationecology of *Orchis galilaea*(Bronm et Schulze) Schltr (Orchidaceae). New Phytol 90: 315-319.

Blanco MA, Barbozoa G 2005. Pseudocopulatory Pollination in *Lepanthes*(Orchidaceae: Pleurothallidinae) by Fungus Gnats. Ann Bot 95: 763-772.

Boberg E 2010. Evolution of spur length in a moth pollinated orchid. Acta Universitatis, Upsaliensis. Digital comprehensive summaries of Uppsala Dissertations from the faculty of Science and Technologies, pp 712-733.

Borba EL, Semir J 2001. Pollinator speci®city and convergence in fly pollinated *Pleurothallis*(Orchidaceae) species: a multiple population approach. Ann Bot 88: 75-88.

Brown JH, Kodric-BrownA 1979. Convergence, competition and mimicry in a temperate community of hummingbird pollinated plants. Ecology 60: 1022-1035.

Brys R, Jacquemyn H, Hermy M 2008. Pollination efficiency and reproductive patternsin relation to local plant density, population size, and floral display in the rewarding *Listera ovata* (Orchidaceae). Bot J Linn Soc 157: 713-721.

Buchmann SL, Nabhan GP 1996. The Forgotten Pollinator. Washington, DC: Island Press.

Burd M 1995. Ovule packaging in stochastic pollination and fertilization environments. Evolution 49: 100-109.

Calvo RN, Horvitz CC 1990. Pollinator limitation,cost of reproduction, and fitness in plants, a transition matrixdemographic approach. Am Nat 136: 499-516.

CaraDonna PJ,Ackerman JD2012.Reproductive assurance for a rewardless epiphytic orchid in Puerto Rico: *Pleurothallis ruscifolia* (Orchidaceae, Pleurothallidinae) Caribbean J Sci 46: 249-257.

Catling PM 1990. Auto-pollination in the Orchidaceae. In: Arditti J, (Ed) *Orchid biology: reviews and perspectives V*. Portland, OR: Timber Press, pp 121-158.

Dafni A, Ivri Y 1981. Floral mimicry between *Orchis israelitica*Bauman and Dafni (Orchidaceae) and *Bellevalia flexuosa* Boiss (Liliaceae).Oecologia 49: 229-232.

Dafni A 1984.Mimicry and deception in pollination. Annu Rev Ecol Evol S 15:259-278.

Dafni A1987. Pollination in *Orchis*and related genera:evolution from reward to deception. In J Arditti (Ed) Orchid Biology: Reviews and Perspectives 4, Cornell University Press, Ithaca, pp. 79-104.

Dafni A, Bernhardt P 1990. Pollination of terrestrial orchids of southern Australia and the Mediterranean region: systematicecological and evolutionary implications. Evol Biol 24: 193-250.

Dafni A 1992. Pollination Ecology: a Practical Approach. Oxford University Press.

Dafni A 2000. Pollen and Pollination. Springer.

Dafni A, Kevan PG, Husband BC 2005. Practical pollination biology.Enviroquest Ltd.

Dangat BT, Gaurav RB2014. Pollination in Habenaria foliosa Var. foetida (Orchidaceae). Richardiana 14: 219-228.

Danks HV, Smith ABT 2009. Insect diversity in the Nearctic Region. In RG Foottit and PH Adler (Eds) Insect Biodiversity.Wiley-Blackwell, pp 35-48.

Darwin C 1862. On the various contrivances by which British and foreign orchids are fertilized by insects. John Murray, Albemarle Street

Darwin C 1875.The variation of animals and plants under domestication. London: John Murray.

Darwin C 1877.Fertilization of Orchids. D Appleton and Company, New York. 2nd edition, revised.

Davies KL, Winters C, Turner MP 2000. Pseudopollen: its structure and development in Maxillaria (Orchidaceae). Ann Bot 85: 887-895.

Davies KL, Roberts DL, Turner MP 2002. Pseudopollen and food-hair diversity in *Polystachya* Hook(Orchidaceae). Ann Bot 90: 477-484.

Davies KL, Turner MP 2004a. Morphology of floralpapillae in *Maxillaria*Ruiz & Pav(Orchidaceae). Ann Bot 93: 75-86.

Davies KL, Turner MP 2004b. Pseudopollen in *Eria*Lindl. Section Mycaranthes Rchbf (Orchidaceae).Ann Bot 94: 707-715.

Davies KL, Turner MP 2004c. Pseudopollen in *Dendrobium unicum* Seidenf. (Orchidaceae):reward or deception? Ann Bot 94: 129-132.

Davies KL, Turner MP 2004. Pseudopollen in Eria Lindl.Section Mycaranthes Rchbf (Orchidaceae). Ann Bot 94: 707-715.

Davis RW 1986.The pollination biology of *Cypripedium acaule* (Orchidaceae). Rhodora 88: 445-450.

Dixon KW, KellSP, Barrett RL, Cribb PJ 2003. Orchid Conservation.Natural History Publications (Borneo), Malaysia.

Dodson CH, Dressler RL, Hills HG, Adams RM, Williams NH 1969. Biologically active compounds in orchid fragrances. Science 164:1243-1249.

Dressler RL 1981. The Orchids, Natural History and Classification. Harvard University Press, Cambridge and London.

Fan XL, Barrett SC, Lin H, Chen LL, Zhou X, Gao JY 2012. Rain pollination provides reproductive assuarence in a deceptive orchid. Ann Bot 110: 953-958.

Fay MF, Chase MW 2009. Orchid biology: from Linnaeus via Darwin to the 21st century. Ann Bot 104: 359-364.

Freudenstein JV, Harris EM, Rasmussen FN2002. Theevolution of anther morphology in orchids: incumbent anthers, superposed pollinia and vancoid complex. Am J Bot 89: 1747-1755.

Galizia G et al. 2004.Relationship of visual and olfactory signal parameters in a food-deceptive flower mimicry system. Behav Ecol16: 159-168.

Galizia CG, Kunze J, Gumbert A, Borg-Karlson AK,Sachse S, Markl C, Menzel R 2005. Relationship of visual and olfactory signal parameters in a food-deceptive flowermimicry system. Behav Ecol 16: 159-168.

Gamisch A, Fischer GA, Comes HP 2014. Recurrent polymorphic mating type variation in Madagascan *Bulbophyllum* species (Orchidaceae) exemplifies a high incidence of auto-pollination in tropical orchids Bot J Linn Soc 175: 242-258.

Gaskett AC 2011. Orchid pollination by sexual deception: Pollinator perspectives. Biol Rev 86: 33-75.

Gogoi K., Borah RL, Sharma GC, Yonzone R 2012. Present Status of Orchid Species Diversity Resources and Distribution in Dibrugarh District of Assam of North East India. Int J Mod Bot2: 19-33.

Goldblatt P, Manning JC, Bernhardt P 2001. Radiation of pollination systems in *Gladiolus* (Iridaceae: Crocoideae) in southern Africa. Ann Missouri Bot Gard 88: 713-734.

Gregg KB 1989. Reproductive biology of the orchid *Cleistes divaricate* (L.) Ames var*bifaria* Fernald growing in a West Virginia meadow.Castanea 54: 57-78.

Gregg KB 1991. Reproductive strategy of *Cleistes divaricata* (Orchidaceae).Am J Bot 78: 350-360.

Hodges SA 1995.The influence of nectar production on hawk moth behavior, self-pollination and seed production in *Mirabilis multiflora* (Nyctaginaceae). Am J Bot 82: 197-229.

Jacquemyn H, Brys R, Vandepitte K, Honnay O, Roldan-Ruiz I, Wiegand T 2007.A spatially explicit analysis ofseedling recruitment in the terrestrial orchid *Orchis purpurea*. New Phytol 176: 448-459.

Jalal JS, Jayanthi J 2012. Endemic orchids of peninsular India: a review. J Threat Taxa 4: 3415-3425

Jersáková J, Johnson SD, Kindlmann P2006.Mechanisms and evolution of deceptive pollination in orchids. Biol Rev 81: 219-235.

Jersáková J, Johnson SD 2006. Lack of floral nectar reduces self-pollination in a fly-pollinated orchid. Oecologia 147: 60-68.

Johnson SD 1993. Carpenter bee pollination of *Herschelianthe graminifoli*a (Orchidaceae) on the Cape Peninsula. Flora 188: 383-386.

Johnson SD 1994. Evidence for Batesian mimicry in a butterfly pollinated orchid. Biol J Linn Soc 53: 91-104.

Johnoson SD 1996. Bird pollination in South African species of *Satyrium* (Orchidaceae). Plant Syst Evol 203: 91-98.

Johnson SD 2000. Batesian mimicry in the non-rewardingorchid *Disa pulchra*, and its consequences for pollinator behaviour.Biol J Linn Soc 71: 119-132.

Johnson SD, Nilsson LA 1999. Pollen carryover, geitonogamy, and the evolution of deceptive pollination system in orchids. Ecology 80: 2607-2619.

Johnson SD, Steiner K 1995. Long-proboscid fly pollination of two orchids in the Cape Drakensberg Mountains, South Africa. Plant Syst Evol 195: 169-175.

Johnson SD, Steiner K 1997. Long-tongued fly pollination and evolution of floral spur length in the *Disa draconis*complex (Orchidaceae). Evolution 51: 45-53.

Johnson SD, Linder HP, Steiner KE 1998. Phylogeny and radiation of pollination systems in Disa (Orchidaceae). Am J Bot 85: 402-411.

Johnson SD, Steiner KE 2000.Generalization versus specialization in plant pollination systems. Trends Ecol Evol 15: 140-143.

Johnson SD, Peter CI, Nilsson LA, Agren J 2003. Pollination success in a deceptive orchid is enhanced by co-occurring rewarding magnet plants. Ecology 84: 2919-2927.

Johnson SD, Brown M 2004. Transfer of pollinariaon birds' feet: a new pollination system in Orchids. Plant Syst Evol 244: 181-188.

Jones DL 1970. The pollination of *Corybas diemenicus* HMRRupp, WH Nichols.Victorian Naturalist 87: 372-374.

Jordan CR, Fauske GM, Harris MO, Lenz D 2006.First record of the Spurge Hawk moth as a pollen vector for the Western Prairie Fringed Orchid. The Prairie Nat 38: 57-62.

Kaiser R1993.The scent of orchids: olfactory and chemical investigations. Amsterdam: Elsevier.

Kelly M, Gaskett A, Toft R 2013. Pollination and insect visitors to the putatively brood-site deceptive endemic spurred helmet orchid, *Corybas cheesemanii*. New Zeland J Bot 51: 155-167.

Kjellsson G, Rasmussen FN 1987. Does the pollination of *Dendrobium unicum* Seidenf. involve 'pseudopollen'? Die Orchidee 38: 183-187.

Koopowitz H, Marchant TA 1998. Post pollination nectar reabsorption in the African epiphyte *Aerangis verdickii* (Orchidaceae). Am J Bot 85: 508.

Kullenberg B 1961.Studies in Ophrys pollination.Zoologiska bidrag från Uppsala34: 57.

Kumar CS, Manilal KS 1994. A catalogue of Indian orchids.Bishen Singh Mahendra Pal Singh.

Kumar P, Rawat GS 2011.Vermiculiphily: Larvae pollinating orchids! Nature Procedings: hdl:10101/npre.2011.6295.1

Lammi A, Kuitunen M 1995. Deceptive pollination of *Dactylorhiza incarnata*: an experimental test of the magnet species hypothesis. Oecologia 101: 500-503.

Li P, Luo YB, Bernhardt P, Yang XQ, Kou Y2006. Deceptive pollination of the lady's slipper *Cypripedium tibeticum* (Orchidaceae). Plant Syst Evol 262: 53-64.

LiP, Luo YB, Bernhardt P, Kou Y, Perner H 2008. Pollination of *Cypripedium plectrochilum*(Orchidaceae) by *Lasioglossum* spp(Halictidae): the roles of generalist attractantsversus restrictive floral architecture. Plant Biol 10: 220-230.

Li P, Pemberton R, Zheng G, Luo Y2012. Fly pollination in *Cypripedium*: a case study of sympatric *C. sichuanense*and *C. micranthum*. Bot J Linn Soc, 170: 50-58.

Little RJ1983.A review of floral food deception mimicries with comments on floral mutualism.In CE Jones and RJ Little (Eds) Handbook of ExperimentalPollination Biology.S & E Scientific and Academic Editions, New York. pp. 294-309.

Lortie CJ, Aarssen LW 1999. The advantage of being tall: higherflowers receive more pollen in *Verbascum Thapsus* L. (Scrophulariaceae). Ecoscience 6:68-71.

Mayer C, Adler WL, Armbruster S et al. (2011). Pollination Ecology in the 21st Century: Key questions for future research. J Poll Ecol 3:8-23.

Mchatton R 2011. Orchid Pollination: Exploring a fascinating world. Orchids 338-349. www.aos.org

Mitchell RJ, Karron JD, Holmquist KG, Bell JM 2004.The influence of Mimulus ringens floral display size on pollinator visitation patterns. Funct Ecol 18: 116-124.

Micheneau C, Johnson SD, Fay MF2009. Orchid pollination: from Darwin to the present day. Bot J Linn Soc 161: 1-19.

Mooney EH, McGraw JB 2007. Effect of self pollination and outcrossing with cultivated plants in small natural populations of American Ginseng, *Panax quinquiefolius* (Araliaceae). Am J Bot 94: 1677-1687.

Neiland MRM, Wilcock CC 1998. Fruit set, nectar reward, and rarity in the Orchidaceae. Am J Bot85: 1657-1671.

Newman E, Anderson B, Johnson SD 2012. Flower colour adaptation in a mimetic orchid. Proc R Soc B 279: 2309-2313.

Nilsson LA 1979. Anthecological studies on the Lady'sSlipper, *Cypripedium calceolus*(Orchidaceae). Bot Notiser 132: 329-347.

Nilsson LA 1983. Anthecology of *Orchis mascula*(Orchidaceae). Nordic J Bot 3: 157-179.

Nilsson LA 1988. The evolution of flowers with deep corolla tubes. Nature 334: 147-149.

Nilsson LA 1992. Orchid pollination biology.Trends Ecol Evol 7: 255-259.

Nosrati H, Hajiboland R, Razban-Haghighi A, Nikniazi M 2011. A comparative assessment of fruit formation in some orchid species from the southern Caucasus region. Turk J Bot 35: 553-560.

Pauw A, Stofberg J, Waterman RJ 2009. Flies and flowers in Darwin's race. Evolution 63: 268-279.

Peakall R 1989. The unique pollination of *Leporella fimbriata*(Orchidaceae) by pseudocopulating winged male ants *Myrmecia urens*(Formicidae). Plant Syst Evol 167: 137-148.

Peakall R, Ebert D, Poldy J, Barrow RA, Francke W, Bower CC, Schiestl FP 2010. Pollinator specificity, floral odour chemistry and the phylogeny of Australian sexuallydeceptive Chiloglottis orchids: Implications for pollinator-driven speciation. New Phytol 188: 437-450.

Peakall R, Handel SN 1993. Pollinators discriminate among floral heights of a sexually deceptive orchid: implications for selection. Evolution 47:1681-1687.

Pedron M, Buzatto CR, Singer RB, Batista JAN, Moser A 2012.Pollination biology of four sympatric species of *Habenaria* (Orchidaceae: Orchidinae) from southern Brazil. Bot J Linn Soc 170: 141-156.

Pemberton RW 2010. Biotic resources needs of specialist orchid pollinators. Bot Rev 76: 275-292.

Perez-Hérnandez H, Damon A,Valle-Mora J, Sanchez-Guillen D2011.Orchid pollination: specialization in chance? Bot K Linn Soc 165: 251-266.

Peter CI, Johnson SD 2009. Autonomous self-pollination and pseudo-fruit set in South African species of Eulophia(Orchidaceae). South African J Bot 75: 791-797.

Primack R, Stacy E 1998. Costs of reproduction in the pink lady's slipper orchid (*Cypripedium acaule*): an eleven-year experimental study of the three populations. Am J Bot 85: 1672-1679.

Reis MG, Pansarim ER, da Silva UF, Amaral MCE, Marsaioli AJ 2004. Pollinator attraction devices (floral fragrances) of some Brazilian orchids. Arkivoc 6: 103-111.

Rodríguez-Robles JA, Meléndez EJ, Ackerman JD 1992.Effects of display size, flowering phenology, and nectaravailability on effective visitation frequency in *Comparettia falcata* (Orchidaceae). Am J Bot 79: 1009-1017.

Rothschild WL, Jordan K (1903). A revision of the lepidopterous family Sphingidae.Hazell, Watson & Viney, Ltd.

Roy BA, Widmer A 1999. Floral mimicry: a fascinating yet poorly understood phenomenon. Trends Plant Sc 4: 325-330.

Sakai S2002.A review of brood-site pollination mutualism: plants providing breeding sites for their pollinators. J Plant Res 115: 161-168.

Salzmann CC, Nardella AM, Cozzolino S, Schiestl FP 2007. Variability in floral scent in rewarding and deceptiveorchids: The signature of pollinator-imposed selection? Ann Bot 100: 757-765.

Schultz CB 2001. Restoring resources for an endangered butterfly. J Appl Ecol 38: 1007-1019.

Scopece G, Musacchio A, Widmer A, Cozzolino S 2009. Contrasting thoughts about deceptive orchids: A response to Sobel and Randle. Evolution 63: 2205-2209.

Scopece G, Cozzolino, S, Johnson SD, Schiestl FP 2010.Pollination efficiency and the evolution of specialized deceptive pollination systems. Am Nat 175: 98-105.

Sheviak CJ, Bowles ML 1986. The prairie fringed orchids: a pollinator isolated species pair. Rhodora 88: 267-290.

Shivanna KR 2014. Biotic Pollination: How plants achieve conflicting demands of attraction and restriction of potential pollination. In Ramawat KG, Merillon JM and Shivanna KR (Eds) Reproductive Biology of Plants. CRC Press, pp. 218-267.

Shivanna KR, Tandon R 2014. Reproductive ecology of flowering plants: a manual. Springer.

Siegel C 2011. Orchids and Hummingbirds: Sex in the fast lane. Orchid Digest 75: 8-17.

Singer RB, Sazima M 1999. The pollination mechanism in the Pelexia alliance (Orchidaceae: Spiranthinae) Bot J Linn Soc 131: 249-262.

Singer RB, Sazima M 2000.The pollination of *Stenorrhynchos lanceolatus* (Aublet) L C Rich(Orchidaceae: Spiranthinae) by humming birds in southeastern Brazil.Plant Syst Evol 223: 221-227.

Southwick EE 1984. Photosynthate allocation to floral nectar a neglected energy investment. Ecology 65: 1775-1779.

SrimuangK, BänzigerH, Pedersen HAE, WatthanaS 2013. Carpenter Bees and the Orchid of a Princess: Natural Pollination of *Sirindhornia monophylla* in Thailand. Taiwania 58: 163-170.

Stockl J, Paulus H, Dafni A, Schulz C, Francke W, Ayasse M 2005. Pollinator attracting odour signals in sexually deceptive orchids of the *Ophrys fusca* group. Plant Syst Evol 254: 105-120.

Stockl J, Brodmann J, Dafni A, Ayasee M, Hansson BS 2010. Smells like aphids: orchid flowers mimic aphid alarm pheromones to attract hoverflies for pollination. Proc Roy Soc B 278: 1216-1222.

Steiner KE, Whitehead VB 1991. Oil flowers and oil bees: Further evidence for pollinator adaptation. Evolution 45: 1493-1501.

Steiner KE 1989. The pollination of *Disperis* (Orchidaceae) by oil collecting bees in South Africa. Lindleyana 4: 164-183.

Stoutamire WP 1967. Flower biology of the lady's slippers (Orchidaceae: *Cypripedium*). Michigan Bot 6: 159-175.

Stoutamire WP 1971. Pollination in temperate Americanorchids. In JG Gorigan (Ed) Procedings of the 6th World Orchid Conference, Sydney Halstead Press, Sydney, pp. 233-243.

Suetsugu K 2013.Autogamous fruit set in a mycoheterotrophic orchid *Cyrtosia septentrionalis*. Plant Syst Evol 299: 481-486.

SugiuraN, GoubaraM, Kitamura K, Inoue K 2002.Bumblebee pollination of *Cypripedium macranthos* var*rebunense* (Orchidaceae); a possible case of floral mimicry of*Pedicularis schistostegia* (Orobanchiaceae). Plant Syst Evol 235: 189-195.

Sugiura N, Fujie T, Inoue K, KitamuraK 2001. Flowering phenology, pollination, and fruit set of *Cypripedium macranthos* var*rebunense*, a threatened lady's slipper (Orchidaceae). J Plant Res 114: 171-178.

Sugiura N 2013. Specialized pollination by carpenter beesin *Calanthe striata* (Orchidaceae), with a review of carpenterbee pollination in orchids. Bot J Linn Soc 171: 730-743.

Sun H, Yi-Bo L, Song G 2003. A Preliminary Study on Pollination Biology of an Endangered Orchid, *Changnienia amoena*, in Shennongjia. Acta Bot Sin 45: 1019-1023.

Sun H, Yi-Bo L 2005. Pollination of a slippery lady slipper orchid in south-west China: *Cypripedium guttatum* (Orchidaceae). Bot J Linn Soc 148: 251-264.

Tremblay RL 1992. Trends in the pollination ecology of the Orchidaceae: evolution and systematics. Can J Bot 70: 642-650.

Tremblay RL, Ackerman JD, Zimmerman JK, Calvo RN 2005. Variation in sexual reproduction in orchids and itsevolutionary consequences: a spasmodic journey to diversification. Biol J Linn Soc 84: 1-54.

Urru I, Stensmyr MC, Hansson BS 2011. Pollination by broodsite deception. Phytochemistry 72: 1655-1666.

van der Cingel NA 1995. An atlas of orchid pollination: European orchids. Rotterdam: A. A. Balkema.

van der Cingel N 2001. An Atlas of Orchid Pollination: America, Africa Asia and Australia. Rotterdam: A. A. Balkeman.

van der Pijl L, Dodson C 1966.*Orchid Flowers: Their Pollination and Evolution.* Coral Gables, FL: University of Miami Press.

van der Pijl L, Dodson CH 1969. Orchid flowers; their pollination and evolution.University of Miami Press, Coral Gables, Florida.

Vereecken NJ, PatinyS 2005. On the pollination of *Ophrys catalaunica* (O Danesch & E. Danesch) by pseudocopulating males of *Chalicodoma pyrenaica* (Lepeletier) (Hymenoptera: Megachilidae). Naturalistes Belges 86:57-64.

Vogel S 1978. Pilzmueck enblumen als Pilzmimeten. Flora 167: 329-398.

Widmer A, Cozzolino S, Pellegrino G, Soliva M, Dafni A 2000. Molecular analysis of orchid pollinaria and pollinariaremains found on insects. Mol Ecol 9: 1911-1914.

Williams NH 1982. The biology of orchids and euglossine bees. In J Arditti (Ed) Orchid Biology: Reviews and perspectives, II. Ithaca: Cornell University Press, pp 119-171.

Woodcock TS, Larson BMH, KevanPG, Inouye DW, Lunau K 2014.Flies and Flowers II: Floral attractants and Rewards. Taiwania 58: 163-170.

Xiaohua J, Dezhu L,Zongxin R,Xiaoguo X 2012. A generalized deceptive pollination system of *Doritis pulcherrima* (Aeridinae: Orchidaceae) with non-reconfigured pollinaria. BMC Plant Biol 12: 67.

Zimmerman JK, Aide TM 1989. Patterns of fruit production in a Neotropical orchid: pollinator vs resource limitation. Am J Bot 76: 67-73.

Zhou X, Lin H, Fan XL, Gao JY 2012. Autonomous self-pollination and insect visitation in a saprophytic orchid, *Epipogium roseum* (D Don) Lindl Austr J Bot 60: 154-159.

Chapter 6

Attractants for Honey Bees

Pradnya V Deshpande and Dattatraya G Naik
pardesh1234@yahoo.co.in; dgnpune@yahoo.co.in

Abstract

Attractants for honey bees have been a subject of research for many years. They are important for inducing the visits of honey bees to plants of interest. Initially clues for developing such formulations were obtained by studying the chemical constituents of attractant pheromone, Nasonov gland pheromone. Interestingly, there were variations in the chemical constitution of Nasonov gland pheromone isolated from different parts of the world. Attractant formulations were developed using the chemical constituents of Nasonov gland pheromone. However the attractiveness of such formulations was found to be concentration dependent. This posed serious limitations to their applications. Attempts were then made to develop attractant formulations using the extracts of plants frequently visited by honey bees. Availability of attractant formulations is helpful not only in improving the quality of honey but also in improving the efficiency of pollination. All these developments indicate the research activities and the application potential in the field. Taking advantage of such research findings certain attractants were developed on commercial scale. However, it was interesting to note that they were not equally effective on all crops. All these findings make development of attractants for honey bees a very active research area.

6.1 Introduction

It is well-known that the largest class in the animal kingdom is of insects. Amongst all the insects we come across, most are harmful but very few are useful to the mankind. Honey bees are amongst these few useful insects. They are Hymenopteran insects and live in social groups, known as 'colonies'. Such colonies are generally found on branches of trees, in cavities of tree trunks and even on some tall buildings. Considering the usefulness of the products obtainable from honey bee colonies, techniques were developed since long

DOI: 10.1201/9781003505471-6

for domesticating them. Colonies of honey bees are now maintained for the production of honey and various beehive products in many parts of the world including India.

The genus *Apis* to which honey bee belong, includes four major well defined species – *A. mellifera*, commonly referred as the European honey bee, *A. cerana*, *A. florea* and *A. dorsata*, commonly referred as the rock bee (Maa 1953). In late sixties *A. mellifera* was also introduced in India and by now all the three species, *A. cerana, A. florea* and *A. mellifera* are found to be domesticated by Indian beekeepers. In India the main objective of beekeeping has always been honey production due to which the important function of pollination carried out by honey bees has attained secondary importance. Although honey bees are the important pollinators of several crops, very little attention has been paid in getting the maximum benefit from their pollination.

Foraging is the important activity carried out by honey bees which is responsible for collection of nectar and transfer of pollen grains. When a honey bee visits a flower, pollens from that plant are collected. If the subsequent visit of the honey bee happens to be to a different plant species, the pollens, even if transferred, get wasted. However if honey bee visits the same plant species, pollination becomes successful (see Chapter 4, this volume). This can result into higher efficiency of pollination directly leading to the higher crop yields! Attempts were, therefore, made to regulate the visits of honey bees so that they can be made to visit only the plants of interest. Availability of honey bee attractants is very important from this point of view.

In order to use honey bees more efficiently and fruitfully, the efforts were made to understand their behaviour in more detail. As in other insects, communication is one of the most important factors determining the pattern of behaviour of honey bees (von Frisch 1967).

Honey bees mainly communicate with each other by means of vision, taste, touch and smell. Pheromones are the chemicals used by insects for the purpose of communication (Karlson and Luscher 1959). Pheromones are normally classified on the basis of their function. Accordingly, pheromones of honey bees, used as attractant, repellent, alarm, trail, sting and queen nuptial pheromones are well known along with their glandular sources (Free 1987). Amongst these, Nasonov gland pheromone is important for attraction.

6.2 Nasanov Gland Pheromone: Attractant Pheromone in Honey Bees

The pheromone secreted by Nasonov gland is very important for the purpose of communication (Snodgrass 1956). The function of Nasonov gland was first described by a Russian Scientist Zoubareff (1883). Sladen (1901, 1902) suggested that the odour from glands of worker bees is dispersed by fanning which attracted other members of their colony. He hypothesized that scenting by the first worker of a colony to discover their hive rapidly initiates similar behaviour in their companions and pointed out that the calling is infectious. It was, then, confirmed that this gland was the origin of odours by which honey bees distinguish their companions (Mclndoo 1914). It was found that if the source of food was odourless, the release of Nasonov gland pheromone was more conspicuous. To understand this, an interesting experiment was done. The bees were allowed to forage on syrup in dishes containing glass balls. The glass balls from the dish at which the bees had foraged for one day were removed, dipped in absolute alcohol and the alcoholic extracts were screened on honey bees. The extracts were found to be very attractive to bees (Lecomte 1957).

The foraging bees were found to mark the food with a scent. Worker honey bees were found to release Nasonov gland pheromone at the hive entrance when virgin queen was about to leave the hive for the nuptial flight (Brian 1983, Kalmus and Ribbands 1952). This was thought to serve the purpose of guiding her to the hive entrance on her return. The conditions under which the foraging bees expose their Nasonov gland were being studied around the same time (Free 1968). Nasonov gland pheromone release was also found to be associated with the collection of water by worker honey bees (Free and Williams 1970). The scout bees were seen using this pheromone in the searching activity (Waller 1973). Worker bees which were denied access to their colony, especially after colony disturbance, were found to expose their Nasonov glands for releasing the pheromone at the entrance to their hive (Ferguson and Free 1981). Pflumm (1970) while studying the factors releasing scent-making behaviour in foraging honey bees found that scent marking with the Nasonov gland was dependent on the 'flux of sugar per unit time'. Subsequently the effect of the composition of the diet on the scent marking behaviour and honey-sack load of a foraging honey bee was also demonstrated (Pflumm et al. 1978). Thus, as per the preliminary observations, it could be hypothesized that the honey bees mainly release the Nasonov gland pheromone during colony movement, reorientation, foraging and collection of water and at the time of nuptial flight of virgin queen. This pheromone was also seen to be released by queenless workers for nest entrance, forage and water source marking and also under artificial circumstances concerned with the reorientation (Free 1987).

Message of attraction is important in order to ensure the visits of honey bees to the desired site. It could be useful in getting unifloral honey, to increase the yield of economically valuable crops and also to attract honey bee swarms.

6.2.1 Chemical Composition

Nasonov scent organ and the physiological role of its secretion were initially studied by Renner (1960). He obtained the scents by wiping the exposed scent glands of bees with small bits of filter paper and demonstrated the strong attractive effect of the substance secreted by scent glands. However, he concluded that it was not specific to a colony. Boch and Shearer (1962) used the wiping method for isolation of pheromone of *A. mellifera* when they found geraniol (Figure 6.1) to be the major constituent of the pheromone. In their study, the chromatography of carbon disulphide extracts of the wipes on a vapour fractometer equipped with a hydrogen flame ionization detector showed characteristic major fractions. To identify these components, gas chromatography (GC) of the extracts was carried out and the effluents were trapped in carbon disulphide (Shearer et al. 1963). Infrared (IR) spectrum of this solution was identical with that of geraniol. As the major chemical constituent of Nasonov gland pheromone, geraniol was expected to possess attractive properties. Accordingly, the attractiveness of geraniol to foraging European honey bees was demonstrated (Free 1962). The geraniol content of bees of different ages was subsequently determined (Boch and Shearer 1963). It was found that the quantity of geraniol in Nasonov gland secretion was more closely related to the foraging activity of bees rather than the chronological age. Formation of two other minor components geranial (Figure 6.2) and neral (Figure 6.3) along with the other major compound nerolic acid (Figure 6.4) was shown by Shearer and Boch (1966). Geranial and neral are double bond isomers of citral. These were shown to enhance the attractiveness of the major component, geraniol. They also found that isomers of citral were not present initially in freshly prepared wipes, and formed when wipes were kept at room temperature. Detailed examination of the extracts of Nasonov gland led to the identification of two isomers of citral, nerolic acid and geranic acid (Figure 6.5) as its constituents (Butler and Calam 1969). Chemical composition of Nasonov gland pheromone isolated from a single *A. mellifera* worker bee was determined by gas chromatography-mass spectrometry (GC-MS) using capillary column (Free et al. 1981, Pickett et al. 1980). The presence of seven terpenoids (Figure 6) was shown in Nasonov gland extracts including the two new compounds, nerol (Figure 6.6) and (E, E)-farnesol (Figure 6.7). The relative proportions of these constituents were also determined (Table 6.1; Picketteet al. 1980).

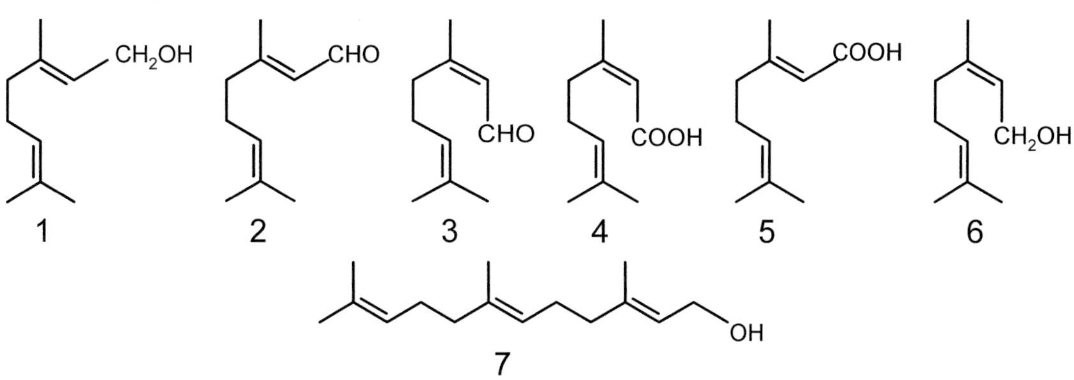

Figure 6.1-6.7 Chemical constituentsof the Nasonov gland pheromone of *Apis mellifera.* (6.1) Geraniol, (6.2) Geranial *i.e.* (*E*)-Citral, (6.3) Neral *i.e.*(*Z*)-Citral, (6.4) Nerolic acid, (6.5) Geranic acid, (6.6) Nerol, (6.7) (*E, E*)-Farnesol

Table 6.1. Relative proportions of the chemical constituents of Nasonov gland pheromone of *A. mellifera* in Europe.

Slno.	Compound	Relative proportion
1	Geraniol	100 parts
2	Nerolic acid	75 parts
3	(E,E)-Farnesol	50 parts
4	Geranic acid	12 parts
5	Geranial	1 part
6	Neral	1 part
7	Nerol	1 part

6.2.2 Bioevaluation of constituents

Identification of the chemical constituents of Nasonov gland pheromone initiated the experiments to examine their attractiveness towards honey bees. These experiments subsequently led to the development of attractant formulations, which involved combination of these compounds.

Although geraniol was the major compound present in the Nasonov gland pheromone of honey bee *A. mellifera*, attractiveness of citral to foraging honey bees was conspicuous (Weaver et al. 1964). It is very interesting to note that even though isomers of citral were minor components of Nasonov gland pheromone, they were found to be more attractive than geraniol. Sedin (1967) studied the influence of some aromatic substances on bees. He found synthetic geraniol, was a less effective marker for 'novice bees' than the natural secretion suggesting the role of other constituents in the process of attraction. Further, he observed that geraniol and nerol, singly or combined, did not influence the behaviour of honey bees unless they were foraging. This showed that the influence of Nasonov gland pheromone was predominant in the foraging activity. Attractiveness of geraniol, citral and anise oil to honey bees on plots of Lucerne (*Medicago sativa*) was studied by Waller (1970). It was interesting to note that geraniol was generally more attractive than citral when it was applied in a solution containing five or ten percent sucrose, but less attractive than citral when applied in water. Studies on responses of honey bees to the chemical constituents of Nasonov gland pheromone continued in 1970s and 1980s also. Ferguson et al. (1979) described the techniques for studying honey bee pheromone involved in clustering based on which the effect of Nasonov gland and queen pheromones were shown. Williams et al. (1981) carried out bioassay of seven individual components of Nasonov gland pheromone. It was demonstrated that each component attracted honey bees foraging on sugar syrup. The mixture of components in the proportions present in the secretion of honey bees was found to be as attractive as natural secretion. They also found that the presence of 'footprint' pheromone enhanced the attractiveness of the mixture of synthetic chemicals constituting Nasonov gland pheromone. Eventually a 'lure' for attracting *A mellifera* bees to a hive or a trap was evolved and patented by Pickett et al. (1982). The synthetic 'lure' contained geraniol, (*E*)-citral or a mixture of isomers of citral, and nerolic acid. Geranic acid and queen mandibular gland pheromone, (*E*)-9-oxo-2-decenoic acid (Figure 6.8) were also present in this 'lure'. A kit consisting of a hive and the 'lure' was described in this patent application.

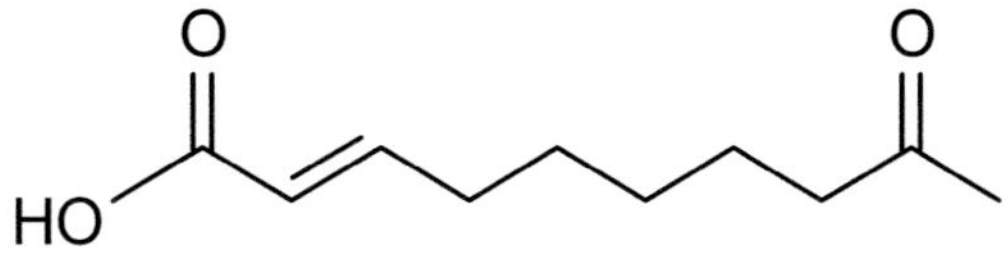

Figure 6.8. (*E*)-9-Oxo-2-decenoic acid

Application of the synthetic lure to stimulate worker honey bees, *A. mellifera* to consume water and artificial forage was shown by Free et al. (1983). Winston et al. (1993) also tested the attractiveness of the constituents of Nasonov gland pheromone and their mixtures. Citral, geraniol plus nerolic acid were found to be the most attractive mixture than citral, geraniol plus queen mandibular gland pheromone followed by citral, geraniol plus hexanal (Figure 6.9) and geraniol, citral alone.

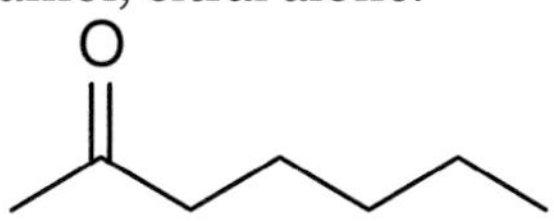

Fig. 6.9 Hexanal

Fine structure of Nasonov gland of the worker honey bees *A. mellifera* was examined using electron microscope and cytochemical techniques. It was proved that the gland secreted volatile terpenoids and protein compounds. The bioassay indicated that the protein fraction enhanced the attractiveness of the volatile fraction to workers (Cassier and Lensky 1994). In the account on honey bee 'lure', Free et al. (1984) have described the development of 'lure' containing components of the honey bee Nasonov gland pheromone. They have also narrated the possible applications of the 'lure'.

6.3 Application of Attractants Based on Nasonov Gland Pheromone for Pollination Services

Applications of attractant properties of chemical constituents for the benefit of agricultural crops were envisaged since long. Some such experiments on attracting *A. mellifera* bees using components of Nasonov gland pheromone on the plots of strawberry are reported in the recent past. The most attractive components were found to be geranial, neral and nerol. It was found that geraniol, nerolic acid and geranic acid were not much effective but other components did exhibit attractant properties. Moreover mixtures of geraniol, geranial, neral, nerol, as well as nerolic acid and geranic acid proved to be more attractive than any one of the individual components alone. The increased number of visits of honey bees to the plots sprayed with Nasonov gland pheromone suggested that it could be used to attract honey bees which in turn are very useful to improve pollination (Pardo and Nates 1994).

Applications of constituents of the Nasonov gland pheromones for attracting honey bees on onion crops were also shown. Application of geraniol and citral was shown to increase onion seed production in Poland (Woyke 1981). Additionally application of Anise oil was also demonstrated for attracting honey bees in this report. Zvedenok (1996) demonstrated that use of geraniol and citral along with other terpenoid limonene (Figure 6.10) as the attractants resulted in improved pollination and yields of onion crop.

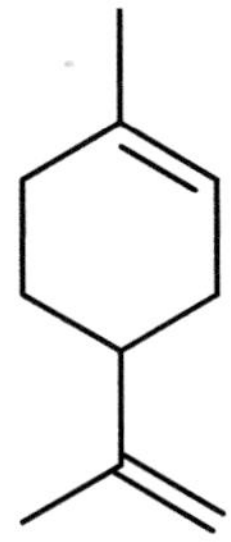

Figure 6.10 Limonene

Attractant properties of constituents of Nasonov gland pheromone were used in yet another way. Geraniol and citral when added to the fungicide, pallinal-endosulfan, masked the repellency of fungicide and improved pollination (Ohe and Praagh 1983).

6.4 Application of 'lures' as Swarm Attractants

6.4.1 Swarming phenomenon

In the flowering season, when food is abundantly available, queen lays more eggs resulting into the increase in the bee population. Due to this the number of workers increases to such an extent that all the bees cannot be accommodated in the original colony. A phenomenon called 'swarming' takes place under such circumstances. In swarming, the original queen flies out of the colony along with a few worker bees and settles elsewhere. The worker bees left behind eventually develop a new queen. Thus, the natural division of colony occurs due to swarming. Nasonov gland pheromone is reported to play an important role of marking the colonies in the process of swarming (Boch and Morse 1974, Mautz et al. 1972, Morse and Boch 1971). It was further demonstrated that swarming bees are guided by queen's substance and pheromone released by worker bees (Avitabile et al. 1975). Beekeepers have no control over the place where the swarm gets settled and usually the swarm is lost. Use of attractants to capture the honey bee swarms is, therefore, an area of active interest since long.

6.4.2 Swarm attractants

Use of "lemon balm" as swarm attractant was made by Parkinson in 1629! Unfortunately his original reports are not available. However reference to Parkinson's work is found in the literature (Burgett 1980). Besides this, Burgett also mentioned that Melissa oil was used as a swarm attractant in 1930s. Witherell (1985) reviewed the scientific literature relating to honey bee bait hives and swarm attractants. Effectiveness of bait hives and 'lures' to attract honey bee swarms were reported as survey tools for use in Africanised honey bee eradication programs. Mixtures of geraniol and citral were found to be the most effective among the 'lures' tested (Witherell and Lewis 1986).

In the bioassays carried out for testing the effects of pure chemicals it was found that synthetic geraniol, (*E*)-citral, nerolic acid, geranic acid, (components of Nasonov gland pheromone of *A. mellifera*) induced worker honey bees to release Nasonov gland pheromone at their hive entrance. Absence of any one of these components stimulated pheromone release less effectively but the lack of nerol or (*E, E*)-farnesol increased effectiveness. These observations were very useful in the development of 'lure' for swarms (Witherell and Lewis 1986, Free et al. 1983a, 1983b).

Use of Nasonov 'lure' to attract Africanised colonies to hives was shown by Ferguson (1985). Kigatiira et al. (1986) carried out experiments using synthetic pheromone 'lure' to attract honey bee colonies in Kenya. They concluded that 'lures' should be placed in hives only at those times of the year when migrating colonies usually settle in the area concerned.

Mixture of geraniol, citral and nerolic acid in equal proportions in the traps was found to be useful in capturing natural swarms of *A. mellifera* (Schmidt and Thoenes 1987, Schmidt 1988, Schmidt et al. 1989). The effect of 'lure' described by Pickett et al. (1982) on attracting the swarms was also studied later (Lesher and Morse 1983). It was found that, of the eight hives occupied by the swarms, seven contained 'lures'. Effectiveness in Nasonov gland pheromone in long distance swarm attraction was shown by Villa and Schmidt (1992). When the efficacy of queen and Nasonov gland pheromones in the capture of *A. mellifera* swarms was compared, it was found that the Nasonov gland pheromone treated traps were more attractive than queen pheromone and queen and Nasonov gland pheromone combinations (Denby and Scott-Dupree 1992).

Detailed investigations on the swarms of both Africanised and European honey bees found that synthetic pheromone blends containing geraniol, citral and nerolic acid in combination were superior to those containing only geraniol or citral. The role of queen pheromone in combination with the constituents of Nasonov gland pheromone was demonstrated in some trials. When tests were performed using bait hives treated with different materials, it was observed that addition of queen pheromone did not increase the attractiveness of Nasonov gland pheromone-treated cavities. However, in the study of short range attractiveness, swarms were found to prefer cavities with Nasonov gland pheromone along with queen pheromone to those with Nasonov gland pheromone alone (Schmidt et al. 1993). Experiments to examine the attraction of reproductive honey bee swarms to artificial nests by Nasonov gland pheromone indicated that it played a vital role in the attraction of honey bee swarms to nest cavities (Schmidt 1994).

It was observed that honey mixed with geraniol and citral was effective to attract honey bees during nectar dearth period and in favourable flight conditions. Interestingly during strong nectar period, this method was found not to be effective (Danka et al. 1990).

An entirely different application of honey bee attractants was shown by Visscher and Khan (1995) in which inexpensive traps which contained pheromone 'lures' and insecticidal soap solution was used for trapping and killing the bees remaining after swarm and colony removal.

Intriguing experiments were carried out to determine if Nasonov gland pheromone served as a true pheromone or it was simply a general attractant for *A. mellifera* swarms by comparing synthetic Nasonov gland secretion with linalool (Figure 6.11), oil of clove, skatole and *Galleria mallonella* sex pheromone.

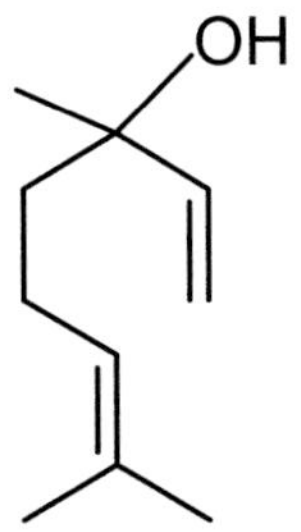

Figure 6.11 Linalool

The results of these experiments confirmed that Nasonov gland pheromone was a true pheromone in the context of attracting swarms to nest cavities and that odours naturally present in the environment or in the cavity play little or no role in nest seeking behaviour (Schmidt 1999).

6.5 Studies on Nasonov Gland Pheromone of Indian Honey Bees

The pheromone chemistry of Indian honey bees had not been studied till late 1980s. The first report indicated the presence of only one major compound, neral, in the extract of Nasonov gland pheromone of the Indian honey bee *A. cerana indica* Fabr. (Naik et al. 1988). This was surprising because the corresponding pheromone in *A. mellifera* was a mixture of seven components (Table 6.1). Moreover neral, found as the major constituent in *A. cerana indica* was one of the minor constituent in the pheromone of *A. mellifera*. A 'lure' to attract honey bees *A. cerana* using citral was eventually developed (Naik et al.1989). Nasonov gland pheromone of other Indian honey bee *A. florea* was studied subsequently. Like the pheromone of *A. cerana*, it also contained only one major component, geranial. Based on this observation a 'lure' to attract the Indian honey bee *A. florea* was also developed and patented (Naik et al. 2001). Occurrence of different but closely related chemicals as the major constituents of Nasonov gland pheromone of the two Indian honey bee species is significant in many ways. It can be seen that geranial and neral are are the geometrical isomers of each other (Figure 6.2 and 6.3). Thus their biosynthesis can be postulated from the same precursors. At the same time the difference in the structure may be responsible in avoiding intermixing of signals. Thus signals given by the workers of *A. cerana* are ineffective for the workers of *A. florea* and *vice versa*.

It is intriguing to note that although Nasonov gland pheromone of European honey bee *A. mellifera* was found to be a mixture of seven compounds in different proportions (Table 6.1), there was only one major constituent in the pheromone of Indian honey bees *A. cerana* as well as *A. florea*. Honey bees *A. mellifera* have been introduced in India in 1960s. They started getting established and being used for improved beekeeping in late 1990s. The Nasonov gland pheromone of *A. mellifera* maintained in India was, then, examined and its chemical constitution was compared with that of *A. mellifera* in Europe. Surprisingly in the preliminary examination it was found that the Nasonov gland pheromone of *A. mellifera* in India was not the mixture of seven compounds. In fact like the Nasonov gland pheromone of *A. cerana indica* and *A. florea, A. mellifera* in India contained only one major component. The component was identified as nerol (Naik et al. 1999). The honey bees studied in these experiments were from the areas where the average temperature was around 25°C, higher than that commonly found in Europe. To examine if the differences in pheromone composition of *A. mellifera* in Europe and in India were due to the differences in temperature of the habitat of honey bees, the pheromone from honey bees foraging in the areas in India having cooler climate, similar to that in Europe, was examined. It was interesting to note that although nerol continued to be the major constituent in this pheromone too, presence of neral was also established (Naik et al. 2001). It is important to note that nerol happens to be the minor component in the Nasonov gland pheromone of *A. mellifera* in Europe. However it is the geometrical isomer of geraniol, the major component of the Nasonov gland pheromone found in Europe (Figure 6). Examination of the chemical structures of geraniol, geranial, neral and nerol easily brings about the similarities in their skeleton which suggests the common precursors in their biosynthesis. The differences in their structures might be the reasons for their specificity.

6.5.1 Variation in Pheromone Composition

Variation in the chemical constitution of Nasonov gland pheromone of *A. mellifera* in Europe and India clearly demonstrated the fact that the pheromone composition is affected by the climatic conditions. Literature search also supported these observations. Neral was reported to be the major constituent of Nasonov gland pheromone of *A. cerana* from India (Naik et al. 1988) but the constituents of the same pheromone of Asiatic honey bee *A. cerana* were identified as geraniol, (E)-citral and (Z)-citral (Abdullah et al. 1990) while occurrence of linalool oxide (Figure 6.12) as a constituent of this pheromone of *A. cerana* from Japan was eventually reported (Matsuyama et al. 1998). It is noteworthy that even the traces of neral were not detected in the bees from Japan and linalool oxide was not found even in minor quantities in the bees from India.

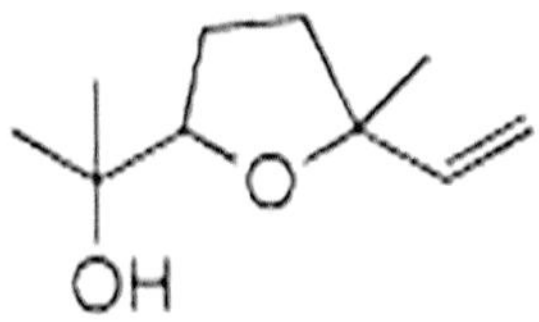

Figure. 6.12 Linalool oxide

6.6 Pheromone Based 'lures' for Honey Bees

'Lures' to attract honey bees were developed using the chemicals constituting Nasonov gland pheromones (Naik et al. 1989). The attractiveness (Δ) of the lure was estimated as the difference between the number of honey bees visiting the dish containing the test formulation and that visiting the dish of the control (liquid paraffin). The attractiveness of the lures was found to be concentration-dependent (Figure 6.13). Formulations of the chemical being used in the lure of lower concentrations possessed marginal attractiveness. The attractiveness (Δ) was found to increase linearly with the concentration of the attractant chemical. However this increase was not uniform. The attractiveness was seen to pass through a maximum and then got reduced. The formulations of higher concentrations than the one having maximum attractiveness were found to be less effective. The formulations of very high concentrations were found to be repellent! These results (Figure 6.13) impose serious limitations to their field applications. Formulations of very lower concentrations are less effective. At the same time the formulations of higher concentrations are also less effective. Furthermore the formulations of the optimum concentration, having maximum attractiveness, are likely to reduce their attractant properties due to volatilisation of the active chemical. These issues make the field application of pheromone based lures almost impractical.

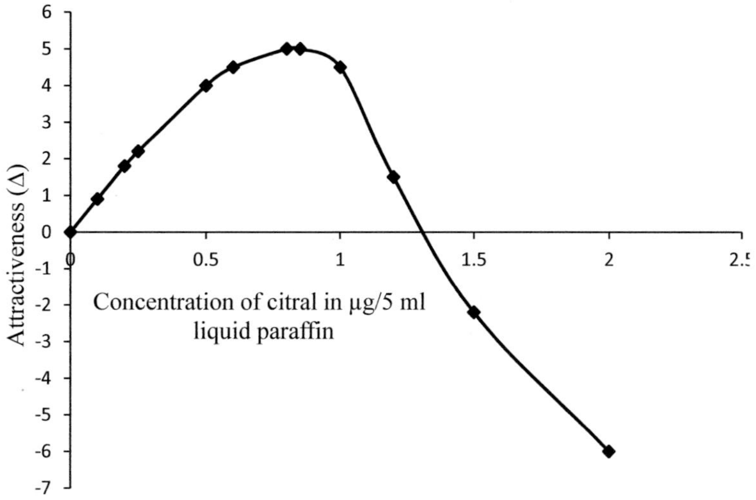

Figure 6.13. Variation of attractiveness of formulations with concentration of citral.

6.7 Search for New 'lures'

Even though there are different varieties of plants and flowers in nature, all of them are not visited by honey bees in the uniform manner. Some of the plants are visited more frequently in preference to others. The volatiles released by the plants were thought to be responsible for this. Investigation of the extracts of such plants was, therefore, considered as a possible alternative to pheromone based lures.

6.7.1 Plant-based attractants

The relation between plants, flowers, nectar and insects are known since ancient times. Insects, particularly honey bees, were known to collect nectar from flowers to produce honey. Amongst the factors that attract honey bees to flowers are the attractive colours of the flowers, but the fragrance of the plant parts can also be considered.

Butler (1951) studied the importance of perfume in the discovery of food by the worker honey bees *A. mellifera*. He carried out the experiments using coloured discs of paper and the perfumes of the white clover (*Trifolium repens*), red clover (*T. pretense*), hawthorn (*Crataegus oxyacantha*), oil of thyme (*Thymus vulgaris*) and Bridal wreath (*Spiraea arguta*) dissolved in odourless paraffin oil. He reported that the untrained bees were dependent upon the sight to find flowers from a distance but on close approach they entered flowers more readily if they were attractively scented.

Wykes (1952) studied the preferences of honey bees for solutions of various sugars, which occur in nectar. From these experiments it was found that the sugars occurring in nectar were not equally attractive to bees. The attention was drawn to the possible biological significance of the selective responses by bees and it was shown that the sugar composition of nectar might be a factor influencing the visits of bees to flowers. Youngken (1956), while studying medicinal plants, observed that certain medicinal plants were selectively visited by honey bees. In his report, he listed the plants that were frequently visited by honey bees (Table 6.2). Anise oil was subsequently used in attracting honey bees towards fruit trees (Mayer and Johansen, 1982).

Table 6.2. Plants frequently visited by the honey bees *A. mellifera*.

Slno.	Common name	Botanical name
1	Rosemary	*Rosmarinus officinalis*
2	Sage	*Salvia officinalis*
3	Frenel	*Foeniculum vulgare, F. dulee*
4	Hyssop	*Hyssopus officinalis*
5	Lavender	*Lavendula spica, L. vera*
6	Hoarhound	*Marrubium vulgare*
7	Peppermint	*Mentha piperita*
8	Spearmint	*Mentha spicata*

While studying the taste and aroma of attractants in honey bee (*A. mellifera)* feeds, Maliszewska and Szymas (1997) made use of geraniol, a plant derived product, in the tests. They found that honey was most attractive to bees, followed by pollen and then geraniol. While studying the *A. mellifera* swarms in Ithaca, NY, Ambrose (1975) found that some substrates such as liliac bushes seemed to be particularly attractive to the swarms.

Use of lemon balm (*Melissa officinalis*) for attracting honey bee swarms was reported by Burgett (1980). The compounds which contribute to the fragrance of the oil in the green leaves of *M. officinalis* were identified; they include geraniol, citral and nerol. Wilhelm and Pflumm (1983) investigated the effect of oil from lily of the valley, rose oil on the exposure of Nasonov glands by honey bees. However, they found the period of Nasonov gland exposure was not significantly affected by the oils being tested.

Honey bees were seen to be attracted to the flowers and fruits of *Evodiahupehensis*. The crude extract of dried fruits was attractive to honey bees when sprayed on Lucerne plots. Two compounds, 2-keto-tridecan-1-yl acetate (Figure 6.14) and 2-keto-tridecanol (Figure 6.15) were identified in the fruit extracts (Gellert et al. 1985).

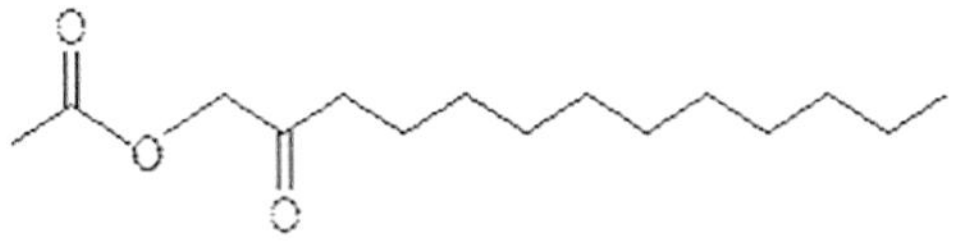

Figure 6.14. 2-Keto-tridecan-1-yl-acetate

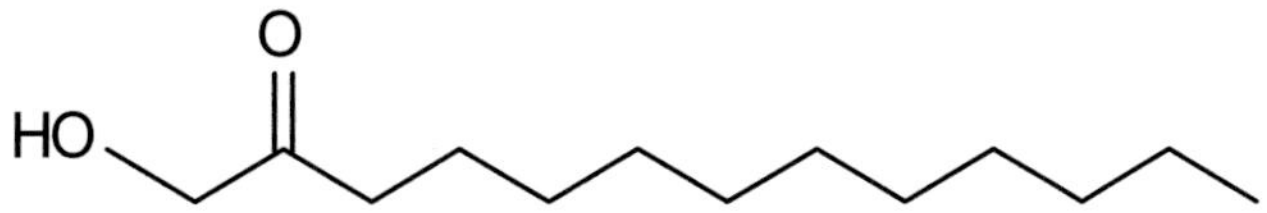

Figure 6.15. 2-Keto-tridecanol

In an interesting experiment Pinzauti (1985) evaluated Beeline® and its mixtures with oil of lemon or oil of orange which were known to attract pronubial insects including honey bees on the plots of safflower (*Carthamus tinctorius*). All of the solutions sprayed made the flowers of *C. tinctorious* more attractive for the honey bees than the untreated controls. The order of the effectiveness of the solutions sprayed was determined. Beeline® (50 g) mixed with 1 L oil of lemon diluted to 50 L with water showed maximum attractiveness; Beeline® (50 g) mixed with 1 L oil of orange diluted to 50 L with water was slightly low in attractiveness than the earlier solution. It was noteworthy that solutions of oil of lemon (1200 ml) and oil of orange (1200 ml) diluted to 50 L were less attractive than the corresponding mixtures with Beeline®. At the same time attractiveness of Beeline® alone was less compared to all the other solutions.

The flower aroma of oriental orchid, *Cymbidium pumilum* was found to attract drones, individual workers as well as swarms of Japanese honey bees*A. cerana japonica;* however, it did not attract *A. mellifera* (Sasaki et al. 1991). Neira and Barriga (1995) studied the pollinating behaviour of honey bees, *A. mellifera*, on flowering raspberry plots. They found that the extract of lavender (*Lavandula latifolia*), when applied to plots resulted in more number of bee visits. However, it did not have any significant effects on yields of fruits (Neira et al. 1994, Neira and Barriga 1995).

Zvedenok (1996) examined the use of secondary attractants such as citral, geraniol, limonene and carrot seed extract to improve the visits of honey bees and pollination of onion crop in the presence of competitors like sunflower or buckwheat. Application of attractants was found to improve the bee visits and consequently the pollination.

Nectars of the flowers of *Angelica gigas* Nakai and *Fagopyrum esculentum* Monech were shown to attract honey bees which resulted in improved pollination. Further examination of these alluring substances resulted in identification of 20 aromatic constituents (Sim and Choi 1999). A formulation was developed using lemon balm oil and clove oil (4:2 by vol) which attracted honey bees to entomophilous plants (Muszynska 2005).

Naik et al. (2005) studied the activity of ethanolic leaf extract of Indian medicinal plant, *Swertia densifolia*, commonly known as Chirayata. The results indicated the dose dependent attractant and repellent properties towards honey bee *A. cerana indica*. Similar results were obtained, when the extract was screened towards *A. florea* (Naik et al. 2007). The essential oil isolated for other experiments and ethanolic extract of de-oiled leaves of *S. densifolia* were also shown to possess dose-dependent attractant-repellent properties towards *A. florea,* similar to pheromones (Naik et al. 2010).

6.8 Commercially Available Honey Bee 'lures'

The fact that Nasonov gland of honey bees secretes a pheromone which gives a message of attraction to the members of the species led to an idea of using attractants for controlling the behaviour of honey bees for better and improved bee keeping. The search for an ideal attractant began and continued with the testing of synthetic components of Nasonov gland in various compositions, plant extracts and randomly selected chemicals. The need of beekeepers induced researchers and manufacturers to develop commercial 'lures'/ attractant formulations for application in the fields which was expected to improve pollination and crop yields.

Commercial 'lures' for attracting honey bees have been developed in the recent past. Their development is based on the observation that foraging honey bees search for nectar or pollen and are attracted towards the food. Along with this it was also known that honey bees are also attracted to certain specific smells. Thus some of the 'lures' are food-based and contain sugar, proteins and pollen supplements while other 'lures' are based on compounds having a peculiar smell like pheromone mimics or the queen's substance.

Various protocols for application of 'lures' have been standardized. They basically use inert supports on which the attractant formulations are loaded. Sometimes the sugar syrup, sugar candy or pieces of sugar cane have also been used.

6.8.1 Results of Applications of Commercial 'lures'

Reports on the application of commercial 'lures' in the field to improve pollination efficiency of crops are available in the literature. Many of the results are encouraging but there are some discouraging results too! The increase in the number of bees visiting the sites marked with 'lures' is observed in most of the cases. However the lures are reported to fail in some instances. The unexpected results of the application of the 'lures' are not yet explained satisfactorily. The details of some of them are as follows:

6.8.1.1 Beeline®

Beeline® is a feeding attractant and contains proteins, sucrose, lactose, fats, minerals and vitamins. Application of Beeline® containing sugar and protein to carrot fields did not attract more insects or enhance pollen transfer between the plants in order to improve pollination (Belietti and Zani 1981). Similarly, Margalith et al. (1984) concluded that

Beeline® was totally ineffective in attracting bees under the experimental conditions. In an investigation on the effect of application of Beeline® on the number and size of raspberry fruits, it was found that the lures did not affect the number of fruits or their weight or size in the fields (Neiraet al. 1994). When Beeline® was tried on plots of flowering raspberry, the number of honey bee visits to flowers and the average time spent by the bees on the flowers were found to be more than on untreated plots with the highest dose of 5 kg/ha (Neira and Barriga 1995) which resulted in improved pollination.

6.8.1.2 Bee-Scent®

Bee-Scent® is marketed by Scentry Biologicals Inc (Montana, USA) as an attractant for honey bees and contains a sugar- pheromone base, the pheromones being a citral-geraniol mixture. Mayor et al. (1989) applied Bee-Scent® on flowering apple, cherry, pear and plum. The treated trees were found to attract significantly more number of honey bees. Moreover, the fruit set was found to be higher in these trees.Bee-Scent® was also applied to five farms of watermelon (Elemstrom and Maynard 1990). The activity of honey bee *A.mellifera* was increased only on a few occasions. Total fruit yield was increased on one farm and there was an apparent increase in early yield on three farms. Also the treatment increased the seed content of fruits of three farms. As against this, Loper and Roselle (1991) showed that there was no significant increase in the yield of watermelon when Bee-Scent® was applied. The 'lure' was found to be ineffective as an attractant to honey bees under the experimental conditions. Neira et al. (1997) treated raspberry field with Bee-Scent®. Although the number of bee visits and the average length of visits increased, there was no significant increase in fruit weight, size and drupelet number compared with freely pollinated controls.

The influence of Bee-Scent® on the number of honey bees visiting apple blossoms and on subsequent fruit production was studied. In these experiments it was found that the effects on fruit set and size distribution varied with season. In the years with light crops and ample fruit size, there were no beneficial effects. However, in years in which there was a significant problems of small fruit set or poor fruit set, applications of Bee-Scent® provided significant advantages in improving fruit size and fruit set. The results were associated with increased bee foraging induced by Bee-Scent® (Tew and Ferree 1998 a).

6.8.1.3 Bee Lure®

Bee Lure® spray on apple trees failed to induce significant increase in the number of foraging bees on treated when compared to untreated plants (Rajotte and Fell 1982). In another experiment, application of Bee Lure® was shown to increase the mean number of foraging honey bees on apple trees. However the increase was not significant. There was no increase in the fruit set also (Tew and Ferree 1984).

6.8.1.4 Bee-Here®

Bee-Here® is used as honey bee attractant and marketed by Hoescht Shering Agrevodo Brasil Ltd. It is generally used by diluting with sugar syrup and water. Its efficiency tested as a honey bee attractant on marrow crop cultivated under greenhouse conditions and reported to be ineffective in attracting honey bees (Ortiz-Sanchez 1993). Tsirakoglou et al. (1997) reported that use of Bee-Here® was one of the ineffective methods out of the methods tested. Although the field evaluation of Bee-Here® was not found to be encouraging, the results of *in vitro* studies were exactly opposite. Bee-Here® was found to be the best honey bee attractant when tested with other commercial 'lures' on the chopped sugarcane (Nogueira-Couto, 1998).

6.8.1.5 Bee-Q®

Bee-Q®, a commercially available honey bee attractant is marketed by M/s Excel Industries, Bombay, India. No increase in the number of honey bee visits or in percentage seed set of hybrid sunflower was found when the plots were sprayed with a commercial attractant, Bee-Q® (Singh and Sinha 1996). Increase in the bee visits were observed when Bee-Q® was applied on farms of radish (Chandrashekhar and Sattigi 2009).

6.8.1.6 Fruit Boost®

Fruit Boost® is a honey bee attractant marketed by Pherotech Inc, Delta, BC, Canada. Ellis and Delaplane (2009) tested Fruit Boost for its efficacy and found that it did not significantly affect the total number of honey bees visiting watermelon flowers, the proportion of honey bee visits to female flowers, fruit set or fruit weight.

6.8.2 Comparative studies on different bee attractants

Reports of comparative evaluation of commercially available bee attractants are available in the literature. It is interesting to note that the effectiveness of these attractants is not the same on all crops.

6.8.2.1 Beeline®and Bee-Scent®

Beeline®and Bee-Scent® were not found to be effective in improving the bee visitation as well as yields of cucumber and watermelon crops (Schultheis et al. 1994, Ambrose et al. 1995). It was, therefore, recommended that instead of using these attractants, additional honey bee colonies should be rented for the crops.

Although the effect of Beeline® and Bee-Scent® was discouraging on watermelon and cucumber, their effect on pear trees was found to be encouraging. The number of bees was found to be significantly more for up to three days between 14-16 hours. Bee-Scent® was found to attract more bees than Beeline® (Neira et al.1996). However the treatment had no effect on fruit yield.

Nakamura (1997) found that Beeline®, a food attractant, and Bee-Scent®, a pheromone mimic attracted honey bees to feeding sites and to wild plants offering nectar. He suggested that these products could be used to attract bees to any area small enough to be sprayed with one of them. Sotomayor (1997) while evaluating different pollination modes in almonds, sprayed the branches with bee attractants Beeline® or Bee-Scent®. He reported that the bee pollination is essential for almond production and the best fruit set was achieved with these attractants indicating the improved efficiency of pollination due to 'lures'.

6.8.2.2 Bee-Scent®, Bee-Q®and Fruit Boost®

Comparative studies were made on the efficacy of three bee attractants, Bee-Scent®, Bee-Q®and Fruit Boost® in pollination of guava (Anita et al. 2012). Based on one week of bee visitations on guava flowers, Bee-Scent® was found to be the most effective bee attractant amongst the three.

6.8.2.3 Bee-Here®and Bee-Q®

Patil et al. (2000) studied the effect of Bee-Q® and Bee-Here® on pollinators and yield of sesamum. They found that both these attractants were effective in attracting higher number of pollinators (*Apis spp.*) and resulted in more yields. The best results were recorded when the concentrations applied were 12.5 g Bee-Q®/l and 4 ml Bee-Here®/l.

6.8.2.4 Fruit Boost®, Bee-Q® and Bee Here®

Orchard evaluation of three commercial honey bee attractants (Bee-Scent®, Bee Here® and Fruit Boost®) for use in pollinating apple blossoms was studied (Tew and Ferree 1998 b). Amongst these, Fruit Boost® was found to be superior; there was highly significant difference between Fruit Boost® and the control. Also there was significant difference between Fruit Boost® and other two treatments.

6.8.2.5 Beeline® and Pollenaid-D®

Screening of Beeline®andPollenaid-D® for their attractiveness to honey bees showed that none of them caused an increase in honey bee activity on the flowering apple trees (Mayer and Johansen 1982).

6.8.2.6 Fruit Boost® and Bee-Q®

Influence of bee attractants on the yield parameters of Ridge Gourd (*Luffa acutangula* L) was studied (Jayaramappa et al. 2011). It was observed that spraying of Fruit Boost® (0.5 ml/l) and Bee-Q® (12.5 g/l) enhanced yield parameters like the number of fruits / plant, fruit length, fruit weight and yield as compared to open pollinated plots. In pumpkin (Jayaramappa and Sivaram 2013) Bee-Q® and Fruit Boost® attracted higher number of foragers than control plots and also enhanced the yield and diameter of fruits. Bee-Q® and

Fruit Boost® on Niger plots significantly increased the number of bee foragers over control plots. They also enhanced the seed set, seed weight and germination of Niger (Sivaram et al. 2013).

6.9 Present Scenario

Availability of attractants for honey bees is helpful to the beekeepers and farmers in many ways. Initially attempts were made to develop attractants based on pheromones. However limitations of pheromone based lures became clear very soon. Lures based on attractive properties of food, sugar and plant based attractants were developed and evaluated, and several attractants such as Fruit Boost®, Bee-Q®, Bee-Here®, Beeline®, Bee-Scent®, Bee Lure® and Pollenaid–D® are now available commercially. However there are contradictory reports about their activities. While many reports comment on their positive effects on honey bee visits and/or yields, a few papers report their ineffectiveness. Reports on plant based attractants show the importance of concentration in deciding the effectiveness of the lure. Thus, search for novel attractants for honey bees and optimization of the concentration of formulations will remain an active field of research at least for a couple of decades.

References

Abdullah N, Hamzah A, Ramli J, MardanM 1990. Identification of Nasonov pheromones and the effects of synthetic pheromones on the clustering activity of the Asiatic honey bee (*Apis cerana*).Pertanika 13: 189–194.

Ambrose JT 1975. A study of honey bee (*Apis mellifera* L.) swarms while in transit to a new homesite. *PhD Thesis,* Cornell University, Ithaca, NY, USA.83 pp. (Cf Apicultural Abstracts. 739/75).

Ambrose JT, Schultheis JR, Bambara SB, Mangum W 1995. An evaluation of selected commercial bee attractants in the pollination of cucumbers and watermelons. Am Bee J 135: 267–272.

Anita M, Sivaram V, Jayaramappa KV 2012. Influence of bee attractants on pollination and yield parameters in Guava (*Psidiumguajava*L.) J Plant Rep Biol 4: 41–46.

Avitabile A, Morse RA, Boch R 1975. Swarming honey bees guided by pheromones. Ann Entomol Soc Am 68: 1079–1082.

Belietti A, Zani A 1981. A bee attractant for carrots grown for seed. Sement Etette 27: 23-27. (Cf Apicultural Abstracts 1998/84).

Boch R, Morse RA 1974. Discrimination of familiar and foreign queens by honey bee swarms. Ann Entomol Soc Am 67: 709–711.

Boch R, Shearer DA 1962.Identification of geraniol as the active component in the Nassonoff pheromone of the honey bee. Nature 194: 704–706.

Boch R, Shearer DA 1963. Production of geraniol by honey bees of various ages. J Insect Physiol 9: 431–434.

Brian MV 1983. Social Insect Ecology and Behavioural Biology. Chapman and Hall, London.

Burgett M 1980. The use of lemon balm (*Melissa officinalis*) for attracting honey bee swarms. Bee World 61: 44–46.

Butler CG 1951. The importance of perfume in the discovery of food by the worker honey bee (*Apis mellifera* L.) Proc Roy Soc B 138: 403–413.

Butler CG, Calam DH 1969.Pheromones of the honey bee – the secretion of the Nassanoff gland of the worker. J Insect Physiol 15: 237–244.

Cassier P, Lensky Y 1994. The Nasonov gland of the workers of the honey bee (*Apis mellifera* L.): Ultrastructure and behavioural function of the terpenoid and protein components. J Insect Physiol 40: 577–584.

Chandrashekhar GS, Sattigi HN 2009. Influence of attractants on bee visitation to radish. Karnataka J Ag Sci 22: 909–911.

Danka RG, Williams JL, Rinderer TE 1990. A bait station for survey and detection of honey bees. Apidologie 21: 287–291.

Denby DLS, Scott-Dupree CD 1992. The efficacy of queen and Nasonov pheromones in the capture of honey bee (*Apis mellifera* L.) swarms in regulatory procedures. Bee Science 2: 120–124.

Elemstrom G, Maynard D 1990. Attraction of honey bees to watermelon with Bee attractant. ProcFl State Hort Soc 103: 130–133.

Ellis A, Delaplane K 2009.An evaluation of Fruit-Boost™ as an aid for honey bee pollination under conditions of competing bloom. Journal of Apicultural Research and Bee World 48: 15–18.

Ferguson AW 1985.The use of pheromones to control honey bee colony behavior in the tropics. In Proceedings of the Third International Conference on Apiculture in Tropical Climates Nairobi, Kenya 5-9 November 1984, pp. 123-126. (Cf. Apicultural Abstracts.146L/86).

Ferguson AW, Free JB 1981.Factors determining the release of Nasonov pheromone by honey bees at the hive entrance. Physiol Entomol 6: 15–19.

Ferguson AW, Free JB, Pickett JA, Winder M 1979. Techniques for studying honey bee pheromones involved in clustering and experiments on the effect of Nasonov and queen pheromones. Physiol Entomol 4: 339–344.

Free JB 1962.The attractiveness of geraniol to foraging honey bees. J Apicult Res 1: 52–54.

Free JB 1968. The conditions under which foraging honey bees expose their Nasonov gland. J Apicult Res 7: 139–145.

Free JB 1987. Pheromones of Social Bees. Chapman and Hall, London.

Free JB, Ferguson AW, Pickett JA 1981.Evaluation of the various components of the Nasonov pheromone used by clustering honey bees. Physiol Entomol 6: 263–268.

Free JB, Ferguson AW, Pickett JA 1983a. A synthetic pheromone lure to induce worker honey bees to consume water and artificial forage. J Apicult Res 22: 224–228.

Free JB, Ferguson AW, Pickett JA 1983b. Effect of the components of the Nasonov pheromone on its release by honey bees at the hive entrance. J Apicult Res 22: 155–157.

Free JB, Pickett JA, Ferguson AW, Simpkins JR, Williams C 1984. Honey bee Nasonov pheromone lure. Bee World 65: 175–181.

Free JB, Williams IH 1970. Exposure of the Nasonov gland by honey bees (*Apis mellifera*) collecting water. Behaviour 37: 286–290.

Gellert M, Rozsa Z, Kovacs Z, Szendrei K, Hussain RA, Reisch G, Reisch J 1985.Honey bee attractants in the fruits of *Evodia hupehensis* Dode. Herba Hung 24: 53–65.

Jayaramappa KV, Pattabhiramaiah M, Bhargava HR 2011. Influence of Beeattractants on Yield Parameters of Ridge Gourd (*Luffa acutangula* L.) (Cucurbitaceae). Appl Sci J 15: 457–462.

Jayaramappa KV, Sivaram V 2013. Some aspects of bee attractants on Pumpkin. J Chem Biol Phy Sci 3: 1801–1807.

Kalmus H, Ribbands R 1952. The origin of the odours by which honey bees distinguish their companions. Proc Roy Soc B 140: 50–59.

Karlson P, Lüscher M. 1959.Pheromones: a new term for a class of biologically active substances. Nature 183: 55–56.

Kigatiira IK, Beament JWL, Free JB, Pickett JA 1986. Using synthetic pheromone lures to attract honey bee colonies in Kenya. J Apicult Res 25: 85–86.

Lecomte J 1957.Sur le marquageolfactif des sources de nourriture par les abeillesbutineuses. Comptus Rendus 245: 2385–2387.

Lesher C, Morse RA 1983. Baits to improve bait hive attractiveness to honey bees. Am Bee J 123: 193–194.

Loper GM, Roselle RM 1991.Experimental use of Beescent to influence honey bee visitation and yields of watermelon. Am Bee J 131: 777.

Maa T 1953. An enquiry into systematics of the tribe apisini or honey bees(Hym). Treubia 21: 525–640.

Maliszewska R, Szymas B 1997. Taste and aroma attractants in feeds for honey bees (*Apis mellifera* L.).Postepyapidologii w Polsce107-116 (Cf. Apicultural Abstracts 943/98).

Margalith R, Lensky Y, Rabinowitch HD 1984.An evaluation of Beeline as a honey bee attractant to cucumbers and its effect on hybrid seed production. J Apicult Res 23: 50–54.

Mastsuyama S, Suzuki T, Sasagawa H 1998. Chemical ecology of honey bees in Asia: Chemical analysis of worker pheromone component. Fourth Asian Apicultural Association International Conference, 23-27 March 1998, Kathmandu, Nepal, pp 13.

Mautz D, Boch R, Morse RA 1972. Queen finding by swarming honey bees. Ann Entomol Soc Am 65: 440–443.

Mayer DF, Johansen CA 1982. Field evaluation of chemical pollinator attractants on tree fruits. Am Bee J 122: 287–289.

Mayor DF, Britt RL, LundenJD 1989.Evaluation of Bee Scent as a honey bee attractant.Am Bee J 129: 41–42.

Mclndoo NE 1914. The scent-producing organ of the honey bee. Proc Soc Nat Sci Phil 66: 542–555.

Morse RA, Boch R 1971.Pheromone concert in swarming honey bees (Hymenoptera: Apidae). Ann Entomol Soc Am 64: 1414–1417.

Muszynska J 2005. Preparation for luring bees to entomophilous plants in bloom.From Polish patent PL 189162 B1 20050630.

Naik DG, Banhatti P, Chawda SS, Thomas D 1999. Chemical constituents of Nasonov gland pheromone of *Apis mellifera* in India. Indian Bee J 61: 59–62.

Naik DG, Banhatti P, Chawda SS, Thomas D 2001. Occurrence of neral in Nasonov gland pheromone of *Apis mellifera* L. Indian Bee J 63: 1–5.

Naik DG, Dandge CN, Deshpande PV 2010.'Pheromone-like' properties of de-oiled leaf extract of *Swertia densifolia* towards *Apis cerana indica* F. J Apicult Res 49: 149–153.

Naik DG, Dandge CN, Puntambekar HM 2007. *Swertia densifolia* leaf extract as a dose-dependent attractant or repellent for *Apis florea*. J Apicult Res 46: 15–18.

Naik DG, Dandge CN, Puntambekar HM, Patil T 2005. Attractant and repellent properties of *Swertia densifolia* leaf extract towards *Apis cerana indica*. J Apicult Res 44: 116–118.

Naik DG, Gadre RV, Kapadi AH, Singh MK, Suryanarayana MC, Kshirsagar KK 1988.

Nasonov gland pheromone of the Indian honey bee, *Apis cerana indica*. J Apicult Res 27: 205–206.

Naik DG, Kapadi AH, Kshirsagar KK 2001.A composition to attract Indian honey bees *Apis florea*. Indian Patent No. 185391.

Naik DG, Kapadi AH, Singh MK, Suryanarayana MC, Kshirsagar KK 1989. Lure development for Indian honey bees *Apis cerana indica* Fabr. Indian Bee J 51: 47–50.

Nakamura J 1997. Bee attractants for pollination. Honey bee Science. 18: 81–85. (Cf. Apicultural Abstracts. 164/98).

Neira CM, Barriga GJ 1995. Honey bee (*Apis mellifera* L., Hymenoptera :Apidae) pollinating behavior on raspberry (*Rubusidaeus* L., cv Heritage) under the effect of two attractant and one repellent treatments. Agro Sur 23: 52–59.

Neira CM, Barriga GJA, Mundiaca BN 1994. Characteristics of *Rubusidaeus* L. fruits, associated with the use of two attractants and a repellent. Agro Sur 22: 77–82.

Neira CM, Lobos SN, Riveros GM, Caprillo LR, Pessot ZR, Mundiaca BN 1997. Insect fauna associated with raspberry flowers (*Rubusidaeus* L.,cvMeeter) and the evaluation of the efficiency of *Apis mellifera* L. as a pollinator under the influence of a pheromonal attractant. Revista Chilena Entomologica 24: 37–44. (Cf. Apicultural Abstracts. 596/00).

Neira CM, Nallar FJ, Pessot ZR, Caprillo LR 1996.The effect of two attractants on the pollination efficiency of honey bee (*Apis mellifera* L.) on pear trees. Agro Sur 24: 167–176. (Cf. Apicultural Abstracts. 597/00).

Nogueira-Couto RH 1998. Effect of attractants and repellents on the behavior of honey bees (*Apis mellifera* L.). Scientia Agricola 55: 388–394. (Cf. Apicultural Abstracts. 987/00).

Ohe von der W, Praagh van JP 1983. Experiment with scent-training in orchards. Nordwestdeutsche Imkerzeitung 35: 100. (Cf. Apicultural Abstracts. 687/84).

Ortiz-Sanchez FJ 1993. Evaluation of the efficiency of Bee Here as a honey bee (*Apis mellifera* L. innaeus, 1758) attractant to marrow crop (*Cucurbita pepo* L 1753) cultivated in green house conditions in southern Spain (Hymenoptera: Apidae). Entomofauna 14: 465–469.

Pardo PR, Nates PG 1994. Increasing flower visits by *Apismellifera* L. (Hymenoptera: Apidae) in crops by the use of synthetic Nasonov pheromone. Revista Colombiana de Entomologia 20: 187–192.

Patil BS, Viraktamath S, Lingappa S, Giraddi RS, Parameshwarappa K, Bhat ARS 2000. Effect of Bee-Q and Bee-Here on pollinators and yield of sesamum. Insect Environ 5: 151–152.

Pflumm W 1970. Auslosende Faktorenbeim Duftmarkieren von Sammelienen. Experientia 26: 318–319.

PflummW,PeschkeC,WilhelmK,CruseH1978.EinflussderineinemFlugraumkontrollierten-Trachtverhaltnisse auf das Duftmarkieren und die Abflugmagenfullung der Sammelbiene. Apidologie 9: 349–362.

Pickett A, Ferguson AW, Free JB, Smith MC 1982. Synthetic bee pheromone and its use in attracting a colony of bees to hive or trap.UK Patent application No.GB 2095 998 A.

PickettJA, Williams IH, Martin AP, Smith MC 1980.Nasonovpheromone of the honey bee, *Apis mellifera* L. (Hymenoptera, Apidae). Part I. Chemical characterization. J ChemEcol 6: 425–434.

Pinzauti M 1985. An evaluation of different substances that attract pronubial insects. In Proceedings of the 29th International Congress of Apiculture, Budapest, 1983. Bucharest, Romania: Apimondia Publishing House pp 327–332. (Cf. Apicultural Abstracts. 149/86).

Rajotte EG, Fell RD 1982. A commercial bee attractant ineffective to enhancing apple pollination. Hortscience 17: 230–231.

Renner M 1960. Das Duftorgan der Honigbiene und die physiologische Bedeutunginres Lockstoffes. Zeitschrift fur Vergleichende Physiologie 43: 411–468.

Sasaki M, Ono M, Yoshida T 1991. Oriental orchid (*Cymbidium pumilum)* attracts drones of the Japanese honey bee (*A. cerana japonica)* as pollinators. Experientia 47: 1229–1231.

Schmidt JO 1988. Swarm trapping. Am Bee J 128: 808.

Schmidt JO 1994. Attraction of reproductive honey bee swarms to artificial nests by Nasonov pheromone. J Chem Ecol 20: 1053–1056.

Schmidt JO 1999. Attractant or pheromone: the case of Nasonov secretion and honey bee swarms. J Chem Ecol 25: 2051–2056.

Schmidt JO, Slessor KN, Winston ML 1993. Roles of Nasonov and queen pheromones in attraction of honey bee swarms. Naturwissenschaften 80: 573–575.

Schmidt JO, Thoenes SC 1987.Honey bee swarm capture with pheromone containing trap boxes. Am Bee J 127: 435–438.

Schmidt JO, Thoenes SC, Hurley R 1989. Swarm traps. Am Bee J 129: 468–471.

Schultheis JR, Ambrose JT, Bambara SB, Mangum WA 1994. Selective bee attractants did not improve cucumber and watermelon yield. Hortscience 29: 155–158.

Sedin IF 1967. The influence of some aromatic substances on bees.Opyt Rab Pchel 1966: 39–45. (Cf. Apicultural Abstracts934/72).

Shearer DA, Boch R 1966. Citral in the Nassonoff pheromone of the honey bee. J Insect Physiol 12: 1513–1521.

Shearer DA, Stone BC, Megugan WA 1963. Collection of gas chromatographic fractions for identification by infrared spectrophotometry. Analyst 88: 147–149.

Sivaram V, Jayaramappa KV, Menon A, Ceballos RM 2013. Use of bee-attractants in increasing crop productivity in Niger (*Guizotia abyssinica*.L). Braz Arch Biol Techn 56: 365–370.

Sladen FWL 1901. A scent organ in the bee. Brit Bee J 29: 151–153.

Sladen FWL 1902. A scent-producing organ in the abdomen of the worker of *Apis mellifera*. Entomol Mag 38: 208–211.

Snodgrass RE 1956. Anatomy of the Honey bee, Constable, London.

Sotomayor SC 1997. Evaluation of different pollination modes in almonds.Cienciae Investigaion Agraria 24: 7–11. (Cf. Apicultural Abstracts.1213/00).

Tew JE, Ferree DC 1984. The effect of bee lure on honey bee (*Apis mellifera*) pollination of apples. Research Circular, Ohio State University, Agricultural Research and Development Centre No 283: 31–32. (Cf. Apicultural Abstracts 117/87).

Tew JE, Ferree DC 1998a. Orchard evaluation of three commercial honey bee attractants for use in pollinating apple blossoms and subsequent fruit set. Progress report for 1998, Ohio State University.

Tew JE, Ferree DC 1998b. The Influence of a Synthetic Foraging Attractant, Bee-Scent TM, on the number of honey bees visiting apple blossoms and on subsequent fruit production infruit crops: A Summary of Research 1998. Research Circular, Ohio State University, Agricultural Research and Development Centre No 299–299.

Tsirakoglou V, Thrasyvoulou A, Hatjina F 1997.Techniques to increase the attractiveness of Kiwi flowers to honey bees. Acta Hort 444: 439–443.

Villa JD, Schmidt JO 1992. Does queen pheromone increase swarm capture in hives baited with Nasonov pheromone? J Apicult Res 31: 165–167.

Visscher PK, Khan AM 1995. A trap for bees remaining after swarm and colony removal. Am Bee J 135: 567–572.

von Frisch K 1967. In the Dance Language and Orientation of Bees. Oxford University Press, London.

Waller GD 1970. Attracting honey bees to alfalfa with citral, geraniol and anise. J Apicult Res 9: 9–12.

Waller GD 1973. The effect of citral and geraniol conditioning on the searching activity of honey bee recruits. J Apicult Res 12: 55–59.

Weaver N, Weaver EC, Law J 1964. The attractiveness of citral to foraging honey bee. Progress Report, Texas A and M University/Texas Agricultural Experimental Station 1–6.

Wilhelm KT, Pflumm WW 1983. Effect of flower- like fragrances on the scent marking behavior of foraging honey bees. Apidologie 14: 183–190.

Williams IH, Pickett JA, Martin AP 1981. The Nasonov pheromone of the honey bee *Apis mellifera* L. (Hymenoptera, Apidae). Part II. Bioassay of the component using foragers. J Chem Ecol 7: 225–237.

Winston ML, Slessor KN, Rubink WL, Villa JD 1993. Enhanced pheromone lures to attract honey bee swarms. Am Bee J 133: 58–60.

Witherell PC 1985. A review of the scientific literature relating to honey bee bait hives and swarm attractants. Am Bee J 125: 823–829.

Witherell PC, Lewis JE 1986. Studies on the effectiveness of bait hives and lures to attract honey bee swarms – A survey tool for use in Africanized honey bee eradication programs. Am Bee J 126: 353-354, 356-357, 359–361.

Woyke HW 1981. Some aspects of the role of the honey bees in onion seed production in Poland. Acta Hort 111: 91–98.

Wykes GR 1952. The preferences of honey bees for solutions of various sugars which occur in nectar. J Exp Biol 29: 511–518.

Youngken HW Jr. 1956.The value of medicinal plants to beekeeping. Glean. Bee Culture 84: 16–18.

Zoubareff A 1883. Concerning an organ of the bee not yet described. Brit Bee J 11: 296–297.

Zvedenok AP 1996. The onion seed crop can be improved. Kartofel'IOvshchi4: 29. (Cf. Apicultural Abstracts. 1076/98).

Chapter 7

Prospects of Honey Bee Attractants in Enhancing Crop Productivity in India

Shashidhar Viraktamath and Anjankumar Naik
shashiv777@gmail.com

Abstract

India with a population of 1.27 billion, growing at 1.54 per cent per year, needs the current food production of 259.27 million tons to be doubled by 2040. Adoption of modern agricultural techniques no doubt has led to green revolution making India self-sufficient, but also has resulted in degraded land and water resources, decline in productivity, climatic changes due to destruction of natural forests and loss of biodiversity. One of the best ecofriendly options for sustainable agriculture is utilization of biological resources to increase the production and attain food security. One third of our food production depends on the pollination services of various pollinators amongst which honey bees are the most dominant. Pollination services is worth of 224 billion US dollars per year globally and 3000 crore rupees in India. Hence, use of honey bees as pollinators is the best, cheap and eco-friendly techniques to enhance the cross pollination and productivity of the crops. However, honey bees visit the crops which reward them with rich nectar and pollen while they tend to ignore the crops with fewer rewards. Pollination of crops at right time is very crucial for optimum yields. Under such situations bee attractants play a promising role in ensuring optimum pollination at right time on preferred as well as less preferred target crops. Bee attractants entice more foraging bees which stay for longer time on the treated flowers ultimately resulting in higher yields. Bee attractants are of food or pheromone or plant based and studies on their role in enhancing crop yields have shown mixed result with outweighing of positive results. At present, a few attractants like Bee-q and Bee-here are being imported and marketed in India. Hence attempts are being made to develop and evaluate indigenous attractants. Prospects of honey bee attractants in enhancing agricultural and horticultural crops are presented and discussed in this paper.

DOI: 10.1201/9781003505471-7

7.1 Introduction

Agriculture is the mainstay of the Indian economy which accounts for 13.7 per cent of India's GDP. India has achieved a record food grain production of 259.29 million tonnes which form about 20 per cent of the world food production (Anonymous 2014). However, India with a population of 1.27 billion, growing at about 1.54 percent per annum, needs the food production to be doubled by 2040. The rapid increase in population and limited land availability act as crucial factors for India's move towards increasing food yield for internal consumption as well as exports.

In the past decade, the agricultural production has either declined or remained stagnant. Several reasons quoted for this stagnation include intensive monoculture, excessive use of chemical fertilizers, and indiscriminate use of pesticides, destruction of natural habitats and lack of pollinators. Adoption of modern agriculture techniques no doubt has led to the green revolution making India a self-sufficient country but at the same time, has resulted in degraded land and water resources, climatic changes due to destruction of natural vegetation and loss of biodiversity. One of the best and ecofriendly options to overcome these problems is utilizing biological resources to increase productivity and attain food security in sustainable manner.

Total cropped area in India is about 160 million hectares out of which 55 million hectares belong to "Entomophilous crops" i.e., crop depending on insects for their pollination requirement (Abrol 2012). Apart from insects several other animals like birds, bats and mammals also contribute to their pollination. It is rightly said that at every bite of food, one must thank a bee for its pollination services. Many fruits, vegetables, oilseeds, plantation, spice, beverage, forage and fibre crops are immensely benefited by the cross pollination carried out by these pollinators.

Use of honey bees as pollinators is considered as one of the best, cheap and eco-friendly techniques to enhance the pollination and thus increase productivity of crops. Honey bees are considered as the best pollinating agents as they have coevolved with the flowering plants with modified morphological and anatomical features. Without bees, there would not have been so much diversity in plant kingdom. The economic value of the bee pollinated crops estimated in different parts of the world is to the extent of US $ 40.0 billion in USA, Aus $ 156 million in Australia, C$ 1.2 billion in Canada (Michener 2000) and in India 3000 crore rupees (Goyal 1993). At the global level the value of the pollination is to the tune of 224 billion US dollars annually (Anonymous 2010).

Studies made on pollination of crops using honey bees revealed an undoubted enhancement in quantitative and qualitative parameters throughout the world (FAO 1995,

Free 1970). However, honey bees have preference to certain flowers, which is mainly guided by the quantity of pollen and nectar available in the flowers. They continue to visit in large numbers such flowers which reward them with rich nectar and pollen (e.g., sunflower) while they tend to ignore the flowers which are less rewarding (e.g. safflower). Many times pollination at right time becomes very crucial for optimum seed or fruit set but weather conditions may not be very conducive for the bee activity. Another problem in pollination management is to attract the bees to target crops and keep them away from more attractive competing blooms (Free 1968). Hence it becomes challenging task to attract bees to the target crop which needs cross pollination but less preferred by bees.

Under such situations bee attractants play a vital role in attracting bees to a less preferred target crop so as to ensure optimum pollination and derive the benefits of cross pollination. Bee attractants when used *boost* the bee visitation and yield parameters.

Honey bees depend entirely on pollen and nectar for their requirement of protein food and energy source. Since flowering plants have coevolved with pollinating bees, depend on bees for their pollination requirement. The plants have evolved various devices to attract the bees by rewarding them with pollen, nectar and by producing showy and aromatic flowers. There is a mutualistic relationship between these plants and bees (see Shivanna, this vol; Kuriakose, this vol). The bees also have adopted elaborate communication devices by using several pheromones (see Deshpande and Naik, this vol). These facts have been exploited to develop bee attractants to enhance the bee pollination and crop productivity.

Based on the source from which they are derived, the bee attractants can be broadly classified into three categories (see also Deshpande and Naik, this vol):

- Attractants based on beefoods, sugars and proteins
- Attractants based on bee pheromones
- Attractants based on plants and other chemicals

Food based attractants mainly comprise of glucose, sucrose, maltose, lactose, protein, fat, minerals, vitamins, gluconic acids in the commercial products (Table 7.1).

Table 7.1: Commercial food based attractants.

Name	Nutritional substances	Others	Company
Api-Fix	6 per centreducing glucose; 25 per centproteins	-----	Coda (France)
Beeline	Saccharose, lactose, protein	Fatty acids; pollen	Custom Chemical (USA)
Beelure	39 per centglucose, 28 per centmaltose, 10 per cent maltotriose, 22 per cent other sugars	Strawberry aroma, red colourant, preservatives	Helena Chemical (USA)
Pollenaid	Gluconic acid	-----	Crop King Chemicals (USA)
Bee-Q	Proteins and sugars	-----	Excel Chemicals

Pheromone-based attractants are further divided into nasonov gland pheromone and queen mandibular pheromone (Table 7.2). These attractants consist of a blend of various components of these two pheromones. Nasonov gland pheromone components of *Apis cerana* and *A. florea,* Citral E and Citral Z have been demonstrated to have attractant properties under laboratory conditions (Naik et al. 1989).

Table 7.2: Commercial Pheromone based attractants.

Name	Nutritional substances	Pheromones	Others	Company
BeeScent	42.5 per cent sugars	9.5 per cent Nasonov	48.5 per cent inert ingredients	Yellowstone Intl. (USA)
Pollinus	------	45 per cent Nasonov	------	Calliope (France)
Beehere	------	Nasonov	------	Troy Biosciences (USA)
Fruit Boost	------	Queen Mandibular Pheromone 1688 ppm	>99% inert ingredients	Phero Tech Inc. (Canada)

Different plant based attractants as well as chemicals have been tried in different parts of the world but none of them has been exploited into a commercial formulation. Some of the plant based and other chemicals evaluated as attractants include alcohols, fatty acids (Woodrow et al. 1965), fruit extract of *Evodia hyphensis* Dode (Gellert et al. 1985), oil of lemon and orange (Pinzauti 1985), carrot seed extract (Zvedenok 1996), leaf extract of cinnamon (Guruprasad 2001, Murasing and Viraktamath 2002), tube rose flower scented water (Guruprasad 2001, Viraktamath and Ganapathi 2008); extracts of dried fruits of *Fagara budrunga* (Naik et al. 2003); leaf extract of *Swertia densifolia* (Naik et al. 2005). Sugar solution and molasses have been evaluated by several researchers.

Experiments conducted worldwide including India for enhancing the honey bee pollination and yield in various crops have shown mixed results and have been summarized in Table 7.3. Higo et al. (1995) has explained the mechanism involved in enhancement of pollination due to application of pheromone-based bee attractants. Application of attractant results in higher recruitment of foraging force to the treated bloom and the foraging bees would stay for a longer time thus ensuring optimum pollination of the flower.

Table 7.3: Effect of bee attractants in enhancement of crop productivity.

Bee attractant	Crop		Effect on productivity	Reference
Beeline	Field crops	Red clover	Negative	Burget and Fisher (1979)
		Soyabean	Negative	Sheppard *et al.* (1979)
		Feverolle	Negative	Pinzanti and Corsi (1981)
	Horticultural crops	Cider apple	Positive	Willams and Jefferies (1982)
		Carrot seed	Negative	Belletti and Zani (1981)
		Cucumber	Negative	Rapp (1981)
		Almond, Lucerne, Apple	Negative	Mayer and Johansen (1982)

Bee attractant	Crop		Effect on productivity	Reference
		Cucumber seed	Negative	Margalith *et al.* (1984)
		Gherkin	Negative	Schultheis *et al.*(1994)
		Raspberry	Positive	Niera and Barriga (1995)
Beelure	Horticultural crops	Apple	Negative	Rajotte and Fell (1982)
		Apple	Negative	Tew and Ferree (1984)
Bee Scent	Horticultural crops	Cherry, Pear, Apple, Plum	Positive	Mayer *et al.* (1989)
		Kiwi	Negative	Pinzauti (1990)
		Watermelon	Positive	Elstorm and Maynard (1991)
		Gherkin and Watermelon	Negative	Schultheis *et al.*(1994)
		Cucurbit and Watermelon	Negative	Ambrose *et al.* (1995); Looper and Rossette 1991
		Strawberry	Positive	Butts (1991)
Pollenaid	Horticultural crops	Pear, Apple	Negative	Mayer and Johansen (1982)
Bee-Q	Field crops	Sesamum	Positive	Viraktamath and Patil (1999)
		Safflower	Positive	Lingappa*et al.* (1999)
		Sunflower	Negative	Singh and Sinha (1996) andSrimathi *et al.* (1999)
		Sunflower	Positive	Sanjivan Kumar *et al.* (2000) Viraktamath and Patil (2002) Manjunath (2003)
		Mustard	Positive	Murasing and Viraktamath (2002)
		Niger	Positive	Sivaram *et al.* (2013)
Bee-Q	Field crops	Niger	Positive	Guruprasad (2001)
	Horticultural crops	Cardamom	Positive	Bhat and Sudharshan (1999)
		Cucumber	Positive	Viraktamath and Anagoudar (2002), Pateel and Sattigi (2007)
		Watermelon	Positive	Lingappa *et al.* (1999)
		Bitter gourd	Positive	Nidagundi (2004)
		Pumpkin	Positive	Jayaramappa and Sivaram (2013)
		Onion	Positive	Patil *et al.*(2011)
Beehere	Field crops	Sesamum	Positive	Viraktamath and Patil (1999) and Patil *et al.* (2000)
		Safflower	Positive	Lingappa *et al.* (1999)
	Horticultural crops	Watermelon	Positive	Maynard *et al.* (1992) Lingappa *et al.* (1999)
		Sweet Orange	Positive	Malerbo-Souza *et al.* (2004)
		Kiwi Fruit	Negative	Tsirakoglou *et al.* (1997)
		Cucurbita pepo	Negative	Ortiz – Sanchez (1993)
		Cucurmis sativa	Positive	Viraktamath and Anagoudar (2002)

Bee attractant	Crop		Effect on productivity	Reference
Fruit Boost (Queen Mandibular Pheromone)	Field crops	Niger	Positive	Guruprasad (2001), Fakkiresh Muddi (2012), Sivaram *et al.* (2013)
		Sunflower	Positive	Viraktamath and Patil (2002), Manjunath (2003), Srikanta Nath (2008)
		Sesame and Niger	Positive	NithyaChandran (2009)
		Cotton	Positive	Viraktamath and Ganapathi (2008), Anjan Kumar Naik (2010), Vinayak Pise (2011)
	Horticultural crops	Cranberry , Blueberry, Apple and Pear	Positive	Currie *et al.* (1992a and b)
		Pear, Cherry, Apple, Cranberry and Blueberry	Positive	Winston and Slessor (1993)
		Pear and Cherry	Positive	Ken-Naumann *et al.* (1994)
		Cranberry	Positive	Mackenzie and Averill (1992)
		Fruit crops	Positive	Higo *et al.* (1995)
	Forage crops	Lucerne	Positive	Chavan (2013)
Other attractants (Citral, Geraniol, Limonene, Carrot seed extract, Rose water spray)	Horticultural crops	Onion	Positive	Zvedenok (1996), AlSahf (2002)

Efficacy of attractants on any crop depends on the following important factors which need to be taken into consideration before employing the attractants.

- Sufficient honey bee population either natural or from apiary should be there in the close vicinity of the target crop
- The weather conditions at the time of application should not be too cloudy and windy or hot to the foraging activity of the honey bees
- More than one application is necessary for the crops which have longer duration of flowering
- Timely application of bee attractants is necessary to obtain optimum benefits

7.2 Advantages and Disadvantages of Bee Attractants

- Increase the production and productivity by increasing the number of bee visits and time spent by the bees on the flowers
- Enhance both quantitative and qualitative parameters of the produce
- Environmentally safe
- Helps to sustain honey bee colonies by increasing the probability of food availability
- Do not cause any phyto-toxicity on the treated plants
- No residue in the products and safer for the consumption

In India, though limited studies have been made on the efficacy of attractants in various crops but most of these studies have demonstrated positive results in enhancing the bee pollination and yield parameters. However, there are several constraints for the adoption of this technology by the farmers, which are given below:

- Lack of knowledge and awareness by the farmers
- Currently commercial bee attractants imported and marketed in India are rather expensive
- These bee attractants are effective only up to two to three days after application. So the crop requires repeated spray of bee attractants during the flowering period
- Crop on which bee attractants are to be sprayed should be free from pesticide sprays
- The treatments are not effective if sufficient honey bee population is not available, which is a limitation in many parts of India
- Beekeeping is not considered as one of the inputs in Indian agriculture

7.3 Indigenous Bee Attractants and their Impact on Crop Yield

Currently all the commercial bee attractants *viz.,* Bee-Q, Bee scent plus, Bee here, Fruit boost, Pollenaid, Bee lure, Pollinus, Api-fix are produced in the countries like France, USA and Canada. Though the Indian companies are involved in importing and marketing some of these, they are cost prohibitive and are practically far away from the reach of majority of Indian farmers. Hence in order to make bee attractants easily available to the Indian farmers, attempts have been made to develop and evaluate a few indigenous bee attractants. These indigenous bee attractants include two plant based (dried fruit extract of *Fagara budrunga* and leaf extract of *Swertia densifolia*) and two nasonov pheromone gland based (Citral E and Citral Z) attractants.

7.4 Case Studies using Commercial and Indigenous Bee Attractants in India

7.4.1 Oil seeds

7.4.1.1 Sunflower

Sunflower being a highly cross pollinated crop, adequate pollination is vital for any significant increase in seed production. The flowers produce abundant quantity of nectar and pollen which attract large number of honey bees. Manjunath (2003) studied the effect of Bee-Q and Fruit boost on yield parameters of sunflower. The crop was raised in the plots of 100 square meters laid out in a randomized block design. There were seven treatments that included application of each attractant once at 10 and 50 per cent and twice at 10 and 50 per cent flowering. The crop that did not receive any spray served as the control. Heavier heads were observed from the treatments that received sprays of Bee-Q and Fruit boost at 10 and 50 per cent flowering (51.17 and 52.26 g). Highest yield of 18.38 q/ha was obtained from the crop sprayed with Fruit boost twice as against 13.98 q/ha in control. In another location, significantly higher yield (11.55 q/ha) was obtained in Fruit boost sprayed treatment followed by Bee-Q (10.02 q/ha) and control (9.23 q/ha). Incremental benefit due to spray of attractants in two locations was to the tune of ₹3205 to ₹5100 and ₹1185 to ₹ 3480 per hectare.

Srikanta Nath (2008) evaluated two pheromone (Citral E and Citral Z) and two botanical-based (leaf extract of *S. densifolia* and fruit extract of *F. budrunga*) bee attractants along with commercial attractant Fruit boost and control (no spray) on sunflower (Morden variety). The experiment was conducted in a randomized block design with plot size of 100 m^2. There were seven treatments each replicated thrice. Sunflower sprayed with *F. budrunga* and *S. densifolia* recorded heavier heads (55.38 and 54.41 g). Percent chaffiness was least in treatments with sugar solution and *F. budrunga* (4.33 and 5.00%). *S. densifolia, F. budrunga* and Citral Z recorded higher number of seeds per head. *S. densifolia* sprayed crop produced highest yield of 19.53 q/ha.

In KBSH-1 hybrid sunflower, higher seed weight was recorded from Fruit boost (5.72 g), *S. densifolia* (5.46 g) and Citral Z (5.44 g) sprayed crop. *Fagarabudrunga* and Fruit boost sprayed crops produced higher number of seeds/head (2.78.75 and 244.17 seeds/ head). Yield obtained from spray of Fruit boost was 12.90 q/ha followed by Citral Z (12.30 q/ha), *S. densifolia* (12.15 q/ ha), Citral E (11.80 q/ha), *F. budrunga* (11.15 q/ha), sugar solution (11.05 q/ha) and control (7.77 q/ha).

7.4.1.2 Mustard

Murasing and Viraktamath (2002) reported that mustard sprayed with Bee-Q @ 15 g/l attracted more pollinators and resulted in more number of pods (61.70 pods/plant), highest test weight (3.51g) and yield (11.81 q/ha). Thus there was an increase of 80.30 and 624.54 per cent over open pollinated and caged plot. Higher number of seeds (16.40 to 17.37 seeds/pod) was recorded in all the plots sprayed with Bee-Q.

7.4.1.3 Seasame and Niger

The flower structure of sesame facilitates cross pollination although it is usually considered to be self-pollinated crop. Honey bees are considered as one of the efficient pollinators of sesame. Patil et al. (2000) reported that spraying of bee attractants Bee-Q @ 12.5 g/l and Bee-here @ 4ml/l attracted more number of bees. These attractants resulted in production of significantly more number of pods (52.29-54.47 pods/plant), more seeds (64.29-66.60 seeds/pod) and heavier seeds (4.26-4.4 g/ 1000 seeds). Highest yield of 8.93 and 8.70 q/ ha was obtained in the treatments with Bee here @ 4 ml/l and Bee-Q 12.5 g/l, respectively.

Niger is a short day-length, photo sensitive plant that is cross-pollinated with bees. Guruprasad (2001) reported that spraying of fruit boost (0.5 ml/l) and tuberose floral-scented water attracted significantly more number of bees. Application of Fruit boost produced more number of seeds (30.72 seeds/head), heavier seeds (3.51g) and higher yields (4.93). There was an increase of 81.25 and 145.27 per cent in yield over the open pollinated and caged crop.

Nithya Chandran (2009) evaluated indigenous bee attractants *viz.*,Citral E, Citral Z, and those from *Fagarabudrunga* and *Swertiadensifolia* and found them to be as good as the commercial bee attractant Fruit Boost in attracting honey bees to sesame and niger flowers (2.12 to 4.23 and 10.18 to 15.35 bees/5 flowers/minute, respectively). Citral Z and Citral E were as good as Fruit Boost producing higher yield in sesame and niger(1.41 to 1.51 in sesame and 0.89 to 1.01 in niger vs 0.95 and 0.62 q/hain control in respective crops). Flowers treated with *F. budrunga* and *S. densifolia* produced next higher yield in both the crops.

Fakkiresh Muddi (2012) observed that though niger crop sprayed with Fruit boost recorded highest yield (1.43 q/ ha), indigenous bee attractants Citral Z and Citral E were the second best treatments in increasing yields (1.16 and 1.08 q/ ha) followed by *S. densifolia, F. budrunga,* tuberose and cinnamon (0.88 to 0.92 q/ ha).

7.4.2 Fodder Crops

7.4.2.1 Lucerne

Fruit boost, sugar solution, *S. densifolia, F. budrunga,* tuberose floral extract and Citral Z successfully enticed more number of bees compared to water sprayed crop (control) to lucerne flowers. The application of bee attractants improved the number of pods per plant (43.20 to 44.80 as against 42.87 pods/plant in control) as well as the number of seeds per pod (5.13 to 5.57 as against 5.03 seeds/pod over control). Highest yield was obtained from Fruit boost-sprayed crop (2.20 q/ ha) which resulted in an increase of 83.33 per cent over control. Sugar solution was the next best treatment in increasing the yield (2.00 q/ ha) followed by *F. budrunga,* tuberose floral scented water, *S. densifolia* and Citral Z (1.45 to 1.80 q/ ha). All the bee attractants enhanced the yield to the extent of 25.00 to 83.33 per cent over unsprayed crop (Chavan 2013).

7.4.3 Commercial Crops

7.4.3.1 Cotton

Ganapathi (2005) and Ganapathi and Viraktamath (2008) observed higher bee visits by *Apis dorsata, A. cerana* and *A. florea* on the flowers of two genotypes of Bt cotton (MECH-184 and RCH-2) treated with bee attractants (Figure 7.1 and 7.2). In MECH-184, spray of Fruit boost recorded higher good opened bolls (26.05/plant), more number of seeds per boll (28.20 seeds/boll), highest seed yield (9.61 q/ha), highest lint yield (5.76 q/ha) and highest kapas yield (15.36 q/ha), which accounted to an extent of 9.08, 8.04, 10.45, 12.28 and 11.30 per cent increase over open pollination, respectively. Bee-Q spray was the second best treatment by recording next higher number of good opened bolls (25.40/plant), more number of seeds (27.10 seeds/boll), higher seed yield (6.89 q/ha), higher lint yield (5.53 q/ha) and higher kapas yield (14.83 q/ha) which accounted to an extent of 6.36, 3.83, 6.89, 7.79 and 7.46 per cent increase over open pollination, respectively. A similar trend was observed in RCH-2 with Fruit boost spray recording higher good opened bolls (24.16/plant), more number of seeds (27.00 seeds/boll), higher seed yield (8.41 q/ha), highest lint yield (8.88 q/ha) and highest kapas yield (13.48 q/ha), which accounted 9.46, 6.92, 8.23, 13.48 and 9.96 per cent increase over open pollination, respectively.

Anjan Kumar Naik (2010) observed that in both BG-I and BG –II, all the four indigenous bee attractants (Citral E, Citral Z, *Fagara budrunga*, *Swertia densifolia)* were equally effective in attracting significantly more number of bees (2.13 to 3.11 and 2.13 to 2.96 bees /10m^2/5 min) which was as good as the Fruit boost, a commercial bee attractant (Figure 7.3). Consequently, these indigenous bee attractants enhanced the yield parameters *viz.,* good opened bolls (6.83 to 16 per cent increase over control), seeds per boll (7.76 to 17.14%), lint yields (6.56 to 12.96%), seed yield (5.80 to 13.33%) and kapas yield (10.02 to 13.51%). At the same time there was a decline in the bad opened bolls to the extent of 19.77 to 32.93 per cent over control due to spraying of attractants.

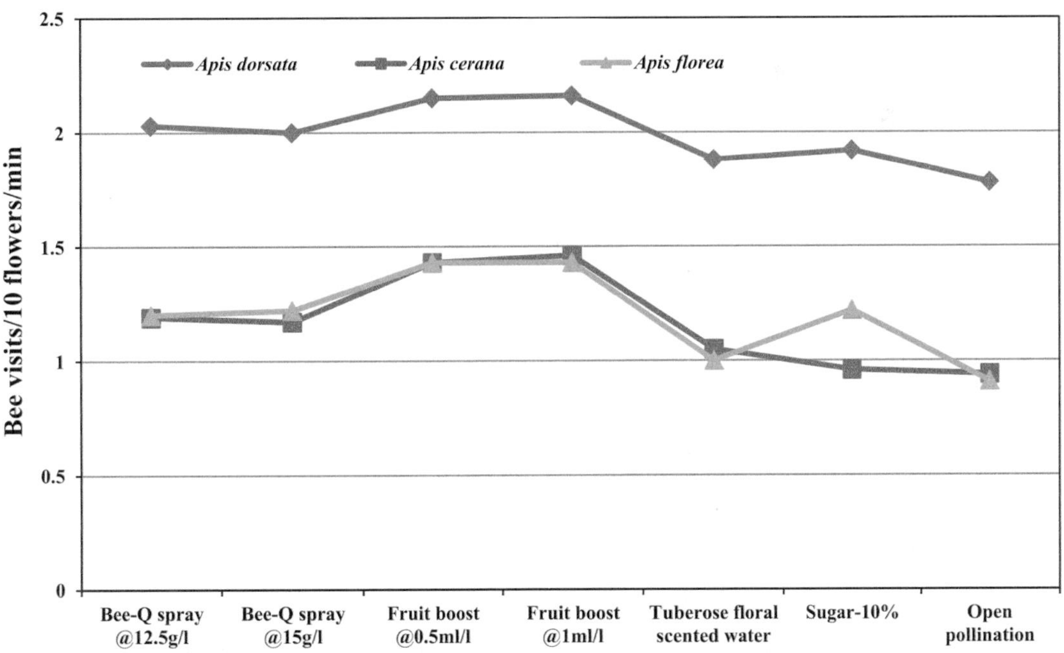

Fig. 7.1. Influence of attractants on bee visitation in MECH-184 Bt cotton

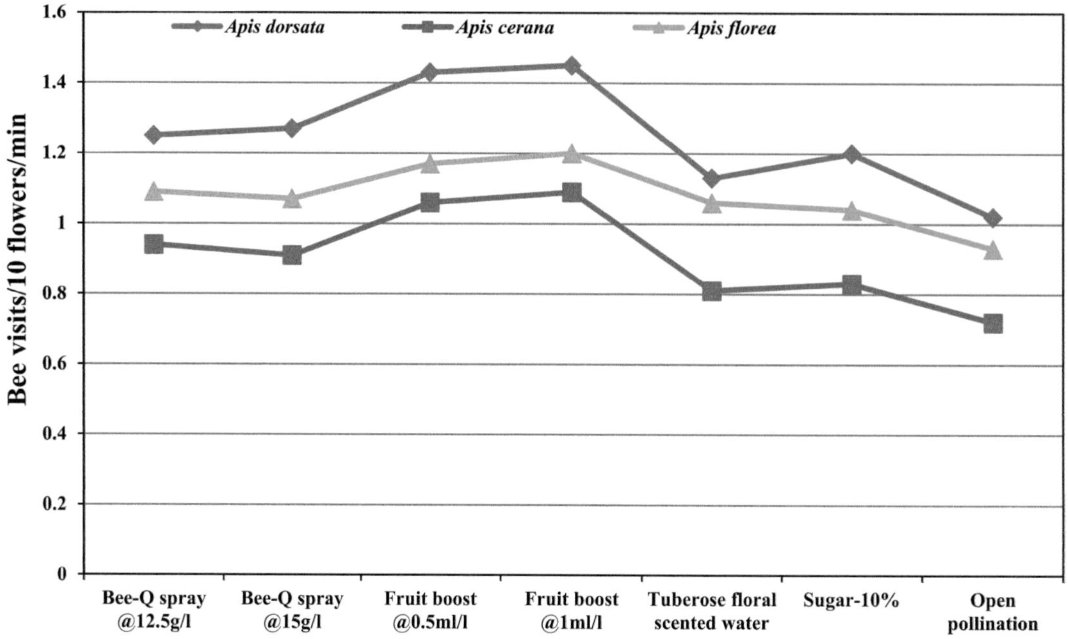

Fig. 7.2. Influence of attractants on bee visitation in RCH-2 Bt cotton

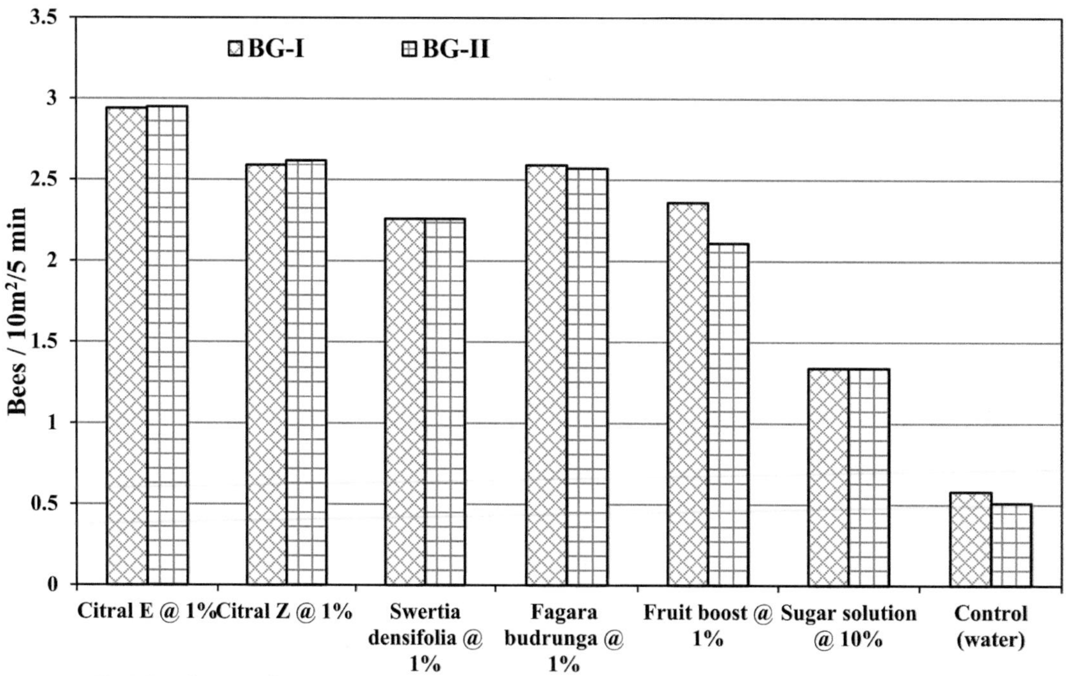

Fig 7.3. Influence of attractants on bee visitation in two genotypes of Bt cotton (BG-I and BG-II)

Vinayak Pise (2011) also observed indigenous bee attractants (Citral E, Citral Z, *Fagara budrunga, Swertia densifolia)* enhancing the bee visitation on Bt and non Bt cotton(2.25

to 3.25 and 2.17 to 3.17 bees /10m²/5 min in Bt and non-Bt cotton, respectively) which was as good as the Fruit boost. Bee attractants also enhanced the yield parameters *viz.,* good opened bolls (8.37 to 9.56 and 7.18 to 8.25% in Bt and non-Bt cotton, respectively), seeds per boll (10.82 to 21.89 and 10.67 to 16.64%), lint yield (5.80 to 7.83% and 4.85 to 5.13), seed yield (11.60 to 14.02 and 7.63 to 11.84%) and kapas yield (10.39 to 15.05 and 8.56 to 12.65%). The extent of increase in the total yield due to spray of indigenous bee attractants was more or less similar both in Bt (10.39 to 15.05%) and non-Bt cotton (8.56 to12.65%) hybrids.

7.5 Concluding Remarks

Although there is disagreement on the efficacy of honey bee attractants in enhancing pollination and crop productivity, the bee attractant arena continues to be of great interest to the bee scientists as well as to the farmers. Survey of the literature indicates that positive results outweigh negative results. However, failure of bee attractants in enhancing productivity needs to be critically analyzed. Presently the efficacy of the bee attractants lasts for a very short period of one to three days. Efforts have to be made to prolong this efficacy by developing techniques to have slow release of the bee attractant. Efforts should be also directed to develop indigenous bee attractants so that they are easily available to farmers.

References

Al-Sahf FH 2002. Effect of planting method, rose water spray on seed production in onion (*Allium cepa* L.). Emirates J AgrSci 14: 14–23.

Anjan Kumar Naik 2010. Evaluation of indigenous bee attractants in Bt cotton. M. Sc. *(Agri) Thesis*, Univ. Agric. Sci., Dharwad (India).

Anonymous 2010.http://www.bbc.co.uk/news/sceince-environment-11418033

Anonymous 2014.http://articles.economictimes.indiatimes.com

Ambrose JT, Schultheis JR, Bambara SB, Mangum W 1995. An evaluation of selected commercial bee attractants in the pollination of cucumbers and watermelons. J Apicult Res 135: 267–271.

Bhat SS, Sudarshan MR 1999. Studies on the efficacy of "Bee-Q" in augmenting pollination, fruit and seed set in cardamom (*Elettaria cardamomum* Moton). A preliminary report. Indian Bee J 61: 49–54.

Belletti A, Zani A 1981. An attractive product for bees on carrot grown for seed. Sementi Elette 27: 23–27.

Burgett M, Fisher GC 1979. An evaluation of Bee line as a pollinator attractant on red clover. Am. Bee J. 119: 356–357.

Butts KM 1991. Bee attractants: improving Strawberry quality. Litrus and vegetable Magazine 55: 16–17.

Chavan Vikas Tukaram 2013. Pollinator fauna, Foraging activity and effect of bee attractants on seed yield of Lucerne. MSc (Agri) Thesis, Univ.Agric.Sci., Dharwad (India).

Currie RW, Winston ML, Slessor KN 1992b. Effect of synthetic queen mandibular pheromone sprays on honey bee (Hymenoptera: Apidae) pollination of berry crops. J Econ Entomol 85: 1300–1306.

Currie RW, Winston ML, Slessor KN, Mayer DF 1992a. Effect of synthetic queen mandibular pheromone sprays on pollination of fruit crops by honey bees (Hymenoptera: Apidae). J Econ Entomol 85: 1293–1299.

FAO 1995. Pollination of cultivated crops in the tropics, FAO Agric Serv Bull, 208 pp.

Elmstrom GW, Maynard DN 1991. Attraction of honey bees to watermelon with bee attractant.Proc of the Florida state Hort Soc 103: 130–133.

Fakkiresh Muddi 2012. Influence of different dates of sowing and indigenous bee attractants on bee pollination and niger yield. MSc (Agri) Thesis, Univ.Agric.Sci., Dharwad (India).

Free JB 1968. Dandelion as a competitor to fruit trees for bee visits. J ApplEcol 5: 169–178.

Free JB 1970. Insect Pollination of Crops, Academic Press, London, 544 pp.

Ganapathi K 2005. Impact of bee attractants on bee activity and yield parameters of Bt cotton. MSc (Agri) Thesis, Univ.Agric.Sci., Dharward (India).

Ganapathi K, Viraktamath S 2008. Effect of bee attractants on bee visits and yield parameters of MECH-184 Bt hybrid cotton. Ad Pol Sp Res 25: 115–127.

Gellert M, Rozsa Z, Kovacs Z, Szendrei K, Hussain RA, Reisch G, Reisch J 1985. Honey bee attractants in the fruits of *Evodia Hyphensis* Dod. Herba Hungaria 24: 53–65.

Goyal NP 1993. Role of honey bees in improving agricultural productivity. In Goyal NP, Shrama RK (Eds) First National Conference on Beekeeping. National Horticultrual Board, Guragaon, India, pp 61–68.

Guruprasad GS 2001. Maximization of productivity of niger through enhancement of pollination. MSc (Agri) Thesis, Uni.Agric,Sci., Dharwad (India).

Higo HA, Winston ML, Slessor KN 1995. Mechanisms by which honey bee (Hymenoptera: Apidae), queen pheromone spray enhance pollination Ann Entomol Soc Am 88: 366–373.

Jayaramappa KV, Sivaram V 2013. Some aspects of bee attractants on Pumpkin. J Chem Biol Phys Sci 3: 1801–1807.

Ken-Naumann Winston ML, Slessor KN, Smirle MJ 1994. Synthetic honey bee (Hymenoptera: Apidae) queen mandibular gland pheromone application affects pear and Sweet cherry pollination. J Econ Ent 87: 1595–1599.

Lingappa S, Viraktamath S, Vastrad AS, Williams R 1999. Utilization of honey bee *Apis cerana* F. for pollination of watermelon and safflower. Proc Api mondia p 235.

Looper GM, Rossette LM 1991. Experimental use of Bee scent to influence the honey bee visitation and yield of watermelon. Am Bee J 131: 177–179

Mackenzie KE. Averill AL 1992. A new honey bee attractant, the queen mandibular pheromone, Cranberries 56: 13–14.

Malerbo Souza DT, Nogueira-Couto RH, Couto LA 2004. Honey bee attractants and pollination in sweet orange, *Citrus sinensis* Osbeck, var. *Pera-rio*.J Venom Anim Toxins incl Trop Dis 10: 362–368.

Manjunath K 2003. Field scale evaluation of bee attractants for their efficacy in sunflower. MSc(Agri) Thesis, Univ.Agric.Sci., Dharwad (India).

Margalith R, Lensky Y, Rabinowitch HD 1984. An evaluation of Beeline as a honey bee attractant to cucumbers and its effect on hybrid seed production. J Apic Res 23: 50–54.

Mayer DF, Johansen CA 1982. Field evaluation of chemical pollinator attractants on tree fruits.Am Bee J 122: 287–289.

Mayer DF, Britt L, Lunden JD 1989. Evaluation of Bee-Scent™ as a honey-bee attractant. Am Bee J 129: 41–42.

Maynard DN, Elmstrom GM, Mc Cuistion FT Jr 1992. Periodicity of watermelon fruit set and effect of bee attractant on yield. Proc Inter Am Soc Trop Hort 36: 81–87.

Michener CD 2000. The bees of the world.The John Hopkins University Press Pub.pp 913.

Murasing S, Viraktamath S 2002. Role of bee attractants in pollination and productivity of mustard (*Brassica juncea* L.).Proc of 6th Asian Apic Assoc (AAA) Intl Conf and World APIEXPO, Feb-24 to March 1, Bangalore, p78.

Naik DG, Banhatti P, Chanwda SS, Thomas D 2003. *Fagara budrunga* fruit extract as an attractant for *Apis cerana*. J Apicult Res 42: 48–49.

Naik DG, Dandge C, Puntambekar H, Patil T 2005. Attractant repllent properties of *Swertia densifolia* leaf extract towards *Apis cerana indica*. J Apicult Res 44: 116–118.

Naik DG, Dandge C, Puntambekar H 2007. *Swertia densifolia* leaf extract as a dose dependent attractant or repellent for *Apis florea*. J Apicult Res 46: 15–18.

Naik DG, Kapadi AH, Singh MK, Suryanarayana MC, Kshirsagar KK 1989. Lure development for Indian honey bees *Apis cerana indica* Fabr. Indian Bee J 51: 47–50.

Nidagundi B 2004. Pollination potentiality of honey bees on yield of bitter gourd (*Momordicac harantia* L.). *MSc (Agri) Thesis,* Univ.Agric.Sci., Dharwad (India).

Niera M, Barriga P, Barriga J 1995. Honey bee (*Apis mellifera* L., Hymenoptera: Apidae) pollinating behaviour on raspberry (*Rubus idaeus* L., cv Heritage) under the effect of two attractant and one repellent treatments. Agro Sur (Chile) 23**:** 52–59.

Nithya Chandran 2009. Evaluation of indigenous bee attractants in enhancing the yield parameters of sesame and niger. *MSc (Agri) Thesis*, Univ.Agric.Sci., Dharwad (India).

Ortiz–Sanchez FJ 1993. Evaluation of the efficiency of Beehere as a honey bee (*Apis mellifera* L) attractant to marrow crop (*Cucurbita pepo*) cultivated in green house conditions in Southern Spain. Entomofauna 14: 465–469.

Patil BS, Viraktamath S, Lingappa S, Giraddi RS, Parameshwarappa K, Bhat ARS 2000. Effect of Bee-q and Bee here on pollinators and yield of sesame. Insect Environ 5: 151–152.

Pateel MC, Sattagi HN 2007. Effect of different attractants on attracting the bees to cucumber (*Cucumis sativa* L.) crop Karnataka J Agric Sci 20: 761–763.

Patil JS, Mokat RB, Mupade RV, Kamate GS 2011.Role of bee attractants in pollination and productivity of onion (*Allium Cepa* L.). J Entomol Res 35: 127–131.

Pinzauti M 1985. An evaluation of different substances that attract pronubial insects. Proc 29th Intl Cong Apic, Budapest, pp 327–332.

Pinzauti M 1990. Kiwi pollination: several ways of increasing the activity of honey bees. Acta Hort 282: 149–150.

Pinzauti M, Corsi M 1981. The use of “Bee line” as an attractant for pollinating insects on flowers of fababean (Viciafaba m.):Informatore Agrario 37:14913–14914.

Rajotte EG, Fell RD 1982. A commercial bee attractant ineffective in enhancing apple pollination. HortScience17: 230–231.

Rapp R 1981. The effect of pollination by honey bees (*Apis mellifera* L.) on cucumber cv. ‘Delila’ seed production. Fac Agric, pp 71.

Sanjivan Kumar, Hari Chand, Singh R 2000. Increasing the attractiveness of sunflower to honey bees pollination. Shashpa 7: 151–154.

Schultheis JR, Ambrose JT, Bambara SB, Mangum WA 1994. Selective bee attractants did not improve cucumber and watermelon yield. HortScience 29:155–158.

Sheppard WS, Jecox ER, Parise SG 1979. Selection and management of honey bees for pollination of soybeans.Proceedings IV International Symposium on Pollination. Maryland Agricultural Experimental Station Special Miscellaneous Publication, pp 123–130.

Singh PB, Sinha SN 1996. Effect of Bee-Q on honey bee and seed yield of hybrid sunflower. Seed Res 24: 151–153.

Sivaram V, Jayaramappa KV, Menon A,Ceballos RM 2013. Use of Bee-attractants in Increasing Crop Productivity in Niger (*Guizotia abyssinica* L). Braz Arch Biol Technol 56: 365–370.

Srikanta Nath 2008. Evaluation of indigenous bee attractants in sunflower, MSc (Agri) Thesis, Univ.Agric.Sci., Dharwad (India).

Srimathi P, Vijaya J, Ananthakalaiselvi A, Krishnasamy V 1999. Bee-Q effect on honey bee visit and seed yield of hybrid sunflower KBSH-1. Madras Agric J 86:338–339.

Tew JE, Ferree DC 1984.The effect of Bee Lure on honey bee (*Apis mellifera*) pollination of apples. Fruit Crops 1984: A Summary of Research. The Ohio State University. Ohio Agric Res and Development Center, Special Circular 283:31–32.

Tsirakoglou V, Thrayavonlou A, Hatjina F, Stakiotakis E, Porlingis J 1997. Techniques to increase the attractiveness of Kiwi flowers to honey bee. *Proc of the Third Intl Symp on Kiwi Fruit*, Greece, 19–22 September.

Vinayak Pise 2011.Comparative performance of indigenous bee attractants in Bt and non-Bt cotton hybrids. *MSc (Agri) Thesis,* Univ.Agric.Sci., Dharwad (India).

Viraktamath S, Anagoudar JA 2002, Influence of bee attractants in enhancing pollination and yield parametersin *Cucumis sativa* L. Indian Bee J 64: 23–27.

Viraktamath S, Patil RK 1999. Preliminary studies on the influence of bee attractants on bee visitation and yield parameters of sesame, *Sesamum indicum* L. Indian Bee J 61:55–58.

Viraktamath S, Patil RK 2002.Effect of bee pollination in maximization of productivity of sunflower.Proc of 6th Asian Apicultural Association (AAA), Int Conf and World, APIXPO, February 24 to March 1, Bangalore, p 62.

Viraktamath S, Ganapathi K 2008. Improving the yield of RCH-2 Bt-cotton hybrid through enhancement of bee pollination. Ad Pol Sp Res 25: 165–168.

Williams RR, Jefferies 1982. Pollination of fruit crops: The evaluation of a pollinator attractant on cider apple blossom. Report of Long Ashton Research Station, *p* 27.

Winston ML, Slessor KN 1993. Application of queen honey bee mandibular Pheromone for bee keeping and crop pollination. Bee World 74: 111–128.

Woodrow AW, Nathagreen TH, Schonhorst MH, Hamilton KC 1965. Evaluation of chemicals as honey bee attractants and repellents. J Econ Ent 58: 1094–1102.

Zvedenok AP 1996. The onion seed crop can be improved. Kartofel I Ovoshchi 4: 29.

Chapter 8

Repellents for Honey Bees

Pradnya V Deshpande and Dattatraya G Naik
pardesh1234@yahoo.co.in; dgnpune@yahoo.co.in

Abstract

Honey bees visit the plants around their colonies for collecting nectar and pollen. An important activity of pollination gets executed in these visits. Although they get pollen from flowers, the bees may get nectar from flowers or any other part of the plant. Thus, honey bees come into intimate contact with plants when foraging for food. If some pesticides are sprayed on the plants, honey bees get exposed to them. This affects not only the honey bees getting exposed to the chemicals, but also the colony as a whole. Sometimes the collected pesticide gets carried into the honey which could be harmful to its consumer. Use of repellents for honey bees, while applying pesticides on plants, offers a solution in this situation. Repellents are important also for keeping the honey bees away from certain areas, as for example, while extracting honey or inspecting a bee colony. Search for repellents for honey bees originated at Agharkar Research Institute from the study of mandibular gland pheromone which is known to be the repellent pheromone. Chemical constitution of the mandibular gland pheromone of European honey bees *Apis mellifera* L. was examined first. Its major constituent was found to be 2-heptanone. When screened 2-heptanone did exhibit repellent properties. Its repellent activity towards the Indian honey bee, *A. cerana indica* Fab. and the dwarf bees, *A. florea* Fab., was eventually demonstrated. Besides, it was also shown to be a repellent towards the stingless bee, *Trigona gribodoi*. This makes 2-heptanone a general repellent for honey bees, as well as the sting-less bees. Since 2-heptanone is a ketone, many other ketones were screened and many possessing repellent activity were identified. While this was being done repellency of many other classes of synthetic and naturally occurring compounds was evaluated. From these studies essential oils emerged as a class of chemicals obtainable from plants which possesses repellent activities towards honey bees. Knowledge of these developments is helpful for developing new repellents for honey bees.

DOI: 10.1201/9781003505471-8

8.1 Introduction

Honey bees are one of the useful insects to mankind. Foraging behaviour of honey bees is one of their unique behaviours. It is a link between honey bee colony and the environment. Various in-colony and out-colony factors affect the foraging activity. Collection of nectar and pollen from the available flora is an important feature of foraging activity. For this purpose honey bees come in intimate contact with the flowering plants. This causes many interactions. Some of them are useful while a few are adverse. A very significant and useful event of pollination is the result of foraging activity of honey bees. Pollen grains get transferred from one flower to another by honey bees. Deodikar and Suryanarayana (1977) gave a comprehensive review on the increased yield in the field and fruit crops due to pollination by honey bees. It has been estimated that 80 per cent or more of pollination of fruit and seed crops is accomplished by honey bees in the United States (Eckert and Shaw 1960).

Organophosphates, carbamates and pyrethroids are extensively used as pesticides since long. Introduction of neonicotinoids has been an addition to the list of pesticides since late 1980s (Kollmeyer et al. 1999). Application of pesticides in the field is certainly useful to protect the plants from pests but it is shown to be harmful to honey bees by many researchers.

If pesticides are sprayed on the plants being foraged by honey bees, foragers get exposed to the pesticides too during the foraging activity. This is a very serious event which is sometimes even detrimental to the life of foraging honey bees (Abou-Shaara 2014).

8.2 Effects of Pesticides on Honey Bees

Pesticide applications under field conditions may produce several direct and indirect effects on honey bees as well as on their colonies. It has been cited as one of the potential causes for global honey bee loss through 'Colony Collapse Disorder,' (CCD) (Frazier et al. 2008, Pettis et al. 2012). Some of the pesticides act as repellents for honey bees which may result in reduced foraging activity. Repellency could also be due to irritant effect on honey bees (Reith and Levin 1987). It is also hypothesized that the so called repellent effect of insecticides can be due to sub-lethal toxicity resulting into transitory inhibition of activity (Reith and Levin 1988). Sub-lethal doses can also influence other bee behaviour patterns *i.e.,* dance rhythm, flight velocity, walking speed, wing beat frequency *etc.* (Brandes 1984, Basu and Chakrabarty, this volume). Abundance of honey bees is reduced in agricultural areas because of regular insecticide use (Crane and Walker 1983). Many such effects are reported in the literature (Kasamatu and Kawachi 1986, Shines et al. 1983, Stoner et al. 1984). At the same time a few pesticides are toxic to honey bees themselves which results into the mortality of a forager. Pesticide poisoning of honey bees has been identified since the advent of synthetic, broad spectrum insecticides. The destruction of honey bee colonies by agricultural insecticides is well documented. First published account of honey bee losses due to insecticides was given in 1881 during an application of Paris Green to a pear tree in blossom (Thompson)! When colonies are disturbed in the field, the use of pesticides becomes an acute problem (Todd and Mc Gregor 1960). After these studies the toxicity of pesticides to honey bees has been studied extensively and it has been observed that almost all synthetic pesticides are hazardous to honey bees (Johnson et al. 2010 and references therein, Karazafiris et al. 2011). Although the use of organochlorines has been banned in Europe for decades, pesticides like DDT, HCH, and HCB were often detected in the honey samples from Portugal and Spain even in the recent past (Blasco et al. 2003). The results obtained could be expected, because those pesticides which were extensively used and are still present in the environment, owing to their high persistence. Organochlorines are lipophilic substances and consequently are soluble and stable in beeswax. Therefore, an amount of these substances gradually migrates from wax into the stored honey. Occurrence of pesticides sprayed on the crops in honey samples was reported from India also (Chaudhary and Sharma 2008). Occurrence of pesticide residues in pollen collected by honey bees is also reported (Chauzat et al. 2006). Neonicotinoid pesticides have been shown to adversely affect pollinator survival. Recently the side effects and risk assessment of neonicotinoids have also been reviewed (Blacquiere et al. 2012).

As a consequence of these observations judicious use of pesticides to avoid losses to honey bees was advocated since 1940s (Anderson and Atkins 1958, Atkins et al. 1954, Atkins and Anderson 1962, Attri and Sharma 1969, Gupta 1988, Hafliger 1949, Johansen 1965). Repeated applications of pesticides can also lead to injuries to honey bees and thereby reduce their longevity (Fries and Wibran 1987, Mackenzie and Winston 1989, Smirle et al. 1984). When honey bees are exposed to pesticide sprayed areas for a long time, collected pesticides gradually accumulate in colonies. Reduced egg laying and brood rearing due to small doses of pesticide application has also been recorded (Fiedler 1986, Lensing 1986, 1987, Nation et al. 1986). Amorphogenic effects in delayed and abnormal development in honey bees were observed in late 1980s (Atkins and Kellum 1987, Czoppelt and Rembold 1988, Ruijter and Steen 1987).

Pesticide contaminated food/nectar may cause honey bees to cease feeding or there may be reduced consumption and collection of nectar (Fiedler 1987). Pesticide application may also change the physiology of nectar or pollen producing plants. Mishra and Sharma (1988) have demonstrated significant effects of growth regulators on pollen production, nectar volume and insect foraging in *Brassica* crops. This results into changes in attraction of honey bees to flowers and consequently affects pollination and yield. Pesticides are shown to affect pollen viability. Reports about effects of organophosphorus insecticides on germination of pollen in tomato (Gentile et al. 1971) and of fenvalerate and phenthoate on apple pollen (Mishra et al. 1987) emphasize this hypothesis.

8.3 Repellents for Honey Bees

Protection of honey bees from the commonly used insecticides has been accepted as one of the important requirements for ensuring efficient pollination in agricultural crops and maintenance of their high yields. In order to reduce the pesticide hazards to honey bees, the repellent additives are useful (Atkins et al. 1977).

The idea of using honey bee repellents to avoid the harmful effects of insecticides by preventing exposure of honey bees to toxic chemicals is not new. The first suggestion to use a chemical addition in insecticide sprays was made in 1900, when use of carbolic acid for repelling honey bees was recommended (Callbreath)! Melander (1921) found carbolic acid, cresol compounds, carbon disulphide, nicotine sulphate and naphthalene to be of use as honey bee repellents. Repellent effects of a number of pesticides viz. BHC, demeton, dieldrin, aldrin, parathion, carbaryl, thiodan etc were studied (Johansen et al. 1957, Lieberman et al. 1954, Linsey and Mac Swain 1947, Palmer-Jones et al. 1954 and 1959). However, all these pesticides were found toxic to honey bees and did not have any significant role to play as honey bee repellents. In 2003, Schmidt et al. studied the olfactory stimulation of Africanized honey bee by three insect repellents, viz. N,N-Diethyl-*meta*-toluamide [DEET], Pyranha*i.e.,* pyrethrins, piperonylbutoxide [(butylcarbityl) (6-propylpiperonyl) ether and related compounds, along with butoxypolypropylene glycol and petroleum distillates], and Repel X [contains pyrethrins, piperonylbutoxide, permethrin (3-phenoxyphenyl) methyl (±) cis/trans 3 - (2,2 dichloroethenyl)-2,2-dimethyl cyclopropane carboxylate as active compounds]. They further observed that none of these pesticides showed any significant repellent activity (Schmidt et al. 2003). Considering these facts many compounds were evaluated for their repellency towards honey bees. Search of other repellent substances for keeping the honey bees away from the areas where pesticides were sprayed was also done (Holtmann 1991).

8.4 Evaluation of Synthetic Compounds as Honey Bee Repellents

Pheromones are the chemicals important in the communication of insects (Deshpande and Naik, Chapter 6, this volume). Mandibular gland pheromone of honey bees is known to possess repellent character. In order to understand the honey bee attractant/repellent compounds, the mandibular gland pheromones of honey bee *A. mellifera* were studied (Butler 1966). It was found that foraging bees *A. mellifera* were strongly repelled by crushed heads and crushed mandibular glands of foraging bees which contain both 2-heptanone and 10-hydroxy-2-decenoic acid. Shearer and Bosch (1969) found 2-heptanone to exhibit repellency to *A. mellifera* L. Morse et al. (1967) reported the absence of 2-heptanone in the mandibular gland secretions of *A. cerana,* but in the same set of experiments they demonstrated elicitation of alarm behaviour due to 2-heptanone in *A. cerana*. An attempt has been made to use 2-heptanone as a repellent in agriculture by controlling volatility by the use of dry PVC resin as an additive (Reith et al. 1986). It is noteworthy that repellency of 2-heptanone towards *A. cerana* (Naik et al. 1997a and 1997b) and *A. florea* (Naik et al. 2002) was subsequently demonstrated. 2-Heptanone was also found to be one of the major constituents of the mandibular gland secretions of the stingless bee, *Trigona gribodoi*. A cephalic extract of these bees was used directly, instead of 2-heptanone, during the experiment to study the repellency (Keeping et al. 1982). These findings show that 2-heptanone is a repellent for all the species of honey bees, as well as stingless bees, and can be regarded as a general repellent for them.

Recently bioassay of the synthetic constituents of the mandibular gland pheromone of *A. florea,* 1-eicosanol, eicosane, heneicosane and 2-heptanol was carried out to study the foraging activity of dwarf honey bees *A. andreniformis* and *A. florea*. The repellent activity of the test chemicals was demonstrated in these experiments (Suwannapong et al. 2011).

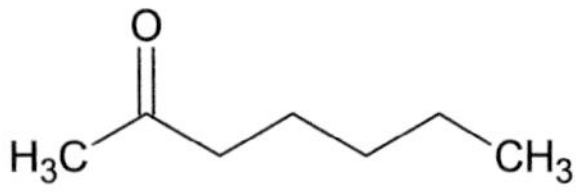

2-Heptanone

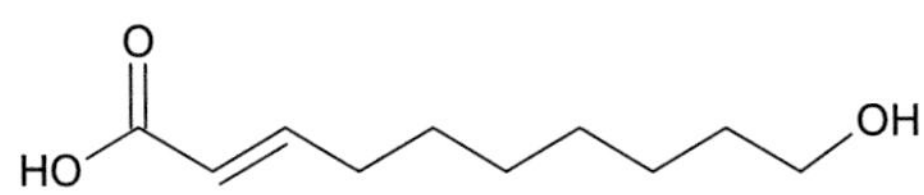

10-Hydroxy-2-decenoic acid

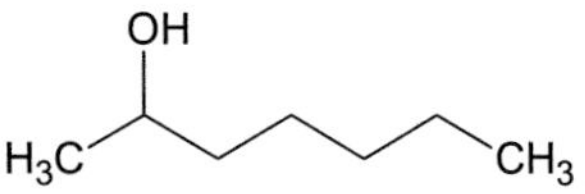

2-Heptanol

1-Eicosanol

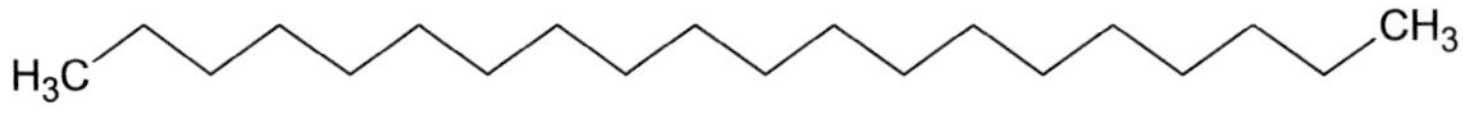

Eicosane

Heneicosane

Evaluation of other carbonyl compounds as repellents for honey bees under laboratory, semi-field and field conditions has been done over the last several decades. The ketones tested can be further classified as follows.

8.4.1 Aliphatic Straight Chain Ketones

(E)-6,10-Dimethyl-5,9-undecadien-2-one was found to exhibit repellent properties towards *A. mellifera* under laboratory conditions (Atkins et al. 1975). Under semi-field conditions 3-octanone, methyl isobutyl ketone, ethyl vinyl ketone and acetyl acetone were found to be good repellents for *A. florea* (Gupta and Kapil 1983). It was reported that amongst the aliphatic straight chain compounds tested for olfacto-gustatory response against *A. florea* under semi-field conditions, 2-octanone showed maximum repellency at lower concentration, followed by methyl isobutyl ketone and ethyl vinyl ketone which also showed good repellency. Surprisingly other carbonyl compounds like acetone, diethyl ketone, 2-hexanone, 3-hexanone and 4-heptanone exhibited moderate repellency. The repellency was found to be dose dependent. It was observed that repellency of all compounds increased with increase in concentration till certain limiting value of concentration. More importantly these compounds did not exhibit oral or respiratory toxicity to the dwarf honey bee (Gupta 1985, Gupta and Mohla 1986). Methyl isobutyl ketone and acetyl acetone exhibited high repellency towards *A. mellifera* under semi-field conditions (Kasana et al. 1997). Under semi-field conditions a few studies have been conducted to assess the efficacy of aliphatic straight chain ketones as *A. florea* repellents. In these studies a concept of duration of significant repellent activity (DSRA) is commonly used. DSRA is defined as the duration for which percent repellency is 60 per cent or above. Ethyl propyl ketone and ethyl butyl ketone were found to exhibit high repellency each showing five hours DSRA whereas ethyl pentyl ketone exhibited four hours DSRA (Singh et al. 1993).

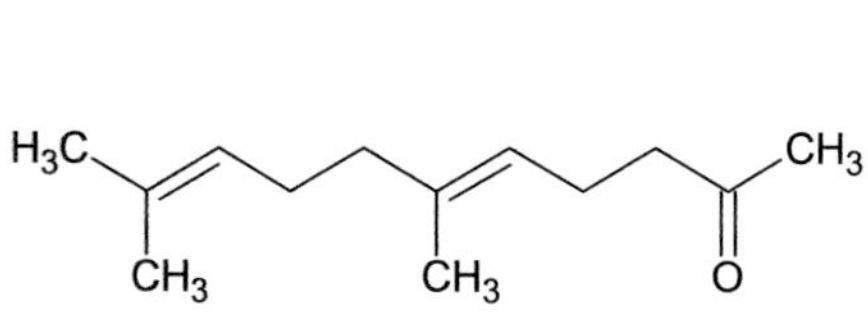
(E)-6,10-Dimethyl-5,9 undecadiene-2-one

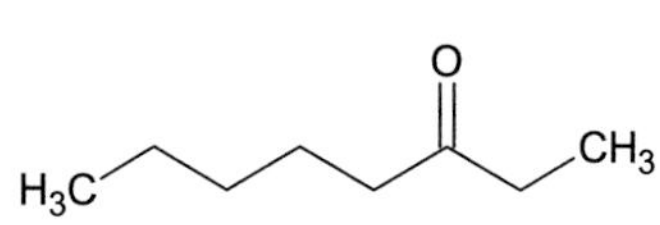
3-Octanone

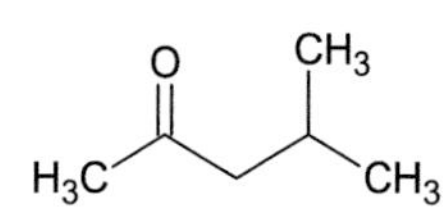
Methyl isobutyl ketone
(4-Methyl-2-heptanone)

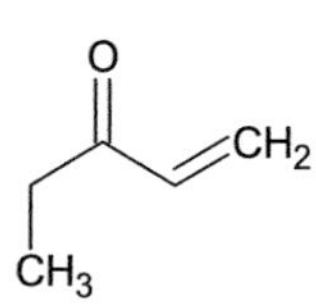
Ethyl vinyl ketone

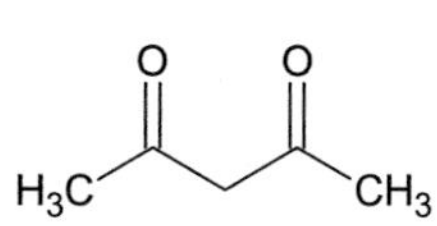
Acetylacetone

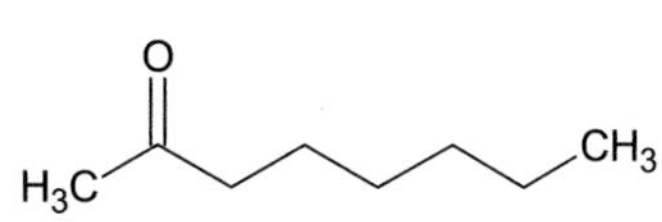
2-Octanone

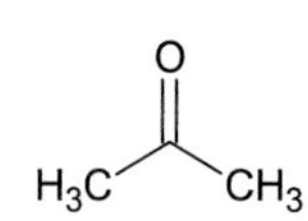
Acetone

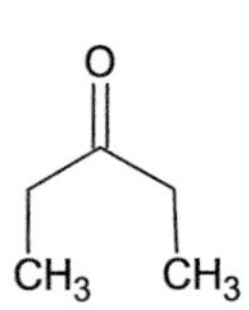
Diethyl ketone

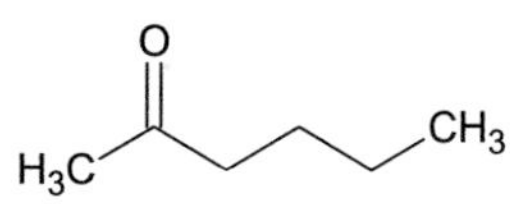
2-Hexanone

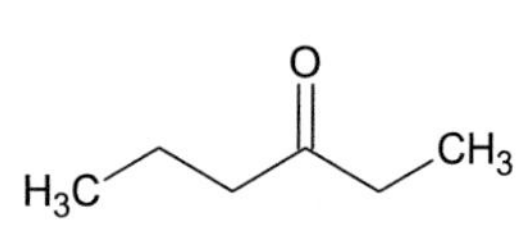
3-Hexanone

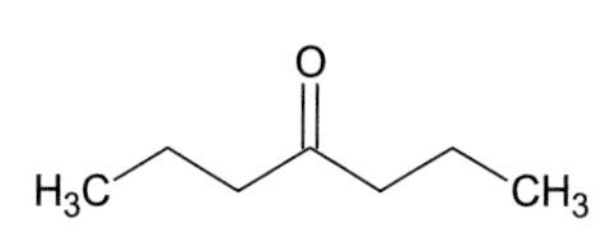
4-Heptanone

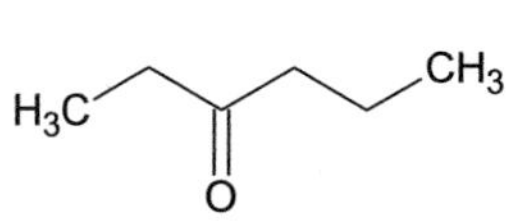
Ethyl propyl ketone (3-Hexanone)

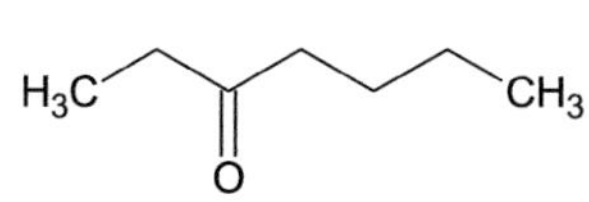
Ethyl butyl ketone (3-Heptanone)

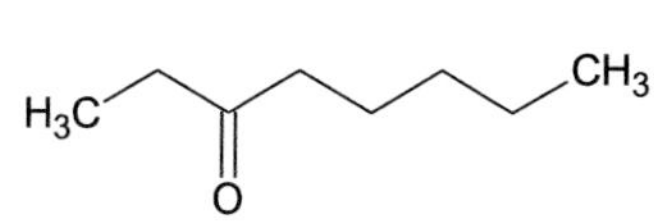
Ethyl pentyl ketone (3-Octanone)

8.4.2 Cyclic Ketones

A number of carbocyclic ketones were reported for their repellency towards *A. florea* F. under semi-field conditions. These findings were of interest since the functional groups and the number of carbon atoms were the same as compared to the acyclic ketones. However the shape of these molecules and their boiling points were different from their acyclic analogues. Their repellency was quantified and the order of repellency was also determined as follows: cyclooctanone>cycloheptanone>cyclopentanone>cyclohexanone. Thus the importance of size of the carbocyclic ring containing carbonyl group is established. In addition to these ketones D-carvone also exhibited significant repellency. Light intensity and temperature were negatively correlated with repellency having significant values of correlation coefficients while relative humidity was found to have positive correlation with repellency with non-significant values of correlation coefficients (Gupta 1986 and 1987a, Gupta and Kapil 1983). Under field conditions, 4-methyl cyclohexanone, when tested for repellency towards *A. florea* F. visiting *Brassica campestris* var. Toria exhibited four hours DSRA (Singh et al. 1993) while cyclopentanone exhibited repellency towards these honey bees visiting *Ziziphus mauritiana* within six hours of DSRA (Singh et al. 1996).

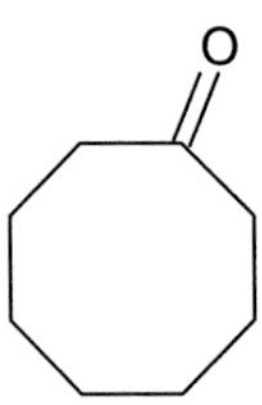

Cyclooctanone

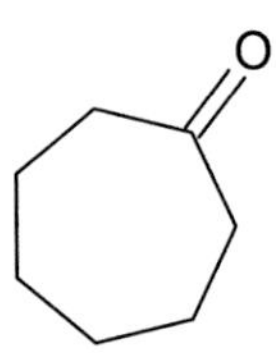

Cycloheptanone

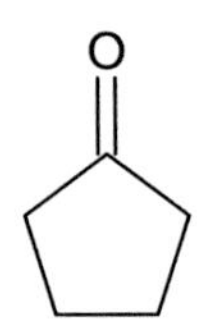

Cyclopentanone

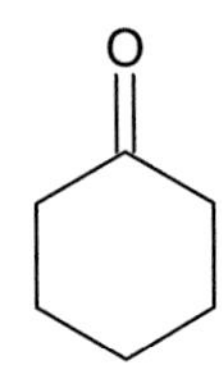

Cyclohexanone

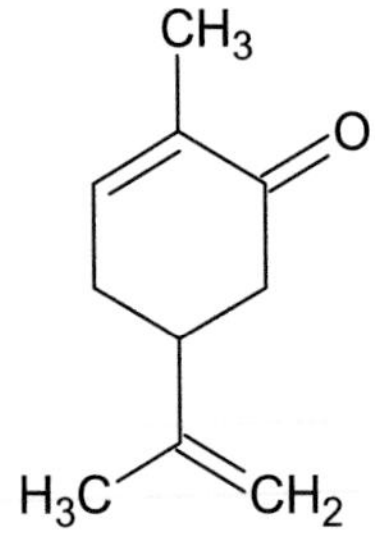

Carvone

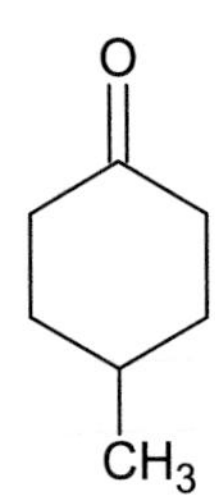

4-Methyl cyclohexanone

8.4.3 Aryl Alkyl Ketones

Acetophenone, the simplest aryl alkyl ketone, showed significant repellent activity to *A. florea* workers in olfactometer tests but its activity was low under field conditions (Chamberlain 1959). Encouraged by these results other aryl alkyl ketones were evaluated for their repellent activity. Under semi-field conditions, these ketones have also been reported to exhibit repellency towards honey bees. *o*-Aminoacetophenone, ethyl benzyl ketone, benzyl butyl ketone, benzyl acetone and 2-methyl butyrophenone were found to exhibit high repellency in semi-field conditions towards *A. florea* (Gupta and Kapil 1983).

1-Phenyl-2-pentanone, 1-phenyl-2-hexanone, ethyl benzyl ketone and benzyl acetone were also tested in semi-field conditions for their repellency towards *A. florea*. At the limiting value of concentration benzyl acetone exhibited maximum repellency while 1-phenyl-2-pentanone exhibited minimum. It was further confirmed that repellency of such compounds was dose dependent up to a certain extent and remained constant beyond that (Gupta 1987b).

Under field conditions, *o*-amino acetophenone and benzyl ethyl ketone were tested with benzophenone and diethyl phthalate as additives. Ethyl benzyl ketone with benzophenone (1:1) or diethyl phthalate (1:1) was effective for six hours. The effectiveness of *o*-amino acetophenone with benzophenone persisted for five hours (Gupta and Kapil 1983). 2-Methyl butyrophenone and 1-phenyl-2-pentanone exhibited significant repellency to *A. florea* visiting *Brassica campestris* var. *Toria* with six hours and four hours DSRA respectively (Singh et al. 1993). 2,4-Dimethyl acetophenone exhibited four hours DSRA while 2-methyl butyrophenone and butyl phenyl ketone exhibited significant activity only for shorter duration i.e., two hours each against *A. florea* visiting *Ziziphus mauritiana* (Singh et al. 1996).

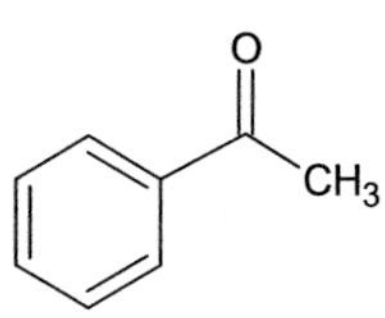

Acetophenone

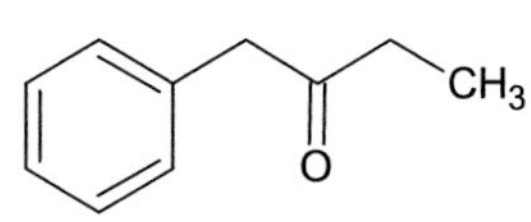

Ethyl benzyl ketone

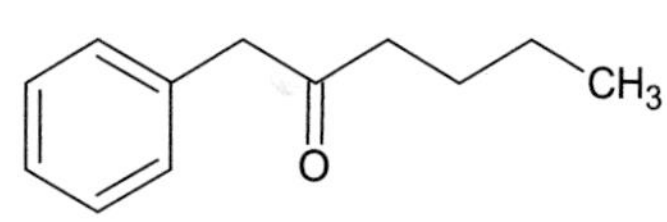

Benzyl butyl ketone

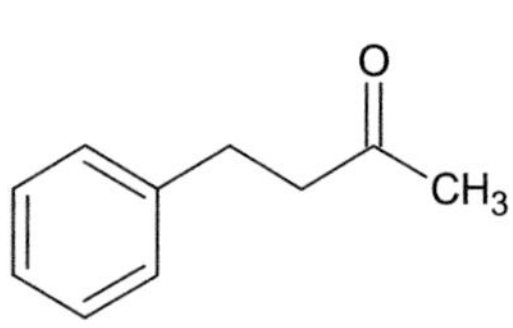

Benzyl acetone

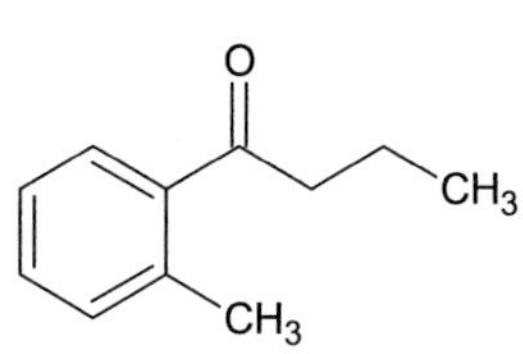

2 -Methyl butyrophenone

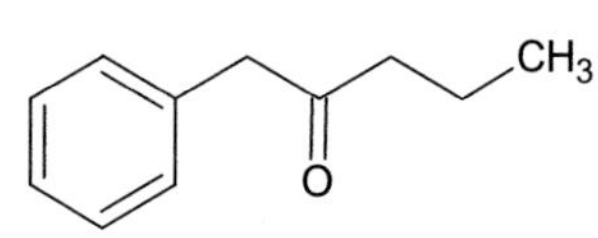

1-Phenyl-2-pentanone

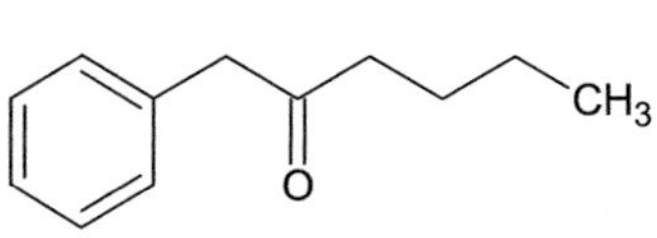

1-Phenyl-2-hexanone

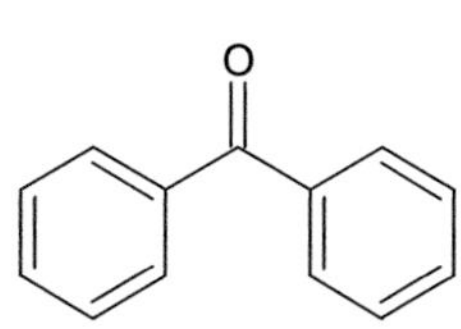

Benzophenone

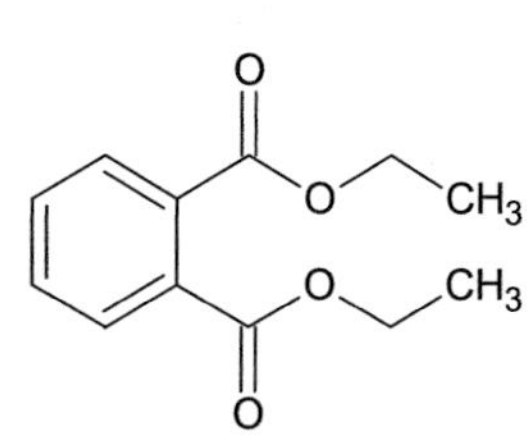

Diethyl phthalate

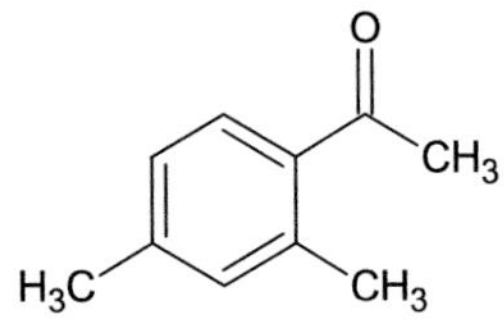

2,4-Dimethyl acetophenone

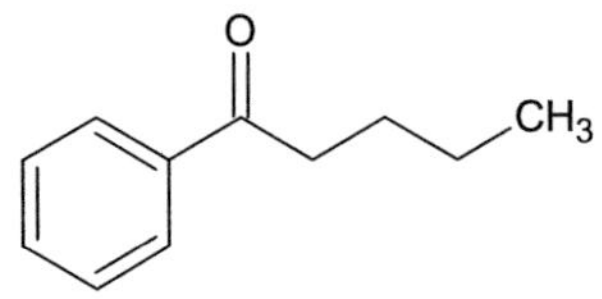

Butyl phenyl ketone (1-Phenyl-2-hexanone)

Considering that ketones are carbonyl compounds aldehydes, a closely related class of other carbonyl compounds, was also evaluated for the repellency towards honey bees.

Townsend (1963) recommended benzaldehyde to repel honey bees from honey combs. Investigations on repellency of *n*-butyraldehyde and *n*-propionaldehyde towards *A. florea* in laboratory conditions at 14 different concentrations have led to the conclusion that repellency of aldehydes is highly concentration dependent. The repellency of propionaldehyde at four per cent and *n*-butyraldehyde at two per cent concentration indicated that the aldehydes are less active than the ketones (Gupta et al. 1986). In field testing, the activity of propionaldehyde was observed for shorter duration (repellency 60.4% and DSRA as one hour) towards *A. florea* visiting *B. campestris* var. *toria* than under laboratory conditions (Singh et al. 1993). *n*-butyraldehyde exhibited five hours DSRA when tested for repellency against *A. florea* foraging on *Ziziphus mauritiana* (Singh et al. 1996).

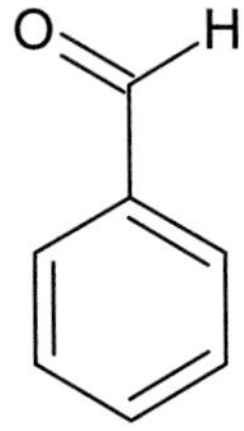

Benzaldehyde

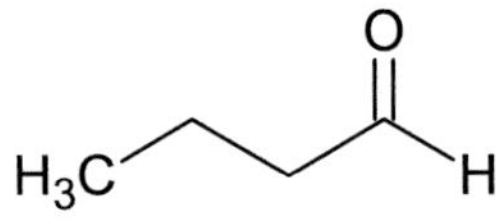

n - Butyraldehyde

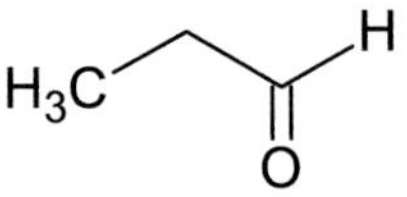

n - Propionaldehyde

8.4.4 Carbonyl Compounds with Additives

Under field conditions, o-amino acetophenone and benzyl ethyl ketone were tested with benzophenone and diethyl phthalate as additives. It was found that ethyl benzyl ketone with benzophenone (1:1) or diethyl phthalate (1:1) was effective for longer time than if used alone (Gupta and Kapil 1983). A comparison of percent repellency and duration of significant repellent activities (DSRA) was made between carbonyl compounds tested with or without additives. Percent repellency to *A. florea* visiting *B. campestris* increased but DSRA decreased on addition of benzyl alcohol as an additive to aldehydes. While in case of ketones DSRA increased significantly but percent repellency either increased or decreased. It may be stated that number of carbonyl compounds tested in combination with benzyl alcohol as additive is not sufficient enough to reach any conclusion. However, certain other carbonyl compounds were also tested on the same crop in combination with benzyl alcohol to find out suitable combination which shows longer DSRA. The ketones *viz.*, cyclopentanone, 2-methyl cyclohexanone and methyl pentyl ketone showed six hours DSRA in these experiments (Anonymous 1987).

Benzyl alcohol responded similarly when its efficacy for longevity of repellency to *A. florea* visiting *Allium cepa* was studied in combination with some carbonyl compounds. It was seen that in case of acetophenone, both repellency as well as durability decreased to a large extent. For diethyl ketone there was decrease in longevity of repellency. In case of *n*-butyraldehyde longevity of repellent activity increased. However, benzyl methyl ketone and methyl pentyl ketone exhibited higher repellency as well as DSRA when tested without additives (Anonymous 1987).

Studies on response of *A. florea* worker honey bees visiting *Brassica juncea* var. *raya* Prakash to three carbonyl compounds *viz.*, diethyl ketone, ethyl propyl ketone and ethyl butyl ketone with or without benzyl alcohol showed that with the additive the percent repellency as well as DSRA increased in case of diethyl ketone while in case of ethyl butyl ketone there was significant decrease in percent repellency as well as DSRA. Ethyl propyl ketone showed increased repellency as well as DSRA in combination with the additive. Thus it could be concluded that efficacy of carbonyl compounds as repellents may increase or decrease with the addition of benzyl alcohol. In other words the effect of addition of benzyl alcohol on the repellency of any carbonyl compound could not be predicted (Anonymous 1987).

Effect of benzyl benzoate as an additive to repellent chemicals for their repellency to *A. florea* visiting *Brassica campestris* var. *Sarson* has been reported (Anonymous 1997). Percent repellency of *n*-butyraldehyde increased in combination with benzyl benzoate while DSRA decreased. It might imply that rate of volatility increased due to which repellency also increased which remained significant only for shorter duration. Percent repellency as well as DSRA decreased when caprylaldehyde was tested in combination with benzyl benzoate. Similar results were obtained with ethyl butyl ketone also. On the other hand propionaldehyde exhibited increased percent repellency as well DSRA when tested in combination with benzyl benzoate (Anonymous, 1987). Some carbonyl compounds were tested alone and in combination with additives for repellency against *A. florea* visiting *Medicago sativa*. Acetophenone alone was found to show maximum repellency for four hours whereas in combination with benzyl benzoate it exhibited repellency with six hours DSRA. Because of variable effects of additives i.e., benzyl alcohol and benzyl benzoate tested in combination with different carbonyl compounds, no hypothesis about their effect on repellency of the active compound could be postulated (Anonymous 1997).

The carbonyl compounds tested alone were found to show a decrease in activity with increase in time. The use of benzyl alcohol as an additive did not show any improvement. As against this, use of benzyl benzoate as additive increased the DSRA of carbonyl compounds. However, it did not show any improvement in activity of acetyl acetone. Dibutyl phthalate was found to increase the degree of repellency and durability of carbonyl compounds. Acetophenone gave the best results with dibutyl phthalate showing six hours DSRA. In the same way diethyl malonate increased repellency as well as DSRA of a few carbonyl compounds like acetophenone, acetyl acetone and cyclohexanone. Comparing the degree of repellency and durability of all combinations studied, the combination, dibutyl phthalate + acetophenone is found to be the best followed by benzyl benzoate + acetophenone > diethyl malonate + acetophenone > diethyl malonate + acetyl acetone > ethyl chloro acetate + acetophenone > methyl ethyl ketone + dibutyl phthalate > cyclohexanone + diethyl malonate. The additives benzyl alcohol and methyl acetoacetate were found to suppress the effectiveness of the repellents (Anonymous 1997).

Two additives, one showing negative effect (benzyl alcohol) and another showing positive effect (dibutyl phthalate) were again studied in combination with different carbonyl compounds for their repellency towards *A. mellifera* visiting *Eruca sativa* TMH-52 for the confirmation of results obtained on *B. campestris* against *A. florea*. It is noteworthy that results of these experiments were in complete agreement with the earlier results (Anonymous 1997).

In addition to the screening of aldehydes and ketones repellency of other classes of carbonyl compounds like carboxylic acids, esters and anhydrides towards honey bees was also determined. About 200 randomly selected carbonyl compounds having different functional groups were tested under laboratory conditions for their repellency towards *A. mellifera* (Woodrow et al. 1965). It was observed that only a few compounds showed moderate to very strong repellency. Among these acids and anhydrides were found to be active. The order of their repellency observed was maleic anhydride > acetic acid > propionic anhydride > isobutyric acid > propionic acid. Combination of propionic anhydride (50%) and propionic acid (50%) was found to exhibit strong repellency. Among the esters, benzyl benzoate exhibited good repellency under laboratory conditions (Chamberlain 1959). Atkins et al. (1975) found the order of repellency of other esters as *p*-tolyl acetate > *o*-tolylchloroacetate > *p*-methoxy phenyl chloroacetate > dimethyl phthalate > *p*-methoxy benzyl propionate > ethyl phenyl acetate > phenyl acetate under the laboratory conditions.

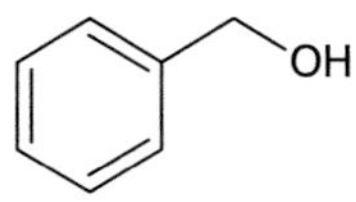

Benzyl alcohol

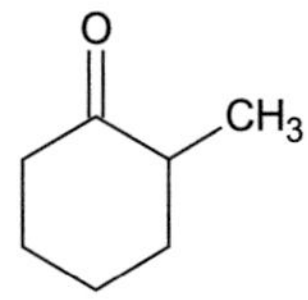

2 - Methyl cyclohexanone

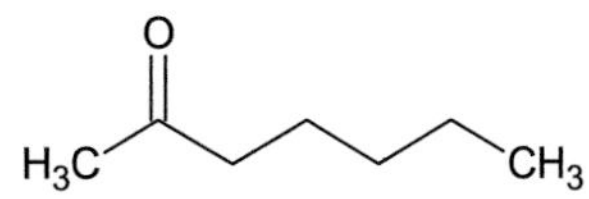

Methyl pentyl ketone (2-Heptanone)

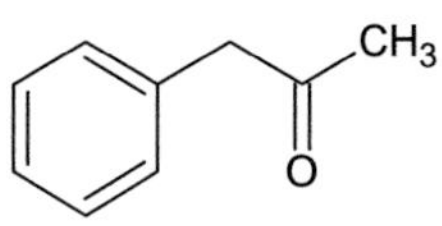

Benzyl methyl ketone
(1-Phenyl-2-propanone)

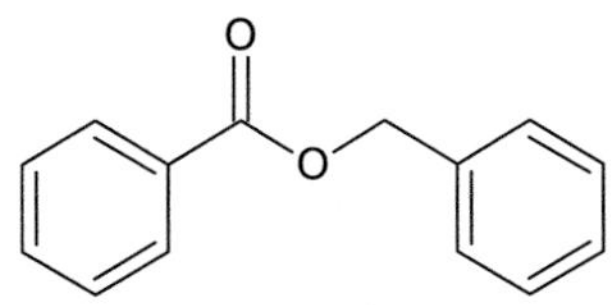

Benzyl benzoate

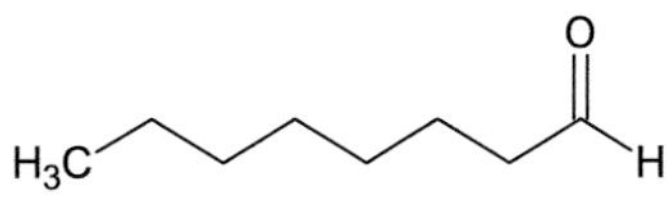

Caprylaldehyde

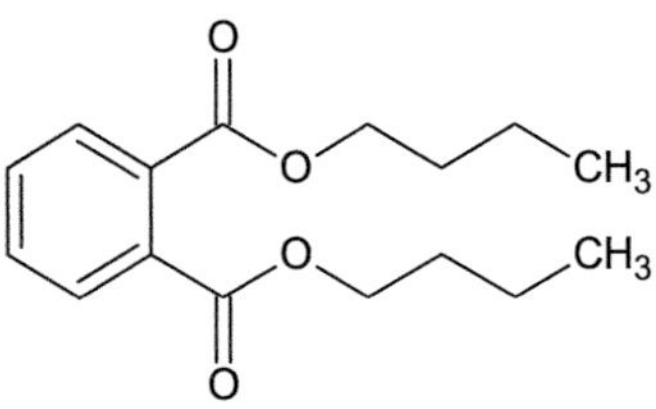

Dibutyl phthalate

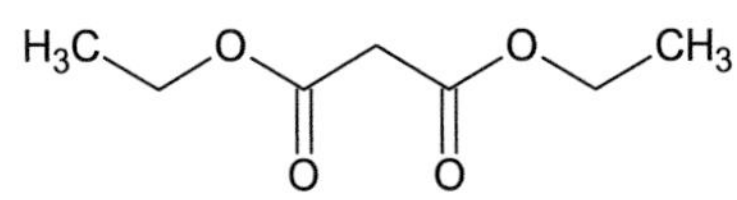

Diethyl malonate

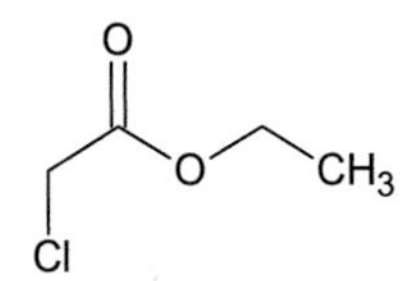

Ethyl chloroacetate

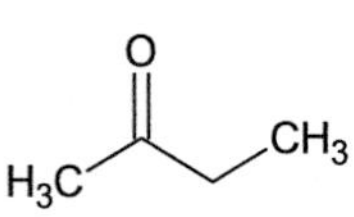

Methyl ethyl ketone
(2-Butanone)

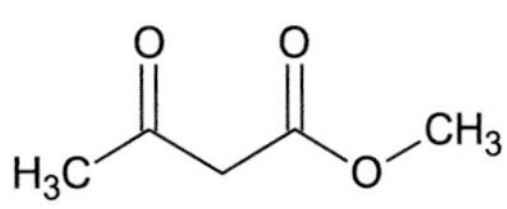

Methyl acetoacetate

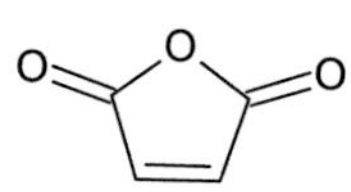

Maleic anhydride

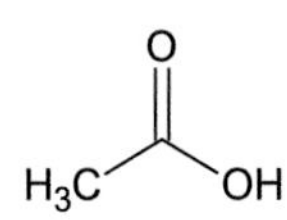

Acetic acid

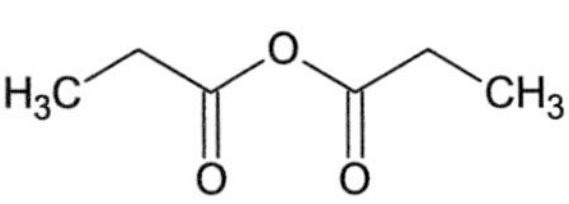

Propionic anhydride

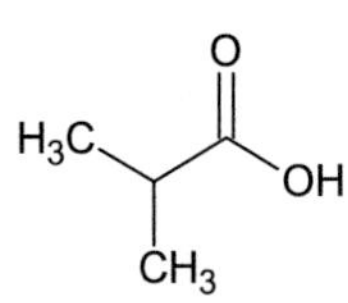

Isobutyric acid

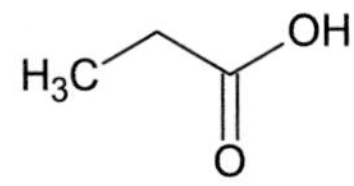

Propionic acid

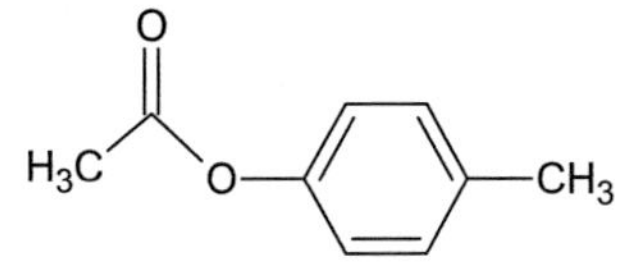

p-Tolyl acetate

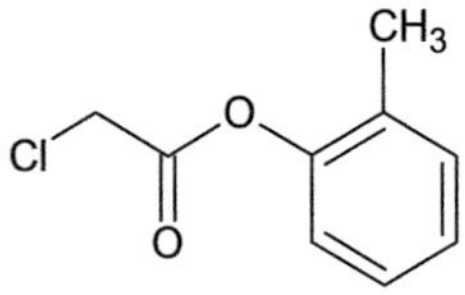

o-Tolyl chloroacetate

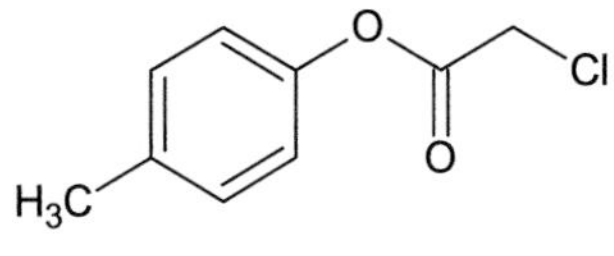

p-Methoxy phenyl chloroacetate

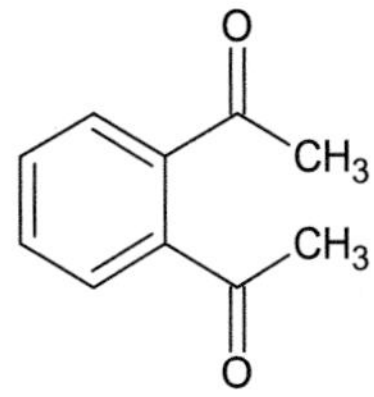

Dimethyl phthalate

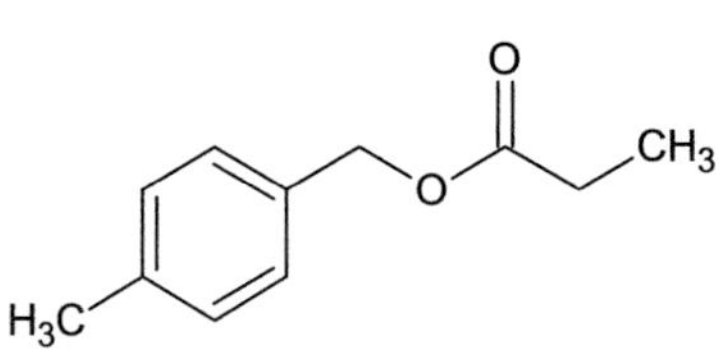

p-Methoxy benzyl propionate

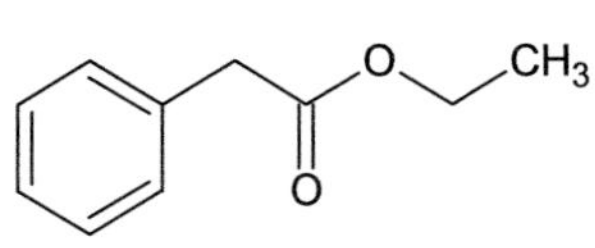

Ethyl phenyl acetate

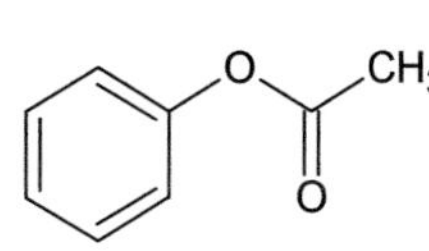

Phenyl acetate

The search for honey bee repellents was also carried out by screening a few selected compounds which are not honey bee pheromones. For this screening of volatile organic compounds belonging to mercaptans were identified as repellents for *A. cerana indica* (Naik et al. 1999). Aliphatic or alicyclic thioethers having six to ten carbon atoms such as ethylthio cyclohexane, found to be effective as repellents, were patented (Naik et al. 2005a). Mayer et al. (2001) conducted field trials of about 240 different compounds, which were not pesticides or pheromones, during 1990-1998 on a variety of crops to evaluate their repellent effect to foraging honey bees. However their results did not lead to effective repellent formulations. Woods et al. (2012) screened a large number of carbocyclic and heterocyclic compounds on foraging honey bees and a patent application has been filed.

Woodrow et al. (1965) reported that the amino compounds also have the potential to repel the honey bees. By screening several amines they found the order of repellency towards honey bees as: 2-ethylhexylamine > N,N-dimethyl,1,1,3,3-tetramethylbutylamine > dibutylamine > isodecylamine > hexylamine > cyclohexylamine > N-*tert*-butyl-1-ethylbutylamine > 1,3-propanediamine > 3-isopropoxypropylamine. 1-Hexanoylpyrrolidine and1-hexanoylpiperidine exhibited high repellency towards *A. mellifera* under laboratory conditions, both being at par with each other (Atkins et al.1975). 2-Ethylhexylamine and isodecyl amine were found to exhibit repellency in the field trials. It was observed that after the application of the compound, the numbers of honey bee visits were reduced. However these compounds were found to exhibit toxicity by damaging considerable number of flowers, hence the reduction in bee visits could not be attributed to a repellent effect of chemical alone (Woodrow et al. 1965). Decylamine, carvacrylamine and 2-(decahydro-2-naphthyloxy) triethylamine were shown to be good repellents when tested under field conditions (Atkins et al. 1977). Among these amines, 2-(decahydro-2-naphthyloxy) triethylamine was found to be the most promising repellent. In complete agreement with the screening results of amines already reported, a detailed study of effect of selected repellent chemicals on the honey bees in canola and alfalfa fields showed that 2-ethylhexyl amine and dibutyl amine acted as honey bee repellent chemicals along with methyl salicylate (Sahebzadeh et al. 2009).

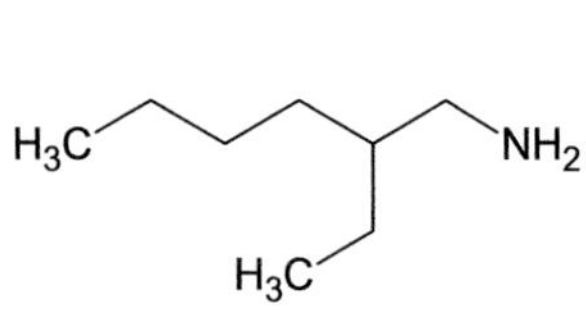

2-Ethyl hexyl amine

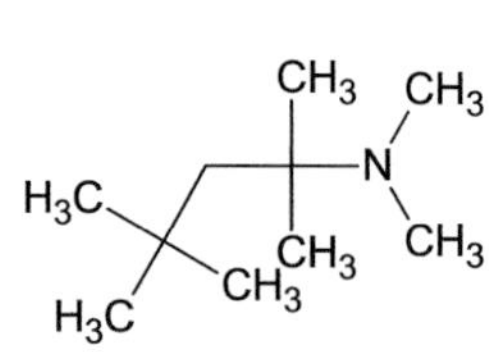

N,N,1,1,3,3-Hexamethyl butyl amine

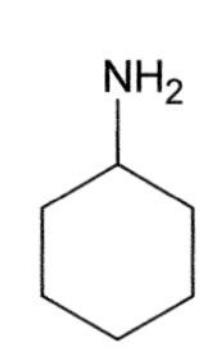

Cyclohexyl amine

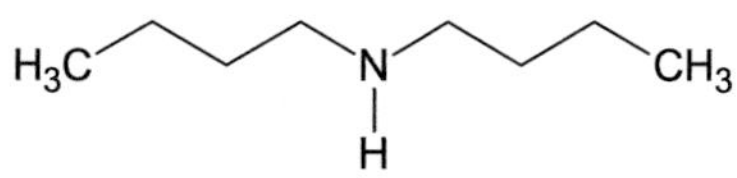

Dibutyl amine

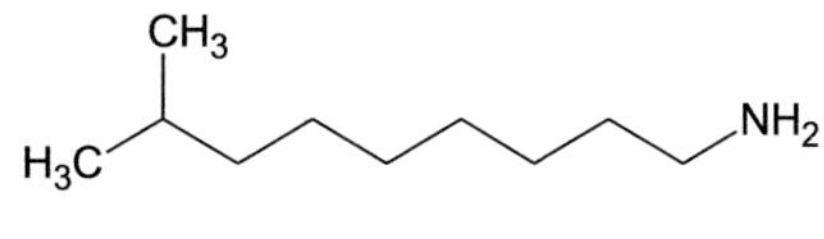

Isodecyl amine (8-Methyl nonyl amine)

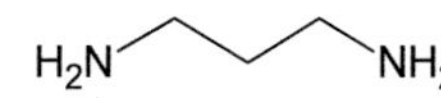

1,3-Propane diamine

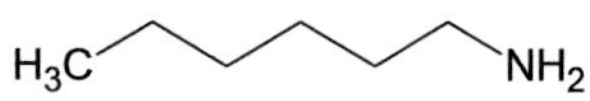

Hexyl amine

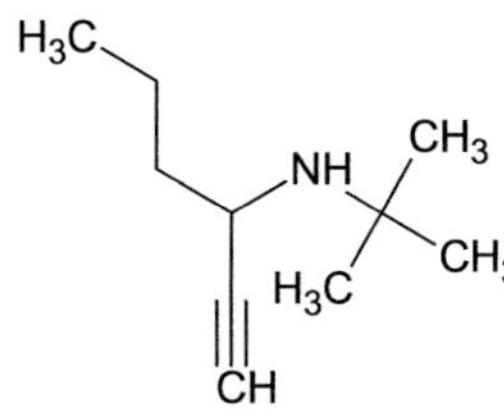

N-*tert*-Butyl-1-ethynylbutyl amine

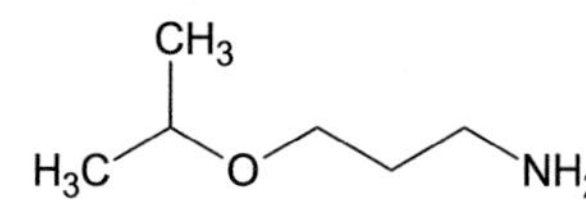

3-*iso* Propoxypropylamine

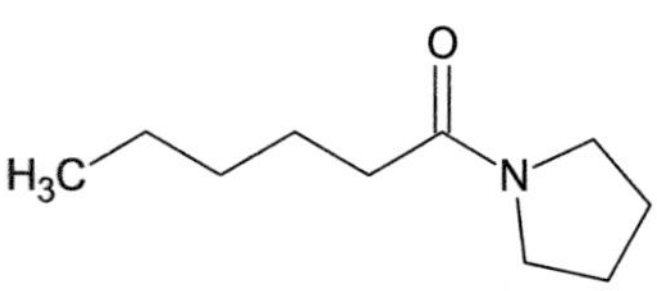

1-Hexanoylpyrrolidine

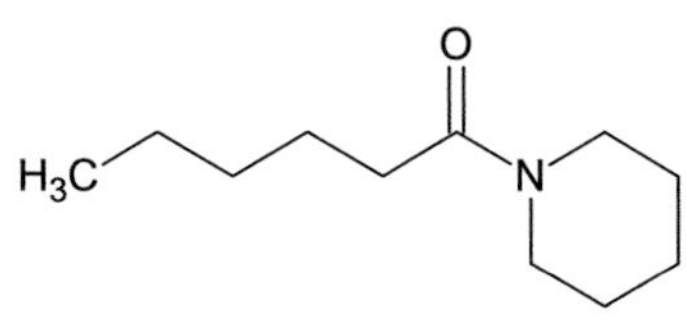

1-Hexanoylpiperidine

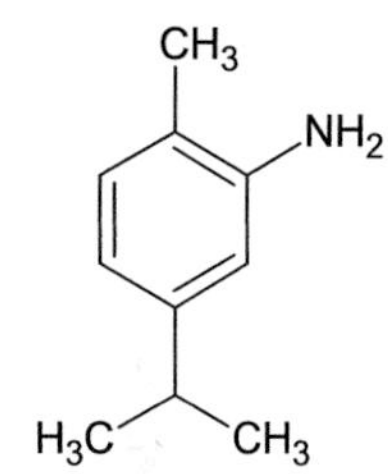

Carvacryl amine

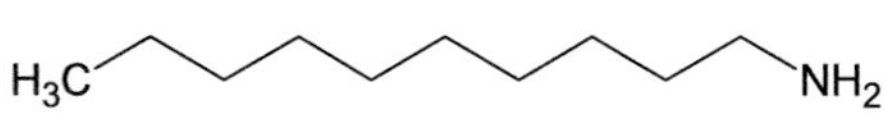

Decylamine

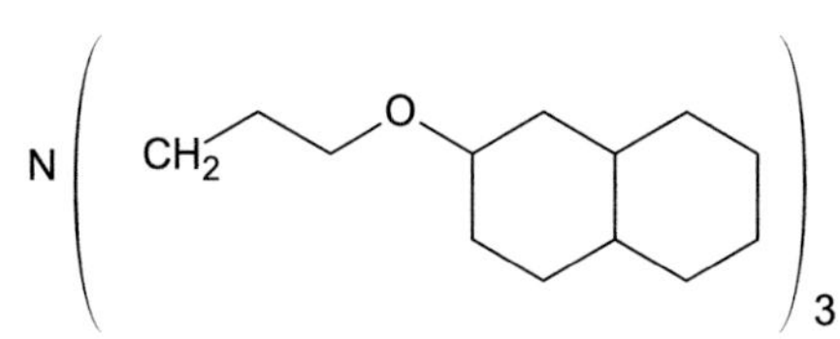

2-(Decahydro-2-naphthyloxy) triethylamine

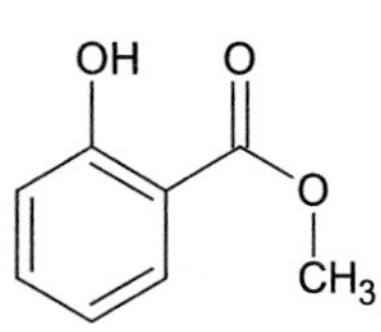

Methyl salicylate

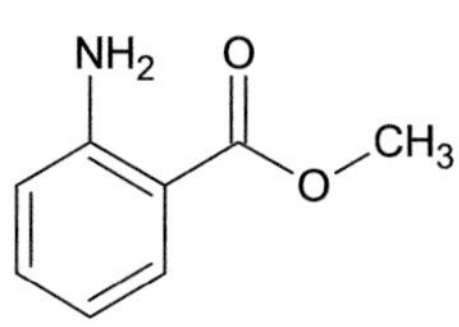

Methyl anthranilate

Alcohols have also been found to show repellent activity under laboratory conditions. Cyclohexylcyclohexanol (Chamberlain 1959), 2-methyl-2-(octylamino)-1-propanol and *p*-mentha-6,8-diene-2-ol exhibited repellency to *A. mellifera* (Atkins et al. 1975).

Applications of repellents for honey bees were demonstrated in yet another way. Methyl anthranilates, a honey bee repellent, was used to protect the beekeeper from honey bee stings. Multiple honey bee stings may result in venom toxicity leading to renal failure and even death. The efficacy of methyl anthranilates as honey bee repellent was tested on Africanised honey bees. Although methyl anthranilate was not totally effective in preventing honey bee stings, it seemed to reduce the number of stings (Pankiw 2009).

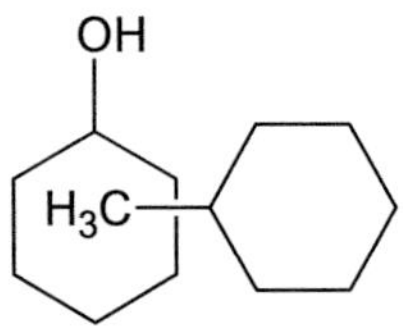

Cyclohexylcyclohexanol

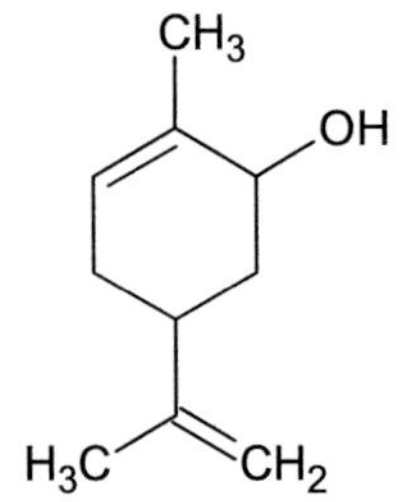

p-Mentha-6,8-diene-2-ol

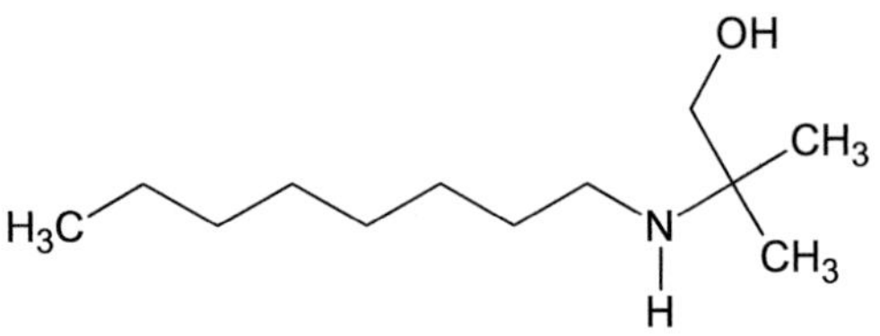

2-Methyl-2-(octylamino)-1-propanol

8.5 Plant-based Repellents

As the evaluation of various organic compounds was being done to determine their repellency towards honey bees, a large number of efforts were made to develop repellent formulations from natural resources. The Negrito Onge tribals of Little Andaman use the sap of *Orophea katschallica* to repel the bees while extracting honey from the hives (Dutta et al. 1983). Further it was reported that the sap of *Amomum aculeatum*, a herb growing in dense tropical forest of South Andamans, acts as a tranquiliser for these bees. With the aid of this sap, honey could be harvested from the hives by natives without protective apparel while the bees remained docile (Dutta et al. 1985). Screening of extracts of plants such as tobacco, rue, garlic, parsley and century showed the efficacy of these plant extracts as honey bee repellents (Nicodemo and Nogueira-Couto 2004). The important group 'essential oils' showed a great promise in this screening.

8.5.1 Repellency of Essential Oils

Essential oil of *Lantana camara* was found to be an effective repellent for honey bees under laboratory conditions. However it was not effective in field trials as it did not show repellency for sufficient duration when applied in field (Attri and Singh 1977).

Gupta and Kapil (1983) tested the response of worker bees of *A. florea* to 16 compounds which included two essential oils in semi-field conditions. In these experiments the maximum repellency was observed for *o*-amino acetophenone and ethyl benzyl ketone, while the essential oils did not show much promise. Only one of the oils tested, that from *Mentha piperata*, showed some repellency but it was the minimum amongst all the test formulations.

Kumar et al. (1986) studied the essential oils of *Citronella*, *Lantana*, *Eucalyptus*, *Chenopodium* and *Trachyspermum* for their gustatory repellency and toxicity. It was shown that *Citronella* oil was the most repellent with least toxicity to *A. cerana indica*. In complete agreement with this observation citronellal, the main component of this essential oil (Wei and Wee 2013), showed significant repellency towards *A. mellifera* (Malerbo-

Souza and Nogueira-Couto 2004). Interestingly an exactly contradictory observation was reported by Abramson et al. (2006) in which they have recorded that citronella oil was not effective in repelling *A. mellifera*. These observations when seen collectively may indicate that essential oils showing repellency for honey bees in one region may not show similar properties towards honey bees in other regions of the world.

Essential oils from *Cinnamomun zeylanicum, C. tamala, Juniperus virginiana, Pinus polustrus, Linum usitatissimum, Syzygium aromaticum, Callistemon lanceolatus, Cuminum cyminum, Lavendula officinalis, Ocimum basilicum, Citrus reticulata, C. aurantium, C. sinensis, Carum capticum, Neolitzea zeylanica* have been proved to be effective repellents to *A. florea* when tested in olfactometer under laboratory conditions (Gupta 1987c). However, their efficacy in the field is not reported. Similarly the essential oil from *Carum capticum* was found to be the best repellent against *A. florea* under laboratory conditions. Repellency of the oil increased directly with increase in its concentration up to a point after which the effect of further increase in concentration was not significant (Gupta 1987d). A report that the essential oil of *Citrus sinensis* was found to show promising repellency towards *A. florea* under semi-field conditions, but high volatility of the oil restricted its use in field (Gupta 1987e) indicating the limitations of laboratory evaluations when extended to the field. In continuation of their studies on essential oils under laboratory conditions twenty four essential oils were tested for repellency to *A. florea* by olfacto-gustatory method. The essential oil from *Lavendula officinalis* showed the maximum repellency (Gupta 1987d) in these experiments. On similar lines the essential oil of bitter almonds and spearmint oil significantly repelled Africanized honey bees (Collins and Hellmich 1988). Perusal of results of studies show that essential oils may be good source of repellents but their efficacy is short lived in field due to their high volatility.

Honey bee repellent compositions comprising tea tree oil, benzaldehyde and alcohol have been patented (Fishman 1998). A comparative study of repellent activity towards honey bee *A. mellifera* of leaf extract and essential oil of *Ocimum sanctum* was carried out (Gill et al. 2002). It was found that both were highly effective, but the essential oil was more active than the leaf extract. A repellent for deterring bees from entomophilous plants was developed using pine oil and eucalyptus oil with water being optionally included (Muszynska 2005).

Recently the essential oil of an Indian medicinal plant, *Swertia densifolia*, commonly known as 'Chirayata', was isolated for the first time and screening of the essential oil and compounds from the essential oil was carried out on *A. cerana* (Naik et al. 2008). As expected, essential oil showed dose dependent repellency and screening of the major compounds of the essential oil, linalool and α-terpineol also exhibited repellent activity, when screened individually. Examination of the essential oil from another Indian medicinal plant, *Terminalia chebula,* was then carried out to test its repellent activity towards *A. florea.* Interestingly the essential oil showed the repellent properties but all the constituents did not show repellent activity. The repellent activity of formulations of the essential oil was found to be dose dependent.

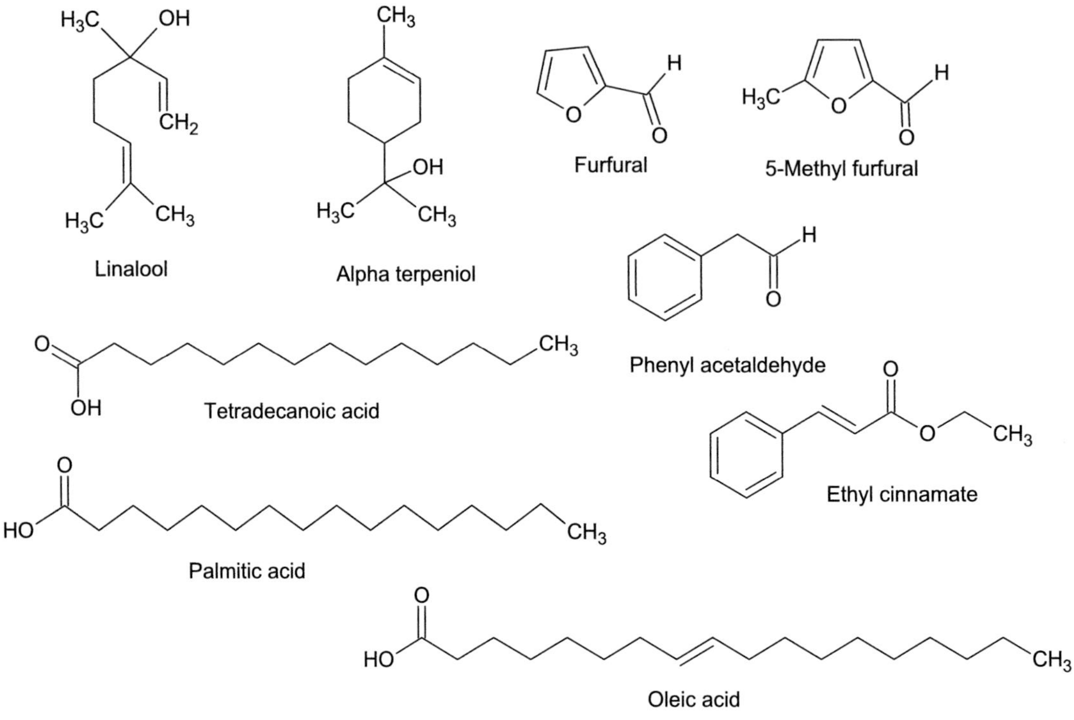

The repellency was found to increase with the concentration of essential oil in the formulations, reached a maximum for the formulation of certain concentration and remained constant beyond that. The screening of formulations of the major essential-oil components showed very interesting results. The formulations of furfural, 5-methylfurfural, tetradecanoic acid, palmitic acid, and oleic acid elicited no response in honey bees! At the same time, the formulations of phenylacetaldehyde were repellent, while those of ethyl cinnamate were attractant (Naik et al. 2010a)!

8.5.1.1 Essential Oils from Propolis

The screening of essential oils was not restricted to the plant essential oil, but was extended to the screening of essential oil of propolis, a beehive product. Propolis, also referred as bee glue, is a sticky dark coloured complex mixture of compounds. These compounds are collected by honey bees from the surrounding flora (Garcia-Viguera et al. 1992, Marcucci et al. 1995). It is applied to internal walls of the hive to repair combs, to fill the crevices in brood frames and for making the entrance of the hive smaller.

The essential oil from Indian propolis was isolated for the first time and its repellency towards *Apis florea* was demonstrated (Naik et al. 2013). It is interesting to note that the essential oil from a substance like propolis, deposited in the bee hive by one species of honey bees themselves, has repellent properties towards other species.

8.5.2 Substances of Natural Origin other than Essential Oils

Under laboratory conditions, substances produced by Bogong moth, *Agrotis infusa*, was found to be a good repellent for honey bees (Bailey 1981). *o*-Aminoacetophenone, a pheromone produced by virgin queen honey bees and released with feces was shown to possess repellent properties. Its effect on groups of ten worker bees was studied in comparison with the effect of octanoic acid and 1-dodecanol which are also present in feces. Among these only *o*-amino acetophenone repelled bees and the effect was comparable with that of feces alone, but other two compounds did not have any repellent effect (Page et al. 1988).

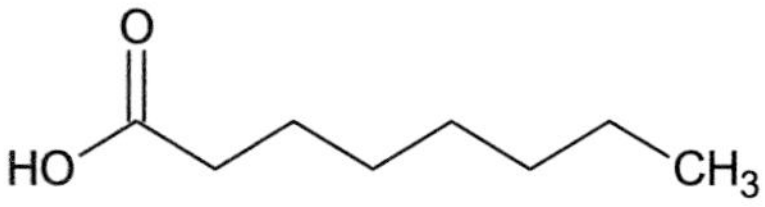

Octanoic acid

1-Dodecanol

In the efforts to screen the plant extracts on Indian honey bees, the total crude extract of leaf of an Indian medicinal plant, *Swertia densifolia* was found to show dose dependent attractant and repellent properties. It was shown that the formulations of low concentrations acted as repellents for *A. cerana indica* (Naik et al. 2005b) and *A. florea* (Naik et al. 2007). It may be recalled that the essential oil of these leaves have been shown to possess repellent properties towards *A. cerana indica* (Naik et al. 2008). Eventual screening of ethanolic extract of de-oiled leaves of *S. densifolia* was carried out on *A. cerana indica*. The extract showed dose dependent attractant properties at lower concentrations and repellent properties at higher concentrations (Naik et al. 2010b). This type of response was similar to that of an attractant pheromone from Nasonov gland (Deshpande and Naik this volume).

8.6 Concluding Remarks

Pesticide poisoning of honey bees which results in their mortality has been an acute problem since long. It has affected crop yield as well as honey bee colonies. Protection of honey bees from hazards of pesticides has been a challenging task. Application of honey bee repellents either in combination with pesticide formulations or separately on the crops to be sprayed with pesticides is one of the alternatives to protect honey bees from these hazards. Although, several repellents and semiochemicals have been identified, only some of them have shown promising repellency in laboratory and semi-field conditions. In field conditions, efficacy of the volatile compounds is limited due to their short persistence on the crop. To increase longevity certain other chemicals have been used as additives and partial success has been achieved. Nevertheless, no compound has yet been found suitable which can be exploited commercially to repel honey bees from pesticide sprayed field for desired duration.

The area of chemical repellency is very rich for further research. There are many opportunities for investigation and possibilities for exploitation. Mechanism of action of perception of repellent compound by honey bees may also be an important area in this field.

References

Abou–Shaara HF. 2014. The foraging behavior of honey bees, *Apis mellifera*: a review. Veterinarni Medicina 59: 1–10.

Abramson CI, Wilson MK, Singleton JB, Wanderley PA, Wanderley MJA, Michaluk LM 2006. Citronella is not a repellent to Africanized honey bees *Apis mellifera* L. (Hymenoptera: Apidae). Bioassay: 1:13.

Anderson LD, Atkins EL Jr. 1958.Toxicity of pesticides to honey bees in laboratory and field tests in Southern California. J Econ Entomol 51: 103–108.

Anonymous 1987.Studies on some new adjuvants and anticoagulating agents (Rodenticides). Final Project Report submitted to Indian Council of Agricultural Research, New Delhi.

Anonymous 1997. Field evaluation of some carbonyl compounds as bee repellents. Technical Yearly Report (1996–97) submitted to Department of Horticulture, Ministry of Agriculture, Govt. of India.

Atkins EL Jr, Anderson LD 1962.DDT resistance in living bees. J Econ Entomol 55: 791–792.

Atkins EL, Kellum D 1987. Comparative morphogenic and toxicity studies on the effect of pesticides on honey bee brood. J Apicult Res 25: 242–255.

Atkins EL, Anderson LD, Tuft TO 1954. Equipment and technique used in laboratory evaluation and of pesticide dusts in toxicological studied with honey bees. J Econ Entomol 47: 965–969.

Atkins EL, McDonald RL, McGovern TP, Beroza M, Greywood Hale EA 1975. Repellent additives to reduce pesticide hazards to honey bees Laboratory testing. J Apicult Res 14: 85–97.

Atkins EL, Kellum D, Neuman KJ 1977. Repellent additives to reduce pesticide hazards to honey bees.Am Bee J 117: 438–9.

Attri BS, Sharma PL 1969. Toxicity of some new pesticides to Indian honey bees *Apis indica* F. Pesticides 3: 27–29.

Attri BS, Singh RP 1977. A note on the biological activity of the oil of *Lantana camara* L. Indian J Entomol 39: 384–85.

Bailey F 1981. Natural bee repellents. Austral Bee Keep 82: 188.

Blacquiere T, Smagghe G, van Gestel CAM, Mommaerts V 2012. Neonicotenoids in bees: a review on concentrations, side effects and risk assessment. Ecotoxicol 21: 973 –992.

Blasco C, Fernaandez M, Pena A, Lino C, Silveira MI, Font G, Picoa Y 2003. Assessment of pesticide residues in honey samples from Portugal and Spain. J Agr Food Chem 51: 8132–8138.

Brades C 1984. Effect of parathion on dance system, sugar consumption, flight velocity, walking speed and wingbeat frequency in *Apis mellifera*. Zoolojischejahrbuches, Abteilung fur Allgemeinezoologic und physiologie der Tiere 88: 343–359.

Butler CG 1966. Mandibular gland pheromone of worker honey bees. Nature 212: 530.

Callbreath JS 1900. The use of carbolic acid during spraying time. Glean Bee Cult 28: 84.

Chamberlain WF 1959. The behavior of agricultural insects towards olfactory repellents in the olfactometer and split–arena tests. J Econ Entomol 52: 286–289.

Chauzat MP, Faucon JP, Martel AC, Lachaize J, Cougoule N, Aubert M 2006. A survey of pesticide residues in pollen loads collected by honey bees in France. J Econ Entomol 99: 253–262.

Choudhary A, Sharma DC 2008. Pesticide residues in honey samples from Himachal Pradesh (India). Bull Environ Contam Toxicol 80: 417–422.

Collins AM, Hellmich RL 1988. Disruption of defensive behavior. Am Bee J 128: 801.

Crane E, Walker P 1983. The impact of pest management on bees and pollination. In Tropical Development and Research Institute, London, College House, Wrights Lane, UK, 199 pp.

Czoppelt C, Rembold H 1988. Effect of parathion on honey bee larvae reared in vitro. Anzeiger fur Schadlingskunde, PflanzenschutzUmweltschutz 61: 95–110.

Dutta TR, Ahmed R, Abbas SR 1983. The discovery of a plant in Andaman Islands that tranquilises *Apis dorsata*. Bee World 64: 158–163.

Dutta TR, Ahmed R, Abbas SR, Rao MKV 1985. Plants used by Andaman aborigines in gathering rock–bee honey. Econ Bot 39: 130–138.

Deodikar GB, Suryanarayana MC 1977. Pollination in the service of increasing farm production in India. Adv Pollen Spore Res 2: 60–82.

Eckert JE, Shaw FR 1960. Beekeeping. Macmillan Co New York.

Fielder L 1986. Results of chronic intoxication of honey bee by acephate. Apidologie 17: 338–340.

Fielder L 1987. Acephate residues after pre–blossom treatments: effect on small colonies of honey bees. Bull Environ Contam Toxicol 38: 594–604.

Fishman Y 1998. Honey bee repellent composition comprising tea tree oil; US Patent No.5030660, filed on April 14, 1998.

Frazier M, Mullin C, Frazier J, Ashcraft S 2008. What have pesticides got to do with it? Am Bee J 148: 521–524.

Fries I, Wibran K 1987. Effects on honey bee colonies following application of pyrethroids, cypermethin and PP 321 in flowering oil seed rape. Am Bee J 127: 266–269.

Garcia–viguera C, Greenway W, Whately FR 1992. Composition of propolis from two different Spanish regions Zeitschriftfür Naturforschung C. Biosciences 47: 634–637.

Gentile AG, Gallaghar KJ, Santner Z 1971. Effect of some formulated insecticides on pollen germination in tomato and petunia. J Econ Entomol 64: 916–918.

Gill K, Malik OP, Kalidhar SV, Mishra RC, Malik MS 2002. Honey bee repellent activity of leaf extract and essential oil of *Ocimum sanctum* and isolation of a new compound. JMAPS 24: 951–956.

Gupta M 1985. Olfacto–gustatory response of *Apis florea* F. to aliphatic straight chain ketones. Indian Bee J 47: 30–33.

Gupta M 1986. Olfactory response of *Apis florea* F. to some cyclic ketones. Indian Bee J 48: 44–46.

Gupta M 1987a. Behaviour of *Apis florea* F. to some ketones and essential oils under semifield conditions. Indian Bee J 49: 29–32.

Gupta M 1987b. Olfactory response of *Apis florea* F. to some aryl ketones. J Anim Morphol Physiol 34: 157–162.

Gupta M 1987c. Essential oils a new source of bee repellents. Chem Ind 5: 162–163.

Gupta M 1987d. Olfactory response of *Apis florea* F. to some essential oils. J Apicult Res 26: 3–10.

Gupta M 1987e. Olfacto–gustatory response of *Apis florea* F. to some essential oils. Indian Bee J 49: 5–10.

Gupta M 1988. Insecticides and bee poisoning. Indian Bee J 50: 42–43.

Gupta M, Kapil RP 1983. Olfactory response of *Apis florea* F. to some repellents in semi–field conditions. In Proceedings of 5th International Symposium on Pollination, Versailles, pp 65–78.

Gupta M, Mohla RS 1986. Aliphatic straight chain ketones as potential bee repellents. Chem Ind 5: 327–328.

Gupta M, Pal HM, Pal B 1986. Olfacto–gustatory screening of n–propionaldehyde and n–butyraldehyde to *Apis florea* F. in laboratory conditions. Indian Bee J 48: 51–52.

Hafliger E 1949. Comparative toxicity of various insecticides to honey bees. J Econ Entomol 42: 523–528.

Holtmann H 1991. Honey bee repellents for repulsion from pesticide treated areas.Ger OffenDE 4012224 A1 19911017.

Johansen C 1965. Bee Poisoning Versus Clover Seed Weevil Control in White Dutch Clover Grown for Seed. J Econ Entomol 58: 857–863.

Johansen CA, Coffey MD, Quist JA 1957. Effect of insecticide treatments to alfalfa on honey bees including insecticidal residue and honey flavour analysis. J Econ Entomol 50: 721–723.

Johnson RM, Ellis MD, Mullin CA, Frazier M 2010. Pesticides and honey bees toxicity – USA. Apidologie 41: 312–331.

Karazafiris E, Tananaki C, Therasyvoulou A, Menkissoglu–Spiroudi U 2011. Pesticide residues in bee products. In Stoytcheva M (Ed) Pesticides in the modern world–risks and benefits. Available from the website http;//www.intechopen.com/books/pesticides–in–the–modern–world–risks–and–benefits/pesticide–residues–in–bee–products.

Kasamatsu K, Kawachi K 1986. Effect of fenvalerate applications on honey bees in flowering plants. In Proceedings of 30th International Conference of Apiculture, Nagoya Japan, Apimondia, pp 233–237.

Kasana VK, Ahlawat DS, Malik MS, Malik OP, Mishra RC 1997. Screening of some chemicals as repellents to *Apis mellifera* L. presented at feeding stations. Indian Bee J 59: 71–73.

Keeping MG, Crewe RM, Field BI 1982. Mandibular gland secretion of the stingless bee, *Trigona gribodoi* Magretti: isolation, identification and compositional changes with age. J Apicult Res 21: 65–73.

Kollmeyer WD, Flattum RF, Foster JP, Powell JE, Schroeder ME, Soloway SB 1999. Discovery of the Nitromethylene Heterocycle Insecticides. In Yamamoto I and Casida J (Eds) *Nicotinoid Insecticides and the Nicotinic Acetylcholine Receptor*, Springer–Verlag, pp71–89.

Kumar J, Palial SK, Mishra RC 1986. Evaluation of some essential oils as repellents to Indian honey bee, *Apis cerana indica* F I Gustatory repellency and toxicity. J Apicult Res 25: 256–259.

Lensing W 1986. Changes in honey bee workers after feeding them with sub–lethal doses of dimethoate. Apidologie 17: 340–343.

Lensing W 1987. Changes in honey bee workers after feeding on sub–lethal doses of dimethoate. Apidologie 18: 353–355.

Liebermann FV, Bohart GE, Knowlton GF, Nye WP 1954. Additional studies on the effect of field applications of insecticides on honey bees. J Econ Entomol 47: 316–320.

Linsey EG, MacSwain JW 1947. The effects of DDT and certain other insecticides on alfalfa pollinators. J Econ Entomol 40: 358–363.

Mackenzie KE, Winston ML 1989. Effects of sub–lethal exposure to diazinon on longevity and temporal division of labour in the honey bee (Hymenoptera:Apidae). J Econ Entomol 82: 75–82.

Malerbo–Souza DT, Nogueira–Couto RH 2004. Efficiency of n–Octyl–Acetate, 2–Heptanone and Citronellal in Repelling Bees from Basil (*Ocimum sellowii*: Labiatae). Braz Arch Biol Technol 47: 121–125.

Marcucci MC 1995. Propolis: Chemical composition, biological properties and therapeutic activity. Apidologie 26: 83–99.

Mayer DF, Lunden JD, Kovacs G, Miliczky ER 2001. Field evaluation of non–pesticide chemicals as honey bee repellents. Colloques INRA 98: 159–168.

Melander AL 1921. Spray poisoning. How bees obtain the poison and how this can be avoided. Glean Bee Cult 49: 210–11.

Mishra RC, Verma AK, Gupta PR 1987. Insecticides for the control of apple blossom thrips. Indian J Hortic 44: 115–120.

Mishra RC, Sharma SK 1988. Growth regulators affect nectar–pollen production and insect foraging in Brassica seed crops. Curr Sci 57: 1297–1299.

Morse RA, Shearer DA, Boch R, Benton AW 1967. Observations on alarm substances in the genus *Apis*. J Apicult Res 6: 113–118.

Muszynska J 2005. Preparation for repellent bees from entomophilous in bloom. Polish patent. PL 189163 B1 20050630.

Naik DG, Katke S, Chawda SS, Thomas D 1997a. Repellency of 2–heptanone to *Apis cerana*. Indian Bee J 59: 80 – 81.

Naik DG, Katke S, Chawda SS, Thomas D 1997b. 2–Heptanone as a repellent for *Apis cerana*. J Apicult Res 36: 151–154.

Naik DG, Banhatti P, Chawda SS, Thomas D 1999. Ethyl–thiocyclopentane as a repellent for *Apis cerana*. Indian Bee J 61: 43–48.

Naik DG, Banhatti P, Chawda SS, Thomas D 2002. 2–Heptanone as a repellent for *Apisflorea*. J Apicult Res 40: 59–61.

Naik DG, Banhatti PR, Thomas D, Chawda SS 2005a. Honey bee repellent. Indian Patent No.:196551 sealed on 31/10/2005.

Naik DG, Dandge CN, Puntambekar HM, Patil T 2005b. Attractant and repellent properties of *Swertia densifolia* leaf extract towards *Apis cerana indica*. J Apicult Res 44: 116 –118.

Naik DG, Dandge CN, Puntambekar HM 2007. *Swertia densifolia* leaf extract as a dose–dependent attractant or repellent for *Apis florea*. J Apicult Res 46: 15–18.

Naik DG, Dandge C, Puntambekar H, Deshpande P 2008. Chemical examination of *Swertia densifolia* leaf essential oil and its repellent activity towards *Apis cerana indica* F. J Apicult Res 47: 63–67.

Naik DG, Puntambekar H, Anantpure P 2010a. Essential oil of *Terminalia chebula* fruit as a repellent for the Indian honey bee *Apis florea*. Chem Biodivers 7: 1303–10.

Naik DG, Dandge CN, Deshpande PV 2010b. Pheromone–like properties of de–oiled leaf extract of *Swertia densifolia* towards *Apis cerana indica*. J Apicult Res 49: 149–153.

Naik DG, Vaidya HS, Namjoshi TP 2013. Essential oil of Indian Propolis: Chemical composition and repellency against the honey bee *Apis florea*. Chem Biodivers 10: 649–657.

Nation JL, Robinson FA, Yu SJ, Bolton AB 1986. Influence upon honey bees of chronic exposure to very low levels of selected insecticides in their diet. J Apicult Res 25: 170–177.

Nicodemo D, Nogueira–Couto RH 2004. Use of repellents for honey bees (*Apis mellifera* L.) *in vitro* in the yellow passion–fruit (*Passiflora edulis*) crop and in confined beef cattle feeders. J Venom Anim Toxins 10: 77–85.

Page RE, Elum MS, Fales HM 1988. o–Aminoacetophenone, a pheromone that repels honey bees (*Apis mellifera* L). Experientia 44: 270–271.

Palmer JT, Bartrum FA, Foster IW, Harrison DL 1954. Effect on honey bees on DDT plus super phosphate applied as a dust to white clover pasture. N Z J Sci Tech A 36: 177–192.

Palmer JT, Bartrum FA, Foster IW, Harrison DL 1959. Effect on honey bees on DDT and Thiodan applied from the air as sprays to *Choumoellier*. New Zeal J Agr Res 2: 481–487.

Pankiw T 2009.Reducing honey bee defensive responses and social wasp colonization with methyl anthranilate. J Med Entomol 46: 782–788.

Pettis JS, van Engelsdorp D, Johnson J, Galen D 2012. Pesticide exposure in honey bees results in increased levels of the gut pathogen *Nosema*. Naturwissenschaften 99: 153–158.

Rieth JP, Wilson WT, Levin MD 1986. Repelling honey bees from insecticide–treated flowers with 2–heptanone. J Apicult Res 25: 78–84.

Reith JP, Levin MD 1987. The pyrethroid insecticide hazard to honey bees. Am Bee J 127: 789–790.

Rieth JP, Levin MD 1988. The repellent effect of two pyrethroid insecticides on the honey bee. Physiol Entomol 13: 213–218.

Ruijter A, van der Steen J 1987. A field study on the effect on honey bee brood of Insegar (fenoxycarb) applied on blooming apple orchard. Apidologie 18: 355–357.

Sahebzadeh N, Ebadi R, Khajehali J 2009. Effect of selected repellent chemicals on honey bees in canola and alfalfa fields. J Apicult Res 48: 29–33.

Schmidt JO, Johnston AN, Ginter DL, Spangler HG 2003. Olfactory stimulation of Africanized honey bee (Hymenoptera: Apidae) attacks by insect repellents. J Med Entomol 40: 275–278.

Shearer DA, Bosch R 1969. 2–Heptanone in the mandibular secretions of honey bee. Nature 206: 530.

Shines SW, Murray A, Deray D, Blanc J Le. 1983. A field study on the effects of a new insecticide, Fastac on honey bees. In 10th International Congress of Plant Protection. Brighton, England. 20–25 Nov 1983 Vol.2 Cogdon UK.

Singh R, Pal HM, Gupta M, Malik OP 1993. Screening of some chemical compounds as repellents to honey bees: Effect of carbonyl compounds on *Apis florea* F. visitor to *Brassica campestris* var. *toria*. Indian Bee J 55: 47–50.

Singh R, Pal HM, Gupta M, Malik OP 1996. Screening of some carbonyl compounds on *Apis florea* F. visitor to *Ziziphus mauritiana* var. *dichotoma*. Korean J Apic 10: 131–134.

Smirle MJ, Mark LW, Kenneth LW 1984. Development of a sensitive bioassay for evaluating sub–lethal pesticide effect on the honey bee (*Apis mellifera*) (Hymenoptera: Apidae). J Econ Entomol 77: 63–67.

Stoner A, Wilson WT, Moffett JO 1984. Effect of long–term feeding of low doses of fenvalerate of fluvalinate in sucrose syrup on honey bees in standardized field colonies. J GA Entomol Soc 19: 490–498.

Suwannapong G, Benbow ME, Chinokul C, Seanbualuang P, Sivaram V 2011. Bioassay of the mandibular gland pheromones of *Apis florea* on the foraging activity of dwarf honey bees. J Apicult Res 50: 212–217.

Thompson G. 1881. Paris green. Am Bee J 17: 85.

Todd FE, McGregor SE 1960. The use of honey bees in the production of crops. Annu Rev Entomol 5: 265–278.

Townsend GF 1963. Benzaldehyde (artificial oil of almonds) for the removal of honey. Glean Bee Cult 91: 464.

Wei LS, Wee W 2013. Chemical composition and antimicrobial activity of *Cymbopogon nardus citronella* essential oil against systemic bacteria of aquatic animals. Iran J Microbiol 2013 5: 147–152.

Woodrow AW, Green N, Tucker H, Schonhorst MH, Hamilton KC 1965. Evaluation of chemicals as honey bee attractants and repellents. J Econ Entomol 58: 1094–1102.

Woods DF, Dimitratos S, Justice R Improved honey bee repellents and uses thereof. WO 2012151556 A2.PCT/US2012/036670. Filed on 4th May 2012.

Chapter 9

Sub-lethal Effects of Pesticides on Pollinators with Special Reference to Honey Bees

Parthiba Basu and Priyadarshini Chakrabarti*
bparthib@gmail.com

Abstract

Pesticides are considered to be a potent driver for loss of important pollinators. Pesticides impose a series of sub-lethal effects on these beneficial insects which are often overlooked. Long term repeated exposure to pesticides can impair behaviour, neurobiology, physiology and various other survival mechanisms of insect pollinator groups and particularly the honey bees. Though these harsh chemicals are targeted towards pests, non-target beneficial insects are often exposed to these and are relentlessly incapacitated. Residues of the pesticides may be brought back to the hive and fed to the brood – thereby imposing an additional threat to the population. The challenge lies in formulating strict pesticide regulations to monitor the indiscriminate use of pesticides. Loss of beneficial pollinators including the honey bees will have serious adverse impacts on the agricultural production and the livelihoods of small and marginal farming communities. This chapterfocuses primarily on the sub-lethal effects that are induced by pesticides on pollinating insects with a special focus on honey bees.

9.1 Introduction

Beneficial insect pollinators like honey bees are an essential component of human agriculture and many crops are completely dependent on the pollination services provided by bees, often over a period as short as a few days (Free 1993). Pollination is essential for the production of a variety of agricultural crops, especially in the fruit growing industry (Heylen et al. 2010). Pollinators also play an important role in the maintenance of biodiversity through the pollination of wild flowers and these in turn provide a food source for many small mammals and birdspecies, either through herbivory or by providing prey

DOI: 10.1201/9781003505471-9

(Williams and Carreck 1994). In recent decades, intensive agricultural practices have led to habitat destruction and increased pesticide use, resulting in a significant reduction in both the numbers and species diversity of wild bees and other beneficial insects. Pesticides by their nature are poisonous and are known to cause significant oxidative stress (Qiao et al. 2005). Hazards arise mainly from the damage resulting from their misuse (Alagh 1988). In 2010, India alone consumed about 20,618.83 tonnes of total insecticides/pesticides out of a total consumption of 98,221.89 tonnes among 23 Asian nations (FAO 2014).

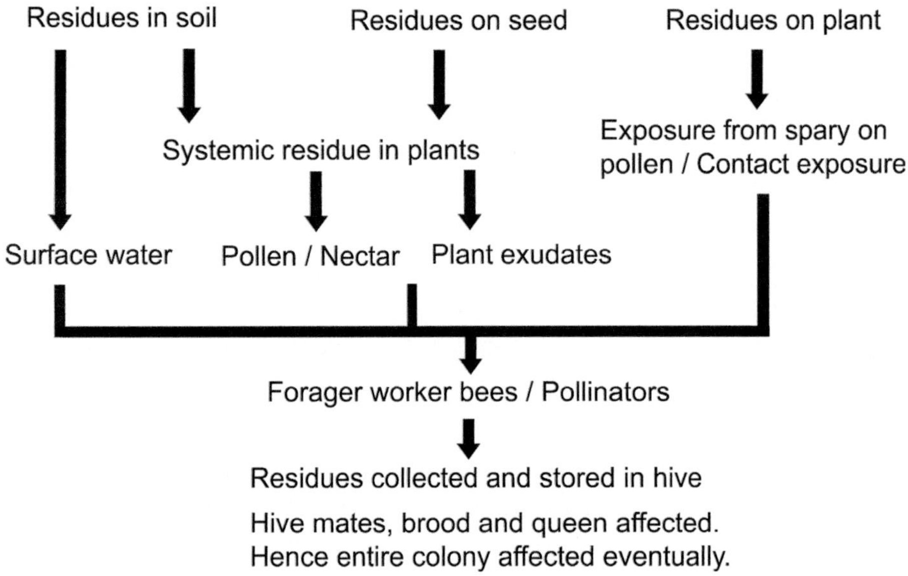

Fig. 9.1. Schematic representation of various routes of pesticide exposure in honey bees (adapted from Scientific Beekeeping.com).

A high dose of pesticide would result in the death of the foragers or would cause an aversion to foraging from the affected patches. However, it is the sub-lethal dose that poses a major threat to the survival of the entire colony. Various studies on sub-lethal effects have revealed new insights into the sub-lethal impacts of pesticides including effects on learning performance, behaviour, and neurophysiology (Desneux et al. 2007). Pesticides adversely affect bees by disrupting their homing behaviour and also affect odour discrimination in individual bees (Thompson 2003). Pesticides are also known to cause a decline in important pollinator populations (Brittain et al. 2010, Mullin et al. 2010, Henry et al. 2012, Whitehorn et al. 2012) and also induce oxidative stress on insect pollinators (Chakrabarti et al. 2014, Qiao et al. 2005). Foraging bees may encounter lethal insecticide levels when pollinating agricultural fields (Colin et al. 2004) or foraging in residential settings (Claudianos et al. 2006) and additional mortality can occur when contaminated nectar and pollen are brought back to the hive (Chauzat et al. 2006, Claudianos et al. 2006) (Figure 9.1).

9.2 Scale of Study

Although a good number of studies have looked into the sub-lethal effects of pesticides on pollinating insects primarily, honey bees and bumble bees, at the laboratory scale; a large information gap exists on the response of naturally occurring native honey bee populations to multiple pesticide exposure in field conditions (Osborne 2012).

Various field and laboratory experiments have determined the LD_{50} values and physiological effects of imidacloprid and other pesticides on honey bees (Choudhary et al. 2009, Decourtye et al. 2004, Gregorc and Bozic 2004, Johnson et al. 2010, Medrzycki et al. 2003, Schmuck et al. 2001). Most honey bee losses from 1966 to 1979 in beekeeping practices in the United States were in fact attributable to pesticide exposure (Atkins 1992, NAS 2007) and pesticides have been identified as a key player for honey bee decline (Mullin et al. 2010, Rortais et al. 2005). However, in laboratory, specific effects of one, two or sometimes even three pesticides are observed individually or in synergy. In field, especially in India, multitude of pesticides is used in varying doses. Therefore, the individual pesticide-centric studies do not reflect the existing field conditions and moreover, the synergistic effects are far more detrimental (Chakrabarti et al. 2014). Native wild pollinators are a bioresource for the nation and importance must be given to them along with the managed colonies of honey bees. Table 9.1 gives a general understanding of the various types of pesticides that are commonly used in the agricultural field. To expand studies on the wild populations along agricultural intensification landscapes, there is a need to first establish the intensification sites by holding extensive surveys of the types and quantities of pesticides used and also by analyses of the soil pesticide residues (Chakrabartiet al. 2014). If possible, pesticide residues from pollinators residing around the particular sites, can also further establish it. The Government data on percentage of cropping intensity and total yield from the regions can further help validate the gradient (Chakrabarti et al. 2014). Laboratory studies can be carried out subsequently to validate the results observed from field samples. This will then complete the picture.

Table 9.1. Major groups of pesticides used (adapted from Abrol 2012).

Botanicals	Inorganics	Organics	Others
Nicotine, derris etc. Can be similar to plant volatiles which naturally act as poison	Arsenicals, fluorides, mercury compounds, sulfur etc. Sulfur	Organochlorine insecticide Organophosphorus insecticide Carbamate insecticide Synthetic pyrethroids Neonicotinoids Ryanoids	Insect Growth Regulators Biopesticides Herbicides Fungicides

9.3 Sub-lethal effects of pesticides

Pesticide use, apart from loss of natural vegetation cover (Winfree et al.2009), has been cited as one of the major drivers of the recent decline in pollinator populations (Brittain et al. 2009, Henry et al. 2012, Mullin et al. 2010,Whitehornet al. 2012). This assertion is underpinned by a number of studies that looked into the toxic impacts of various pesticides on important pollinators like *Apis mellifera* or bumble bees through dose or exposure–mortality response relationships in controlled ecotoxicological experiments (Brittain et al. 2010, Choudhary et al. 2009, Decourtye et al. 2004, Johnson et al. 2010). Apart from dose-mortality responses, a number of biomarkers including fecundity (Cresswell and Laycock 2012, Henry et al. 2012), behavioural markers like proboscis extension response/reflex (Decourtye et al. 2005, Han et al. 2010) or homing (Henry et al. 2012) and physiological markers (Badiou-Bénéteau et al. 2012, Chakrabartiet al. 2014) have also been used to study the sensitivity of honey bees to pesticide exposure in controlled experiments. More in depth studies of the sub-lethal effects have been conducted mainly on honey bees as their biological system is quite well documented.

9.3.1 Physiological effects

9.3.1.1 Stress markers

Recent studies have established pesticide induced oxidative stress on the honey bees (Carvalho et al. 2013, Chakrabarti et al. 2014). However, acetylcholinesterase has been particularly prominent among the physiological markers for xenobiotics studies and as has been suggested, a combination of other different physiological response markers would provide a more precise diagnosis (Badiou-Bénéteau et al. 2012).Antioxidant enzymes such as superoxide dismutase (SOD) and catalase (CAT) are of vital importance in an organism's defense against oxidative stress (Chakrabarti et al. 2014, Fridovich 1982, Khessiba et al. 2005, Mamidala et al. 2011, McCord and Fridovich 1969, Schriever et al. 2008) and have both been associated with pesticide toxicity in insects (Landa et al. 1991). A number of anti-oxidant enzymes, such as SOD, CAT, glutathione *S*-transferase, glutathione peroxidase and glutathione reductase, have been reported to occur in insects and are known to be affected by pesticides (Ahmad et al. 1991, Chakrabarti et al. 2014, Felton and Summers 1995, Joanisse and Storey 1996).Unlike younger workers, who mainly perform housekeeping tasks inside the colony, various mixed function oxidases get activated in older forager workers and it has been suggested that this is a biochemical adaptation against environmental toxins such as pesticides and plant allelochemicals (Smirle and Winston 1988). Recent studies have also reported, for the first time, an increased level of enzyme xanthine oxidase in pesticide treated honey bee populations (Chakrabarti et al. 2014).

On the one hand, an increased level of antioxidant enzymes may help the pollinating insect to cope in a pesticide laden environment. On the other hand, free radicals act on important macromolecular structures of the organism and can cause severe damage to the physiology of the organism by interfering with cell components, including proteins, lipids, and DNA (Akhgari et al. 2003, Ranjbar et al. 2005).

9.3.1.2 Neurophysiology

The honey bee model system is well studied to test the effects of pesticides on the neurophysiology of the insect (Desneux et al. 2007). Acetylcholinesterase activity is reported to be most affected by pesticides (Alix et al. 2001, Chakrabarti et al. 2014, Desneux et al. 2007). Acetylcholinesterase, being a major neurotransmitter, inhibition of its activity indicates major concern for synaptic processes (Kreissl and Bicker 1989). Mushroom body, a major component of brain involved in honey bee olfaction, was also reported to be affected by pesticides (Decourtye et al. 2004, Palmer et al. 2013). Palmer et al. (2013) used recordings from mushroom body Kenyon cells in acutely isolated honey bee brains and showed that imidacloprid, clothianidin and coumaphosoxon cause a depolarization block of neuronal firing and inhibit nicotinic responses. These effects were observed at field realistic concentrations and were additive with combined applications. Their study predicted that exposure to multiple pesticides, which target cholinergic signaling, will cause enhanced toxicity to pollinators.

9.3.2 Behavioural impairments

9.3.2.1 Navigation / Orientation / Learning

Navigation can be controlled by various cues – either chemical or visual. How an insect pollinator navigates is essentially based on its memory of the location of hive and the food resource. Learning abilities of the pollinators are in fact the memory acquisition of the various cues and learning to respond to these cues. With respect to insect pollinators, honey bee is the best studied model to test the sub-lethal impacts of pesticides on insect navigation and orientation.

9.3.2.2 Olfaction

Olfaction plays a crucial and determining role in mediating most of the chemical responses including discovering food resources and ultimately survival of the foragers. A number of early studies have upheld the 'odour-search hypothesis' where the studies assumed olfactory cues to be extremely important for nectar and pollen search by honey bees as the honey bees were reported to use their antennae to locate food rewards (Vadas 1994, Wenner 1971). Caste-specific pheromonal communication in honey bees determines mating, alarm, defense, orientation, self-colony recognition and incorporation of all conducts within the hive (Blum 1992, Katzav-Gozansky et al. 2002, Winston 1987). Hence the importance of olfaction in the recognition and sustenance of the various tasks within the hive is undeniable and its subsequent damage by pesticides is alarming for the entire colony.

Honey bees, like other insects, are constantly exposed to numerous random odor stimuli which they use to synthesize consequential information. The principle factor to locate food is to identify and process the floral odours (Hussaini et al. 2007). Such types of associative learning help an organism to relate various environmental odour cues of biological importance to its own benefit. The simplest form of associative learning is called first order conditioning (Hussaini et al. 2007). Additionally associative learning can be indirect or of second order (Pavlov 1927, Hussaini et al. 2007). Any particular cue, which acts as the primary reinforcer or elicits a response without any prior training is called the unconditioned stimulus (US). The response to US is called unconditioned response (UR). Additionally, there can be another stimulus in addition to US which, initially in combination with US and later on by itself, can elicit a response. This is called conditioned stimulus (CS) and its response is called conditioned response (CR). If a third stimulus is now paired with CS and the organism, being trained to respond to these two stimuli, elicits a response to the new stimulus in the absence of CS, this is then called second order conditioning (Hussaini et al. 2007). Measurement of proboscis extension response/reflex (PER) (Figure 9.2) based on this Pavlovian milieu has been used to identify the extent of memory acquisition and retention in honey bees and has also been greatly used to assess the adverse effects of pesticides on honey bees (Aupinel et al. 2005, Candolfi et al. 2001, Decourtye et al. 2005, Desneux et al. 2007, Han et al. 2010).

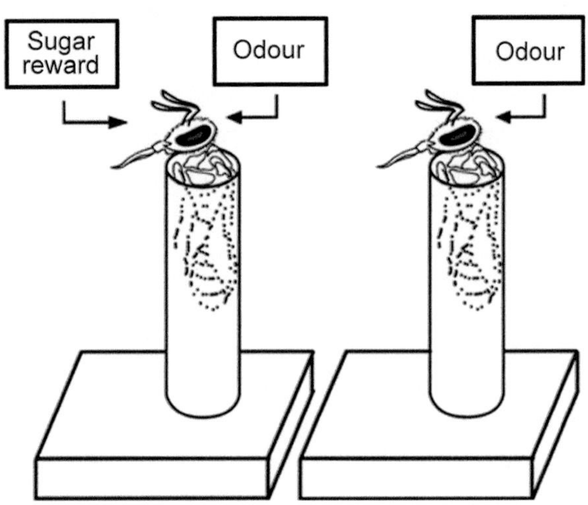

Fig. 9.2. A simple representation of proboscis extension reflex where the honey bee learns the odour and extends its proboscis in response to the odour. This happens following a Pavlovian milieu where the odour was initially administered with a sugar reward and the bee learns to associate the odour with the sugar reward.

Exposure to pesticides can be done before, during or after PER conditioning (Abramson et al. 2004, Desneux et al. 2007, Mamood and Waller 1990, Stone et al. 1997). Feeding on contaminated food can lead to learning inability in developing bees and thereby affecting its foraging later on. Studies have reported reduced learning performance in bees surviving 11 days of oral exposure to imidacloprid, 5-hydroxy-imidacloprid, fipronil, deltamethrin, endosulfan, and prochloraz (Decourtyeet al. 2003, Decourtyeet al. 2005,

Desneux et al. 2007). Hence, PER approach essentially helps to understand if the pesticides can interfere with the memory and learning process in insect pollinators and thereby affecting their ability to memorize a food resource and return back to it.

9.3.2.3 Vision

The floral resource colours and the landmarks required to return home (homing) constitute the visual cues that a pollinator is exposed to. It is extremely important to learn these visual cues for spatial orientation and homing.A honey bee exposed to pesticides in any of its foraging trip or a bee being reared on contaminated food is shown to be easily be affected such that it acquires the visual data incorrectly and also interprets it wrongly. This not only causes wrong information being sent to the nest mates to locate a food resource through an imprecise bee dance but also in extreme cases causes disorientation and loss of the individual forager honey bee (Desneux et al. 2007). A few important studies have reported how sub-lethal exposure affects the honey bees and reduces their homing ability (returning to hives) using radio-frequency identification tag (Henry et al. 2012) or coloured number tags (Vandame et al. 1995).

Another set of studies have taken the help of complex mazes to test the visual learning capacity of pollinators through the honey bee model. These mazes associatedcolour with sugar reward. The bees learn the maze by making correct or incorrect decisions. Figure 9.3 shows a simple Y-maze arrangement where two or one colour choice(s) can be given and any colour can be associated with sugar reward. Pesticide treated bees are often found to make a higher proportion of incorrect choices (Desneux et al. 2007, Han et al. 2010) thereby further asserting the trepidation that pesticides indeed affect the learning abilities in pollinators.

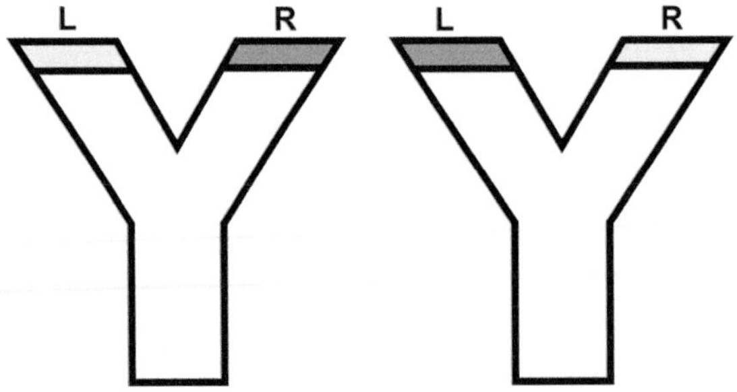

Fig. 9.3. A simple colour memory acquisition in relation to food reward can be done with the help of a Y – maze. Two arms left (L) and right (R) can be differentially coloured or only one arm can be coloured. Any of the colours can be trained to a sugar reward and memory retention can then be checked.

9.3.2.4 Feeding

Pesticides may affect the feeding behaviour by having a general repellent effect. Some pesticides may be formulated to incorporate antifeedant properties and pesticides in general may interfere with the olfactory, learning and foraging abilities of pollinators, thereby reducing feeding (Desneux et al. 2007). These insecticides, specifically intended for pests, can easily affect non-target insects like pollinators and other beneficial arthropods. Pesticide exposures have been reported to decrease foraging activity in bees (Haynes 1988, Taséi et al. 1994). Incorrect dances by foragers exposed to pesticides may cause a reduction in forager recruitment and thereby less visitation to floral resources and less feeding for the entire hive (Dechaume-Moncharmont 2003). A recent study by Gill and Raine (2014) reported that neonicotinoid exposure has both acute and chronic effects on overall foraging activity in bumble bees. Whilst foragers from control colonies improved their pollen foraging performance as they gained experience, the performance of bees exposed to imidacloprid became worse indicating a chronic behavioural impairment. Their study also suggested that pesticide exposure can change forager preferences for the flower types from which they collect pollen. Thus feeding of entire hive or nest mates can be largely dependent on the exposed individual's foraging abilities.

9.3.3 Immunology

Insecticides are reported to affect insect immunity. Though targeted at pests, the on target beneficial pollinator group can be easily affected. It has been reported that the number of plasmatocytes in the hemolymph of a particular insect (*Rhynocoris kumarii*) was decreased by 16% and 13% when monocrotophos and methylparathionwere applied, respectively at one tenth of the LC50 values. It was reported that decreases in the number of plasmatocytes were associated with an increase in thenumber of granular hemocytes, which play a role in detoxification through phagocytosis (George and Ambrose 2004). Possible reason suggested was that plasmatocytes are transformed into granular hemocytes duringthe detoxification process, indicating that the tested pesticides acted on the predator's immunological response indirectly by mobilizing immunity cells for detoxification tasks (George and Ambrose 2004, Desneux et al. 2007).

9.3.4 Development

Development of pollinators can be easily interrupted by pesticides meant to target and imbalance the pest development. Insect growth regulators (IGR) are hormones commercially produced and mimic developmental hormones, thereby disrupting the functions of ecdysone and juvenile hormones and affect the endocrine system (Dhadialla et al. 1998). This poses a major threat because loss of individual forager to pesticide can be counteracted by colony plasticity of division of labour to convert more nurses and juveniles to new foragers depending on the colony size (Thompson 2003). However if all the growing honey bees are developmentally impaired, there would be none to successfully replace the existing forager. The colony would starve without efficient foragers and remain unprotected without strong soldiers. A number of developmental abnormalities have been reported in honey bees when exposed to pesticides like reduction in brood surface area (Chandel and Gupta 1992), malformed larvae or pupae (DeRuijter and Vander Steen 1987), reduced weight gain and suppressed formation of hypopharyngeal glands (Gupta and Chandel 1995), impaired vitellogenin synthesis in the haemolymph (Pinto et al. 2000) etc. All these in turn show repercussion in brood care and development of the next batch of larvae.

9.3.4.1 Adult longevity

Pesticides are extremely toxic to pollinators when administered in high doses. Chronic exposure to pesticides may reduce adult longevity in the insects as a consequence of sub-lethal toxicity. In honey bees, as mentioned before, contaminated food may be stored in hive for longer term (Chauzat et al. 2006, Claudianos et al. 2006). Hence, death in honey bees due to acute toxicity is only a partial measure as chronic sub-lethal effects may be further critical.

Discerning long term pesticide effects on honey bee survival is however not easy and there are statistical issues of analyzing survival data (Desneux et al. 2007) and there has been several approaches to tackle this. Often the end result is stated in chronic toxicity tests (Schmuck 2004). Some reports graphically state how mortality rate varies during the time of pesticide exposure (Suchail 2001, Taséi et al. 2000). Recent studies by Chakrabarti et al. (2014) have reported lethal time 50 (LT50) instead of LC50 of a particular pesticide treatment group in *Apis cerana* and *A. dorsata* laboratory populations exposed to pesticides. This study has also shown the cumulative survival functions of these two species exposed to the various pesticide treatment groups. There are numerous reports (National Geographic 2013,Sandrock et al. 2014) from many bee keepers all over the world as to how their colony faces extinction due to severe adult honey bee death exposed to pesticides in the neighbouring crop fields.

9.3.4.2 Fecundity

Reduced fecundity could be due to the combined effects of behavioural impairments and / or physiological aberrations. When contaminated food is brought back to the hive, the entire colony mates feed on it including the queen. This can greatly damage the reproductive capacity of the queen and as reported by Idris and Grafius (1993) there are effects on the fertilization of ova, as well as differential survival of sexes in juveniles exposed to pesticides in haplodiploid insects like bees. The effects of four insect growth-regulating (IGR) insecticides on honey bee (*Apis mellifera*) colony development, queen rearing and drone sperm production have also been reported (Thompson et al. 2001). (Describe the observations in brief). Imidacloprid has been reported to reduce reproductiveoutput in *Bombus terrestris* (Laycock et al. 2012, Mommaerts et al. 2010, Tasei et al. 2000). Reduced fecundity can in turn affect the survival of the entire colony.

9.3.5 Community Ecology

Physiological effects of pesticides include effects on the population, because pesticides have been reported to induce deformations of ovaries and testes in insects (George and Ambrose 2004, Medina et al. 2004). Especially in haplodiploid species, like honey bees, in which the female chooses the fertilization of ova while laying eggs, effects on the fertilization of ova, as well as differential survival of sexes in juveniles exposed to pesticide (Idris and Grafius 1993), can cause a serious threat to population survival. In the above context, a resultant population imbalance could affect honey bee colony survival, population structure and perhaps the inter-specific behaviour of affected species. In reality, the field scenario is complex, with multi-species interactions and multiple pesticide usage (Chakrabarti et al. 2014). Because sub-lethal effects of pesticides interact with numerous life-history traits involved in reproduction, they are also likely to have an impact on insect communities (Desneux et al. 2007). However, further investigationsare necessary in a larger trialsto discern the impacts of such stresses on their population dynamics (Chakrabarti et al. 2014). Loss in foragers and hive workers cause a decline in not only the concerned colony but also thus affect the fine intra and inter specific balance (Figure 9.4)

Fig. 9.4. Huge losses of foragers in the colony pose a serious threat to the colony propagation. Picture Courtesy Mother Nature Network (2013).

9.4 Challenges Ahead

In India, multitude and magnitude of pesticide use poses serious threats to the pollinator populations. However, most of the work done on pollinators is based on the honey bee or bumble bee model particularly *Apismellifera* and *Bombus sp*. Therefore serious information gap exists on the effects of pesticides on the host of other wild pollinators including the wild bees. Even though the mainstream agricultural establishment is becoming increasingly aware and sensitized towards the deleterious impact of pesticides on important elements of biological diversity like the pollinating insects, effective regulation of pesticide application in field is still strikingly amiss. More studies in various agro-ecological regimes of the country on various groups of pollinating insects would be crucial for not only raising our overall understanding of the extent of damage but will also go a long way towards making the case stronger for more stringent pesticide regulatory regime.

References

Abramson CI, Squire J, Sheridan A, Mulder, PG Jr 2004. The effect of insecticides considered harmless to honey bees (*Apis mellifera*): proboscis conditioning studies by using the insect growth regulators tebufenozide and diflubenzuron. Environ Entomol 33: 378–388.

Abrol DP 2011. Foraging. In Hepburn HR, Radloff SE (Eds) Honey bees of Asia. Springer, New York

Ahmad S, Duval DL, Weinhold LC, Pardini RS 1991. Cabbage looper antioxidant enzymes: Tissue specificity. Insect Biochem 21: 563–572.

Akhgari M, Abdollahi M, Kebryaeezadeh A, Hosseini R, Sabzevari O 2003. Biochemical evidence for free radical induced lipid peroxidation as a mechanism for subchronic toxicity of Malathion in blood and liver of rats. Hum Exp Toxicol 22: 205 – 211.

Alagh YK 1988. Pesticides in Indian agriculture. Econ Polit Weekly 23: 1959–1964.

Alix A, Cortesero AM, Nénon JP, Anqer JP 2001. Selectivity assessment of chlorfenvinphos reevaluated by including physiological and behavioral effects on an important beneficial insect. Environ Toxicol Chem 20: 2530–2536.

Atkins EL 1992. Injury to honey bee by poisoning. In: JM Graham (Ed.), *The Hive and the Honey bee*. Dadant and Sons, Hamilton, pp 1153–1208.

Aupinel P, Fortini D, Dufour H. et al 2005. Improvement of artificial feeding in a standard in vitro method for rearing *Apis mellifera* larvae. B Insectol 58:107–11.

Badiou–Bénéteau A, Carvalho SM, Brunet J–L, Carvalho GA, Bulete A et al 2012. Development of biomarkers of exposure to xenobiotics in the honey bee *Apismellifera*: Application to the systemic insecticide thiamethoxam. Ecotox Environ Safe 82: 22–31.

Blum MS 1992. Honey bee pheromones. In: Graham JM (Ed.), *The hive and the honey bee*, Dadantand sons Hamilton, Illinois, pp 373–400.

Brittain CA, Vighi M, Bommarco R, Settele J, Potts SG 2010. Impacts of a pesticide on pollinator species richness at different spatial scales. Basic ApplEcol 11: 106–115.

Candolfi MP, Barrett KL, Campbell P, Forster R, Grandy N et al 2001. Guidance document on regulatory testing and risk assessment procedures for plant protection products with non–target arthropods. In SETAC/ESCORT 2 Workshop Rep. 21–23, March 2000, Wageningen

Carvalho SM, Belzunces LP, Carvalho GA, Brunet JL, Benateau AB2013. Enzymatic biomarkers as tools to assess environmental quality: a case study of exposure of the honey bee *Apis mellifera* to insecticides. Environ ToxicolChem 32: 2117–2124.

Chakrabarti P 2014 *The vanishing honey bees*. www.livelystories.com, http://livelystories.com/2014/06/24/vanishing–honey–bees/

Chakrabarti P, Rana S, Sarkar S, Smith B, Basu P 2014.Pesticide induced oxidative stress in laboratory and field populations of native honey bees along intensive agricultural landscapes in two Eastern Indian states. Apidologie 46: 107–129.

Chandel RS, Gupta PR 1992. Toxicity of diflubenzuron and penfluron to immature stages of *Apis cerana indica* F. and *Apis mellifera* L. Apidologie 23: 465–473.

Chauzat MP, Faucon JP, Martel AC, Lachaize J, Cougoule N et al 2006. A survey of pesticide residues in pollen loads collected by honey bees in France. J Econ Entomol 99: 253–262.

Choudhary A, Sharma DC, Badiyala A 2009. Relative safety of some pesticides against honey bees, *Apis cerana cerana* Fab. And *Apis mellifera* L. on mustard (*Brassica juncea* L. Czern). Pest Res J 21: 67–70.

Claudianos C, Ranson H, Johnson RM, Biswas S, Schular MA et al 2006. A deficit of detoxification enzymes: pesticide sensitivity and environmental response in the honey bee. Insect Mol Biol 15: 615–636.

Colin ME, Bonmartin JM, Moineau I, Gaimon C, Brun S et al 2004. A method to quantify and analyze the foraging activity of honey bees: relevance to the sublethal effects induced by systemic insecticides. Arch Environ Contam Toxicol 47: 387–395.

Cresswell JE, Laycock I 2012. Towards the comparative ecotoxicology of bees: the response–response relationship. Julius–Kuhn–Archiv, pp 437.

Dechaume–Moncharmont FX 2003. *Butinagecollectif chez l'abeille* Apis mellifera *L.: etude th´eoriqueetexp´erimentale*. PhD thesis. Univ. Pierre–et–Marie–Curie, Paris, pp 309.

Decourtye A, Armengaud C, Renou M, Gauthier M, Pham–Delegue MH 2004. Imidacloprid impairs memory and brain metabolism in the honey bee (*Apis mellifera* L.). Pest Biochem Physiol 78: 83–92.

Decourtye A, Devillers J, Genecque E, Le Menach K, Budzinski H et al 2005. Comparative sublethal toxicity of nine pesticides on olfactory learning performances of the honey bee *Apis mellifera*. Arch Environ ContamToxicol48: 242–50.

Decourtye A, Lacassie E, Pham–Del`egue MH 2003. Learning performances of honey bees (*Apis mellifera* L.) are differentially affected by imidacloprid according to the season. Pest ManagSci 59: 269–278.

DeRuijter A, VanderSteen J 1987. A field study on the effect on honey bee brood of insegar (fenoxycarb) applied on blooming apple orchards. Apidologie 18: 356–357.

Desneux N, Decourtye A, Delpuech JM 2007. The sublethal effects of pesticides on beneficial arthropods. Ann Rev Entomol 52: 81–106.

Dhadialla TS, Carlson GR, Le DP 1998. New insecticides with ecdysteroidal and juvenile hormone activity. Annu Rev Entomology43: 545–569.

FAO 2014. http://faostat.fao.org/site/424/default.aspx#ancor as on 18.02.2015.

Felton GW, Summers CB 1995. Antioxidant systems in insects. Arch Insect Biochem Physiol 29: 187–197.

Free JB 1993. Insect pollination of crops. Academic Press, London, UK.

Fridovich I 1982. Superoxide dismutase in biology and medicine. In: Autor AP (Ed.), *Pathology of Oxygen*. Academic Press, New York, pp 1–19.

George PJE, Ambrose DP 2004. Impact of insecticides on the haemogram of *Rhynocoriskumarii* Ambrose and Livingstone (Hem., Reduviidae). J Appl Entomol 128: 600–604.

Gill RJ, Raine NE 2014. Chronic impairment of bumblebee natural foraging behaviour induced by sublethal pesticide exposure. Funct Ecol 28: 1459–1471.

Gregorc A, Bozic J 2004. *Is honey bee colonies mortality related to insecticide use in agriculture?* Sodobno Kmetijstvo 37: 29–32.

Gupta PR, Chandel RS 1995. Effects of diflubenzuron and penfluron on workers of *Apis cerana indica* F. and *Apis mellifera* L. Apidologie26: 3–10.

Han P, Niu C–Y, Lei C–L, Cui JJ, Desneux N 2010. Use of an innovative T–tube maze assay and the proboscis extension response assay to assess sublethal effects of GM products and pesticides on learning capacity of the honey bee *Apis mellifera* L. Ecotoxicology 19: 1612–1619.

Haynes KF 1988. Sublethal effects of neurotoxic insecticides on insect behavior. Annu Rev Entomol 33: 149–168.

Henry N, Béguin M, Requier F, Rollin O, Odoux J F et al2012. A common pesticide decreases foraging success and survival in honey bees. Science 336: 348–350.

Heylen K, Gobin B, Arckens L, Huybrechts R, Billen J 2010. The effects of four crop protection products on the morphology and ultrastructure of the hypopharyngeal gland of the European honey bee, *Apis mellifera*. Apidologie 42: 103 – 116.

Hussaini SA, Komischke B, Menzel R, Lachnit H 2007. Forward and backward second–order Pavlovian conditioning in honey bees. Learn Mem 14: 678–683.

Idris AB, Grafius E 1993. Pesticides affect immature stages of *Diadegmainsulare* (Hymenoptera, Ichneumonidae) and its host, the diamondback moth (Lepidoptera, Plutellidae). J Econ Entomol 86: 1203–1212.

Joanisse DR, Storey KB 1996. Oxidative stress and antioxidants in overwintering larvae of cold–hardy goldenrod gall insects. J ExpBiol 199: 1483–1491.

Johnson RM, Ellis MD, Mullin C et al 2010. Pesticides and honey bee toxicity – USA. Apidologie 41: 312–331.

Katzav–Gozansky T, Soroker V, Hefetz A 2002. Honey bees Dufour's gland – idiosyncrasy of a new queen signal. Apidologie 33: 525–537.

Khessiba A,Roméo M, Aïssa P 2005. Effects of some environmental parameters on catalase activity measured in the mussel (*Mytilusgalloprovincialis*) exposed to lindane. Environ Poll 133: 275–281.

Kreissl S, Bicker G 1989. Histochemistry of acetylcholinesterase and immunocytochemistry of an acetylcholine receptor–like antigen in the brain of the honey bee. J Comp Neurol 286: 71–84.

Landa V, Sula J, Marec F, Matha V, Soldan T 1991. In: Tardiffand RG, Goldstei B (Eds.), Methods for assessing exposure of insects. In: Methods for Assessing Exposure of Human and Non–Human Biota.John Wiley and Sons Ltd., Hoboken

Laycock I, Lenthall KM, Barratt AT, Cresswell JE 2012. Effects of imidacloprid, a neonicotinoid pesticide, on reproduction in worker bumble bees (*Bombus terrestris*). Ecotoxicology 21: 1937–1945.

Mamidala P, Jones SC, Mittapalli O 2011. Metabolic resistance in bed bugs. Insects 2: 36–48.

Mamood A, Waller G 1990. Recovery of learning responses by honey bees following a sublethal exposure to permethrin. Physiol Entomol 15: 55–60.

McCord JM, Fridovich I 1969. Superoxide dismutase: an enzymic function for erythrocuprein (heinocuprein). J BiolChem 244: 6049–6055.

Medina P, Budia F, Del Estal P, Vinuela E 2004. Influence of azadirachtin, a botanical insecticide, on Chrysoperlacarnea (Stephens) reproduction: toxicity and ultrastructural approach. J Econ Entomol 97: 43–50.

Medrzycki P, Montanari R, Bortolotti L, Marcelino J, Maini S et al 2003. Effects of imidacloprid administered in sub–lethal doses on honey bee behaviour: Laboratory tests. B Insectol 56: 59–62.

Mommaerts V, Reynders S, Boulet J, Besard L, Sterk G et al 2010. Risk assessment for side–effects of neonicotinoids against bumblebees with and without impairing foraging behavior. Ecotoxicology 19: 207–215.

Mother Nature Netwrok (2013) http://www.mnn.com/earth–matters/animals/stories/new–studies–provide–clues–into–colony–collapse–disorder–and–other–bee as on 18.02.2015

Mullin CA, Frazier M, Frazier JL, Ashcraft S, Simonds R, et al2010. High levels of miticides and agrochemicals in North American apiaries: implications for honey bee health. Plos One 5: e9754.

NAS (National Academy of Sciences) (2007) Status of pollinators in North America. In: National Academy Press, Washington, DC.

National Geographic 2013. http://news.nationalgeographic.com/news/2013/13/130510–honey bee–bee–science–european–union–pesticides–colony–collapse–epa–science/ as on 18.02.2015

Osborne JL 2012. Bumble bees and pesticides. Nature 491: 43–45.

Palmer MJ, Moffat C, Saranzewa N, Harvey J, Wright GA et al. 2013. Cholinergic pesticides cause mushroom body neuronal inactivation in honey bees. Nat Commun 4: 1634.

Pavlov IP 1927. *Lectures on conditioned reflexes*, International Publishers, New York, pp. 103 –114.

Pinto LZ, Bitondi MMG, Simoes ZLP 2000. Inhibition of vitellogenin synthesis in *Apis mellifera* workers by a juvenile hormone analogue, pyriproxyfen. J Insect Physiol 46: 153–160.

Qiao D, Seidler FJ, Slotkin TA 2005. Oxidative mechanisms contributing to the developmental neurotoxicity of nicotine and chlorpyrifos. Toxicol Appl Pharmacol 206: 17–26.

Ranjbar A, Solhi H, Mashayekhi FJ, Susanabdi A, Rezaie A et al 2005. Oxidative stress in acute human poisoning with organophosphorus insecticides; a case control study. Environ Toxicol Pharmacol 20: 88–91.

Rortais A, Arnold G, Halm M–P, Touffet–Briens F 2005. Modes of honey bee exposure to systemic insecticides: estimated amounts of contaminated pollen and nectar consumed by different categories of bees. Apidologie 36: 71–83

Sandrock C, Tanadini M, Tanadini LG, Fauser–Misslin A, Potts SG et al. 2014. Impact of Chronic Neonicotinoid Exposure on Honey bee Colony Performance and Queen Supersedure. PLoS One 9(8): e103592.

Schmuck R 2004. Effects of a chronic dietary exposure of the honey bee *Apis mellifera* (Hymenoptera: Apidae) to imidacloprid. Arch Environ Contam Toxicol 47: 471–478.

Schmuck R, Schoning R, Stork A, Schramel O 2001. Risk posed to honey bees (*Apis mellifera* L., Hymenoptera) by an imidacloprid seed dressing of sunflowers. Pest Manag Sci 57: 225–238.

Schriever CA, Callaghan A, Biggs JP, Liess M 2008. Freshwater biological indicator of pesticide contamination. In: Environment Agency, Rio House, Bristol.

Scientific bee keeping (2012). http://scientificbeekeeping.com/neonicotinoids–trying–to–make–sense–of–the–science/as on 18.02.2015

Smirle MJ, Winston ML 1988. Detoxifying enzyme activity in worker honey bees: an adaptation for foraging in contaminated ecosystems. Can J Zool66: 1938–1942.

Stone J, Abramson C, Price J 1997. Task–dependent effects of dicofol (Kelthane) on learning in the honey bee (*Apismellifera*). B Environ ContamToxicol58: 177–183.

Suchail S, Guez D, Belzunces LP 2001. Discrepancy between acute and chronic toxicity induced by imidacloprid and its metabolites in *Apismellifera*. Environ ToxicolChem 20: 2482–2486.

Tas´ei JN, Lerin J, Ripault G 2000. Sub–lethal effects of imidacloprid on bumblebees, *Bombus terrestris* (Hymenoptera: Apidae), during a laboratory feeding test. Pest ManagSci 56: 784–788.

Taséi JN, Sabik H, Pirastru L, Langiu E, Blanche JM et al 1994. Effects of sublethal doses of deltamethrin (Decisce) on *Bombus terrestris*. J Apicult Res 33:129–135.

Thompson HE 2003.Behavioural effects of pesticides in bees – their potential for use in risk assessment. Ecotoxicology 12: 317–330.

Thompson HM 2001. Assessing the exposure and toxicity of pesticides to bumblebees (*Bombus* sp.). Apidologie 32: 305–321.

Vadas Jr RL 1994. The Anatomy of an Ecological Controversy: Honey–Bee Searching Behavior. Oikos 69: 158–166.

Vandame R, Meled M, Colin ME, Belzunces LP1995. Alteration of the homing flight in the honey bee *Apis mellifera* L. exposed to sublethal dose of deltamethrin. Environ ToxicolChem14: 855–860.

Wenner AM 1971. The bee language controversy: an experience in science. – Biological Sciences Curriculum Study, Educational Programs Improvement Corp., Boulder, CO.

Whitehorn PR, O'Connor S, Wackers FL, Goulson D 2012. Neonicotinoid pesticide reduces bumble bee colony growth and queen production. Science 336: 351–352.

Williams IH, Carreck NL 1994. Land use changes and honey bee forage plants. In: Forage for bees in an agricultural landscape. International Bee Research Association, Cardiff, pp 7–20.

Winfree R, Aguilar R, Vázquez DP, LeBuhn G, Aizen MA 2009. A meta–analysis of bees' responses to anthropogenic disturbance. Ecology 90: 2068–2076.

Winston ML 1987. The Biology of the Honey Bee, Harvard University Press, Cambridge, Massachusetts, London, England.

Chapter 10

Pollen-Pistil Interaction and Fertilization

K R Shivanna
shivanna@atree.org

Abstract

Following pollination pollen grains have to complete a series of sequential events in the pistil, termed pollen-pistil interaction, before they can discharge male gametes into the embryo sac for fertilization. Successful completion of pollen-pistil interaction is essential for development of fruits and seeds, the end products of sexual reproduction. Pollen-pistil interaction is dynamic and there is a continuous dialogue between the pollen grain and later the pollen tube, and the saprophytic tissues of the pistil. During this interaction, pollen grains are recognized; if compatible, the pistil facilitates post-pollination events and if incompatible, it inhibits pollen germination or pollen tube growth. Pollen-pistil interaction also creates intense competition amongst even the compatible pollen and only the most vigorous pollen grains are able to achieve fertilization. Less vigorous pollen grains are eliminated from this competition. Pollen competition has been shown to improve the fitness of the progeny. Considerable progress has been made in identifying the pollen and pistil components involved in this interaction, their nature and location. Extra-cellular components present in the pollen wall, and on the surface of the stigma and in the intercellular spaces in the style have been shown to be involved in pollen-pistil dialogue. In a few species some of the components involved in this dialogue have been identified. Impressive advances have been made in understanding the components responsible for guiding the pollen tube from the placenta in the ovary into the micropyle of the ovule. Synergids and the filiform apparatus present at the micropylar tip of the synergids play a crucial role in this guidance. In a few species the genes and gene products involved in this guidance have been identified.

Double fertilization is unique to flowering plants; of the two sperm cells discharged in one of the synergids, one of them fuses with the egg to give rise to the zygote and the

DOI: 10.1201/9781003505471-10

other fuses with the central cell. The zygote develops into the embryo and the central cell into the endosperm. As fertilization occurs deep inside the ovule, accessibility has been a major problem for effective experimentation and has remained an enigma since long. In recent years the success in achieving in vitro fertilization using isolated egg and sperm cells combined with the application of the techniques of molecular genetics and high resolution imaging to study fertilization in living cells have started unrevealing the details of double fertilization. With continued application of these advanced approaches to study pollen-pistil interaction and fertilization more rapid progress are likely to be made in the coming years leading not only to a better understanding of these processes but also in their effective manipulation for practical benefits.

10.1 Introduction

In flowering plants both the male and female gametophytes are highly reduced. Male gametophyte is represented by the pollen grain. They develop in anthers and are released to the atmosphere following anther dehiscence. The pistil, made up of the stigma, style and ovary, represents the female partner in sexual reproduction. Ovary contains the ovules in which embryo sac, the female gametophyte is located. For successful fertilization and seed set, pollen grains have to land on the stigma (pollination) through a pollinating agent (*see* Chapter 2, this volume). Following successful pollination pollen grains germinate on the stigma and the resulting pollen tubes grow through the tissues of the stigma and style and enter the embryo sac where they discharge the two sperm cells for fertilization. Another unique feature of flowering plants is double fertilization. It involves two fusion events, one of the sperm cells discharged by the pollen tube fuses with the egg and the other fuses with the central cell of the embryo sac. The zygote develops into the embryo and the central cell develops into the endosperm which nourishes the embryo; the ovule develops into the seed and the ovary develops into the fruit. Thus, successful fertilization and seed formation depends not only on effective pollination but also successful completion of pollen-pistil interaction and fertilization. Pollen-pistil interaction and fertilization were not the topics of intensive studies by earlier investigators. This was largely due to lack of suitable techniques to study these aspects. In recent years, however, a number of advanced techniques have become available and extensive studies have been carried out on both structural and functional aspects pollen-pistil interaction and fertilization. This review gives a brief account of pollen and pistil biology, pollen-pistil interaction and fertilization. More details are available in Stanley and Linskens (1974), Shivanna and Johri (1985), Raghavan (1999), Shivanna (2003) and in several reviews (Lord and Russell 2002, Raghavan 2003, Ma 2005, Suzuki 2009, Higashiyama 2010, Chapman and Goring 2010, Berger 2011, Berger et al. 2008, Shivanna 2010, Kawashima and Berger 2011, Hepler et al. 2013, Dresselhaus and Franklin-Tong 2013, Guan et al. 2013, Williams et al. 2014).

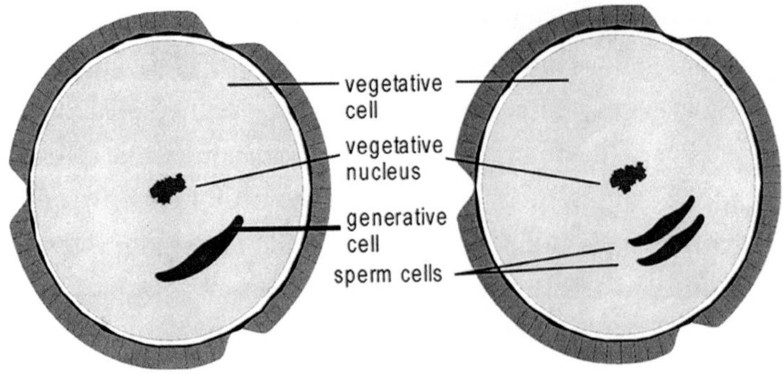

Fig.10.1. Pollen cytology. Diagrammatic representation of 2-celled (left) and 3-celled (right) pollen.

10.2 Pollen Biology

Pollen biology can be studied under three phases: developmental phase, free-dispersed phase and pollen-pistil interaction phase. In this review the dispersed phase and pollen-pistil interaction phases are covered. Free-dispersed phase of pollen covers the events from anther dehiscence until the pollen lands on the stigma. During this phase pollen grains act as independent functional units and are affected by the prevailing environment. At the time of shedding, pollen grains are made up of a vegetative cell and a generative cell (2-celled pollen present in about 70% of the species) or a vegetative cell and two sperm cells (3-celled pollen present in 30% of the species) formed by the division of the generative cell (Figure 10.1, *see* Ma 2005, Williams et al. 2014).Pollen size generally varies between 15-35 μm. The smallest pollen grains are reported in members of Boraginaceae (ca 5 μm) and the largest pollen of over 300 μm has been recorded in a member of Anonaceae. Pollen grains resemble seeds in several features; similar to seeds, pollen grains are desiccated and dormant units, and also the units of dispersal and gene flow. Their moisture content is generally less than 20% and their metabolic activity is very low. Desiccation of pollen sets in before the division of the generative cell in 2-celled pollen, whereas desiccation is delayed until the completion of the generative cell division in 3-celled pollen. The cytology of the pollen (2- or 3-cells) at the time of shedding is correlated with several functional aspects of pollen grains (Table 10.1).

Table 10.1: Correlations between pollen cytology (2/3-celled) and their function.

Functional features	**2-celled pollen**	**3-celled pollen**
Viability	Maintained for longer period	Lost rapidly
Storage requirements	Less stringent	More stringent
Requirements for in vitro germination	Less stringent	More stringent
Lag phase (the period from pollen hydration to pollen tube emergence)	Long/short	Short
State of mitochondria	Differentiated/undifferentiated	Differentiated
Controlled hydration for in vitro germination	Not Required	Required
Stigma surface	Wet/dry	Dry
Genetics of self-incompatibility	Gametophytic	Sporophytic
Zone of inhibition of self-pollen	Style	Stigma surface

Pollen grain wall is one of the most complex structures known in the plant kingdom. It has three domains: the intine, exine and pollenkitt/pollen coat. An important feature of the pollen wall is the presence of extracellular components (ECM) in all the three pollen wall domains. The ECM of pollen wall plays an important role in pollen-pistil interaction. Pollen exine is made up of sporopollenin, one of the most resistant organic materials known. It can withstand high temperature, high acidity and alkalinity and is insoluble in most of the organic solvents. Chemical analysis of sporopollenin has indicated that it is a mixed polymer with a large amount of aliphatics with additional compounds such as phenols (*see* Brooks and Shaw 1971, Shivanna 2003). Pollen grains are well preserved as fossils because of the presence of sporopollenin.

10.2.1 Pollen Fertility and Viability

Pollen fertility refers to the development of functional pollen grains. In most of the species, pollen grains are fertile at the time of shedding. However, in species of hybrid origin and in plants grown under environmental stresses, particularly of water and temperature, a proportion of pollen grains that develop may be sterile. The degree of sterility may vary depending on the extent of reproductive isolation of the parent species of the hybrid and the severity of environmental stresses. Pollen viability refers to their being alive at a given time. Viable pollen grains are able to deliver functional sperm cells in the embryo sac for fertilization. At the time of shedding, fertile pollen grains are also viable. However, fertile pollen grains do not remain functional forever; they lose their viability over time. The period for which they remain viable is highly variable; it depends partly on intrinsic nature of the pollen and partly on the prevailing environmental conditions. Pollen grains of some species, particularly the members of Poaceae lose their viability within 30-60 min after shedding while pollen of several other species may remain viable for months under ambient conditions. Pollen grains of a vast majority of the species fall between these two extremes; they remain viable for several days.

Pollen fertility is generally assessed through acetocarmine staining. Fertile pollen grains take up dense uniform stain whereas sterile pollen grains are generally shrivelled without or with partial contents and hence do not take up or take up partial and faint stain. There are several tests to assess pollen viability. Two of the most authentic and routinely used tests are in vitro germination test and fluoresceine diacetate (FDA) test (Heslop-Harrison and Heslop-Harrison 1970, Shivanna and Rangaswamy 1992). It is necessary to standardize an optimal in vitro germination medium to perform in vitro germination test. Viable pollen grains germinate in vitro in a suitable germination medium while nonviable pollen grains fail to germinate. It is difficult to achieve optimal in vitro germination of pollen of several species, particularly of those with 3-celled pollen; this test, therefore, cannot be performed on such pollen systems. FDA test is much simpler and works with pollen of most of the species; it can be performed in a few minutes (Shivanna and Rangaswamy 1992, Shivanna and Tandon 2014). FDA is a non-polar and non-fluorescent compound. When pollen grains are mounted in a drop of FDA, it readily enters the pollen cytoplasm as its plasma membrane does not act as a barrier. If esterases are active in the pollen cytoplasm, they hydrolyze FDA into fluorescein and acetate molecules. Fluorescein is polar and fluorescent; it does not move out of pollen grains readily if their plasma membrane is functional and thus accumulates in the pollen cytoplasm. When observed under a fluorescence microscope using blue filter combination, viable pollen grains fluoresce brightly (Figure 10.2) and non-viable pollen grains do not fluoresce. Thus, the FDA test assesses two properties of the pollen – the activity of esterases in the pollen cytoplasm and the presence of functional plasma membrane. When esterases are inactive, FDA is not hydrolyzed and pollen grains do not fluoresce. When plasma membrane is not functional, fluorescein, formed as a result of active esterases, leaks out of the pollen cytoplasm giving a mild background fluorescence rather than intense fluorescence of the pollen grains.

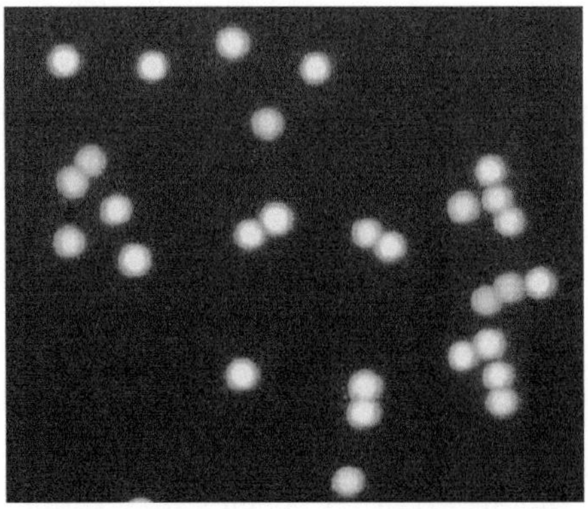

Fig. 10.2. FDA test for pollen viability. All pollen grains are viable and show bright fluorescence.

Table 10.2: Morphological diversity of the stigma (based on Heslop-Harrison and Shivanna 1977, Shivanna 2003).

Dry Stigma (without apparent fluid secretion)
Group I: Plumose – receptive surface dispersed on multiseriate branches
Group II: Receptive surface localized to well-defined stigma
A. Stigma surface non-papillate
B. Stigma surface distinctively papillate
1. Papillae unicellular
2. Papillae multicellular
a. Papillae uniseriate
b. Papillae multiseriate
Wet stigma (fluid secretion present on the stigma surface)
Group III: Receptive surface papillate
Group IV: Receptive surface non-papillate
Group V: Receptive surface covered with copious exudates in which detached secretory cells of the stigma are suspended

10.3 Pistil Biology

The pistil is the seat of all post-pollination events – pollen-pistil interaction, fertilization, and fruit and seed development. To understand these events, it is necessary to understand the structural details of the pistil. The pistil is made up of the stigma, style and ovary. The stigma is the terminal part of the pistil, the style is the elongated structure connecting the stigma and the ovary and the ovary is the swollen basal part of the pistil which houses the ovule(s). The ovules contain the female gametophyte (embryo sac) containing the egg cell along with two synergids, the central cell and three antipodal cells. The stigma is the recipient of the pollen grains during pollination. The stigma is classified on the basis of the distribution of the receptive surface and the presence or absence of the stigmatic exudates (a secretion product) on the surface of the stigma (Table 10.2). It is divided further based on the presence or absence of the papillae on the stigma (Heslop-Harrison and Shivanna 1967, Shivanna 2003). One of the important general features applicable to all types of stigmas is the presence of extracellular matrix (ECM) on the surface of the receptive stigma. The ECM is heterogeneous and contains enzymic and non-enzymic proteins, carbohydrates, phenolics and a variety of other components (*see* Shivanna 2003). It is rich in polysaccharides, glycoproteins and glycolipids (see Pereira et al. 2015). These components are secreted by the cells of the stigma. In wet type of the stigma, the ECM is present as a part of the stigmatic exudates and in the dry type it is present as a thin lining, termed pellicle, on the surface of the cuticle (Figure 10.3).

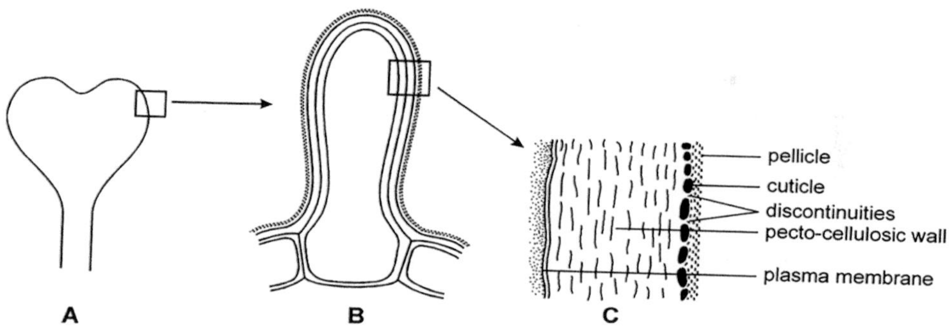

Fig. 10.3. Diagrammatic representation of the dry stigma to show the location of the pellicle. A. Outline of the stigma. B. One of the papillae from the blocked region in A enlarged. C. A part of the wall of the stigmatic papilla enlarged to show the details.

The style is basically of two types, the solid and the hollow. In solid style a strand of transmitting tissue connects the stigma with the ovarian cavity (Figure 10.4). The transmitting tissue is made up of narrow, elongated cells, connected to each other at their ends through plasmodesmata. However, their side walls are separated from each other by intercellular spaces filled with ECM secreted by the cells of the transmitting tissue. Thus, there are no plasmodesmatal connections on their longitudinal walls. In hollow styles the transmitting tissue is replaced by a continuous canal, termed stylar canal, that connects the stigma with the ovarian cavity. The stylar canal is bordered by one or a few layers of glandular cells, the canal cells. In hollow styles also the cells of the canal cells secrete ECM which accumulates on the inner surface of the canal cells (facing the canal) or fills up the stylar canal. The transmitting tissue/canal cells continue into the ovary along the placenta on which the ovules are located. Thus a major highlight of the structural features of the pistil is the presence of ECM on the surface of the stigma which receives the pollen grains and along the path of the pollen tube growth in the pistil until they reach the vicinity of the ovules.

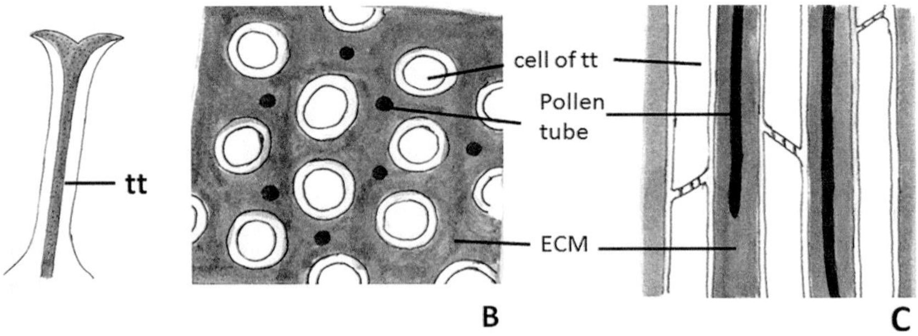

Fig. 10.4. Diagrammatic representation of the solid style. A. Longisection of the stigma and the style to show the location of the transmitting tissue (tt). B. Transection of the transmitting tissue. C. Longisection of the transmitting tissue. Massive intercellular spaces filled with extra cellular matrix (ECM) through which pollen tubes grow, are clear in B and C.

10.4 Significance of Pollen-pistil Interaction

Sexual reproduction in any organism involves fusion of male and female gametes. This fusion has to be specific between the gametes of the same species to maintain species integrity. The specificity is mediated through the ability of the gametes to recognize each other. In the Cryptogams, male and female gametes are released into the surrounding water and they come in contact with each other. If the gametes are of the right type (conspecific and compatible), they recognize each other and fuse. If the gametes are of the wrong type (heterospecific or incompatible) fusion is not permitted. In Bryophytes and Pteridophytes, the female gametes is retained inside the archegonium and the male gametes are released into the aqueous medium. Male gametes enter the archegonium, move through the neck and eventually come in contact with the female gamete; if the male and female gametes belong to the same species and are compatible, they recognize each other and fuse. Heterospecific and incompatible gametes fail to fuse. Thus in these groups of plants also the function of gamete recognition is bestowed with the gametes themselves; no other cell or tissue has any role in gamete recognition. In Gymnosperm, male gametes are not released into the aqueous medium but are formed and retained inside the pollen grain, the male gametophyte. Female gamete develops inside the ovule. Pollen grains released from the microsporangia are deposited through abiotic or biotic agents in the pollination drop, present at the tip of the ovules, where they germinate. A few layers of sporophytic nucellar cells intervene between archegonia and the site of pollen germination. Pollen grains/tubes have to interact with these layers before the sperm reaches the female gamete for fertilization. We still do not know clearly whether these nucellar layers have any role in pollen recognition (Owens et al. 1990, Kormutak 1999, see Shivanna 2003, Little et al. 2014).

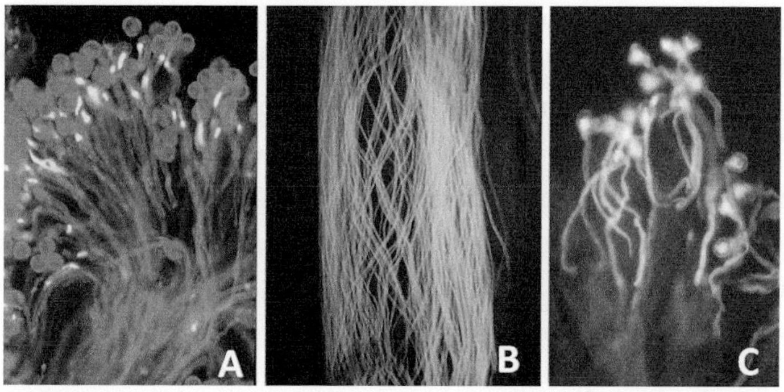

Fig. 10.5. Pollen-pistil interaction. Fluorescence micrographs of pollinated stigma and style stained with aniline blue. A, B. Profuse pollen germination on the stigma (A) and pollen tube growth in the style (B) following compatible pollination. C. Inhibited pollen tubes just below the stigma following incompatible pollination.

In flowering plants, a number of evidences indicate that the function of gamete (pollen) recognition has been taken over by the tissues of the pistil. Pollen-pistil interaction involves a massive interaction between the gametophyte (pollen grain and pollen tube) and sporophytic tissues of the pistil (the stigma and the style). During pollen-pistil interaction compatible pollen grains are recognized by the pistil and facilitate pollen germination and pollen-tube growth while incompatible pollen grains are inhibited in the pistil before they reach the ovules (Figure 10.5), where the female gamete is located (de Nettencourt 2001, Shivanna 2003). When pollen grains are deposited directly on cultured ovules (bypassing the intervening tissues of the pistil), even incompatible pollen grains are able to achieve fertilization and result in seed set (see under double fertilization) indicating the inability of the gametes to recognize each other and inhibit the fusion of the wrong type of gametes. More recent studies in achieving in vitro fertilization using isolated sperm and egg cells have also shown that gametes in conspecific as well as heterospecific combinations readily fuse, further emphasizing the inability of gametes to establish recognition (*see* under double fertilization). Thus the pistil screens the pollen grains before they reach the ovule; incompatible pollen grains are inhibited while compatible pollen grains are promoted to complete pollen-pistil interaction and fertilization.

The function of screening of pollen grains by the pistil has important implications on the breeding system of flowering plants. Self-incompatibility (SI) is an important outbreeding mechanism in almost all groups of plants. In flowering plants, it is determined by multiple alleles named S_1, S_2, S_3 . . . S_n at one locus in most of the species. When there is allelic matching at the S locus between the pollen and the pistil, pollen grain is recognized and inhibited. In S_1 and S_2 plant for example, self-pollen grains contain either S_1 or S_2 allele. As the sporophytic tissues of the pistil contain both S_1 and S_2 alleles, both types of pollen grains produced by the same plant are recognized and inhibited. However, SI in lower groups of plants, in which the recognition is established at the level of the gametes, it is able to prevent the fusion of only 50% of the gametes. For example, in a sporophyte with S_1 and S_2 alleles, half the gametes are of S_1 and the other half are of S_2. SI prevents fusion of S_1 with S_1 as well as S_2 with S_2 gametes, but cannot prevent fusion of S_1 with S_2 gametes, although they are produced by the same sporophyte. Thus, unlike lower groups of plants, evolution of SI has made flowering plants complete outbreeders (*see* Shivanna 2003). Enormous information is available on the operation of SI (Chan et al. 2010, de Nettencourt 2001, Qin et al. 2006, Rea et al. 2010, Takayama and Isogai 2005, Wheeler et al. 2010). This review covers the details of only compatible pollinations.

The screening of pollen also takes place for the quality. Many evidences indicate that the pistil is able to select the most vigorous pollen grains for fertilization. Under optimal pollination conditions the number of pollen grains deposited on the stigma is many times more than the number of ovules available for fertilization. As the pollen tubes grow through the pistil for considerable length, there is ample opportunity for vigorous pollen tubes to overtake less vigorous pollen tubes and achieve fertilization. Thus, the presence of pistil creates intense competition amongst even the compatible pollen. Those pollen grains that germinate early and those pollen tubes which are more vigorous are able to reach the ovules early and effect fertilization and less vigorous pollen grains are eliminated in this competition (Mulcahy 1979). There are many reports to show that pollen competition increases the vigor and fitness of the progeny (Pannell and Labouche 2013, Sari-Gorla and Frova 1997, Skogsmyr and Lankinen 2002). In *Petunia* for example, the seeds obtained from intense pollen competition (by pollinating the stigma with many times more pollen grains than the number of ovules) showed increased seed germination and seedling growth, enhanced competitive ability, greater reproductive output and less variation in phenotypic characters when compared to those obtained from no competition (by pollinating the stigma with less number of pollen grains than the number of ovules (Mulcahy 1984). Evolutionary success of angiosperms over other groups of plants is well-recognized. Although angiosperms were the last to evolve about 100 million years ago, they soon became the most dominant group. Evolution of the pistil along with the elaboration of the flower for efficient biotic pollination which enabled efficient cross-pollination, is considered to be the main causative factor for the evolutionary success of angiosperms (Endress 2011, Shivanna 2014, Willmer 2011).

10.5 Pollen-Pistil Interaction

Pollen-pistil interaction involves a series of sequential events starting with pollination (see Chapter 2, this vol) until the pollen tube enters the ovule (Figure 10.6). As pointed out earlier, pollen-pistil interaction plays a crucial role in sexual reproduction of flowering plants. However, earlier investigators had taken pollen-pistil interaction for granted. Although extensive structural studies were conducted on the development of pollen grains, embryo sac, embryo and endosperm, there were hardly any studies on pollen-pistil interaction (see Maheshwari 1950, 1964). The realization of the significance of pollen-pistil interaction and development of suitable techniques to study pollen germination and pollen tube growth in the pistil have stimulated extensive studies on this aspect during the last 30-40 years and considerable information is now available (*see* Shivanna 2003).

Pollination
↓
Pollen adhesion
↓
Pollen hydration
↓
Pollen germination
↓
Pollen tube entry into the stigma
↓
Pollen tube growth through the style
↓
Pollen tube entry into the ovule

Fig. 10.6. Sequential steps involved in pollen-pistil interaction.

Pollen-pistil interaction is a dynamic process and involves continuous communication and coordination between the pollen and the pistil. An understanding of these interactions involves: a) identification of the pollen and pistil components involved in this dialogue, b) their location in the pollen grains and the pistil and c) details of their interactions leading to pollen recognition followed by physiological and biochemical processes associated with promoting or preventing pollen germination and/or pollen tube growth in the pistil. Recent studies have highlighted the role of various components present in the ECM of the pollen wall and on the stigma surface, and later the pollen tube and the ECM of the transmitting tissue or the stylar canal during pollen-pistil interaction. In recent years, a number of studies have highlighted the role of arabinogalactan proteins (AGPs) during pollen-pistil interaction (*see* Pereira et al. 2015). AGPs are a large family of heavily glycosylated, hydroxyproline-rich proteins and have been shown to be present in almost all reproductive cells and tissues of flowering plants, although their expression varies temporally and spatially.

10.5.1 Events on the Stigma Surface

For initiation of pollen-pistil interaction, pollen grains that land on the stigma have to be viable and compatible, and the stigma has to be receptive. Stigma receptivity refers to the ability of the stigma to facilitate post-pollination events of compatible pollen. Although appearance of some of the enzymes such as esterases and peroxidases on the stigma surface have been reported to indicate stigma receptivity (Dafni et al. 2005, *see* Shivanna 2003), it is not conclusive. The only authentic method to assess stigma receptivity is through pollination. In recent years several arabinogalactan proteins (AGPs) have been localized in the stigma and also along the path of pollen tube growth in the style in several angiosperms including early divergent (basal angiosperms) lineages (see Losada et al. 2014 and references therein). Earlier studies implicated AGPs in some of the functions of pollen-stigma interaction such as pollen adhesion on to the dry stigma surface (Knox et al. 1986, *see* Shivanna 2003,) and in the synthesis of pollen wall material of the growing pollen tubes in the style (Dardelle et al. 2010). In a recent detailed study on *Magnolia virginiana*, a protogynous member in which the stigma receptivity lasts only for 4 h after flower opening, two AGPs have been shown to be closely associated with stigma receptivity (Losada et al. 2014). Pollen grains fail to germinate until the appearance of AGPs on the stigma and the AGPs could no longer be detected on the stigma surface after completion of pollen germination and pollen tube growth or after the stigma loses its receptivity. Thus the presence of AGPs acts as a marker for stigma receptivity. As AGPs have been reported to be present on the stigma surface of a number of species (*see* Losada et al. 2014), their role in stigma receptivity is likely to be widespread.

Pollen adhesion, hydration, germination and pollen tube entry into the stigma are the major events that take place on the surface of the receptive stigma. As pollen grains are desiccated systems, pollen hydration is a pre-requisite for subsequent events. The moisture needed for pollen hydration has to come from the stigma for which effective adhesion of the pollen on to the stigma surface is required. Adhesive carbohydrates, lipids and proteins present on the surface of the pollen and the stigma play a critical role in these events (Chapman and Goring 2010, Sanchez et al. 2004, Shivanna 2003). In some species such as *Gladiolus* and *Brassica,* characterized by dry stigma, washing of the stigma with a detergent or enzymic digestion of pellicle components have been shown to remove the ability of the stigma to support pollen germination and/or pollen tube entry into the stigma (Clarke et al. 1977, Knox et al. 1976, Shivanna et al. 1978) indicating that some components of the pellicle are involved in pollen germination and others in the entry of pollen tubes.

In dry stigma types, pollen adhesion is a critical step as it is a pre-requisite for pollen hydration and germination. Many evidences indicate that pollen coat substances especially lipids play an important role in pollen adhesion and hydration. Pollen coat substances flow on to the stigma surface and come in contact with the pellicle components present on the surface of the dry stigma and form a meniscus; this meniscus promotes not only pollen adhesion but also its hydration by facilitating the movement of water from the stigma to the pollen. In *Brassica* pollen grains in which pollen coat substances have been removed, fail to adhere, hydrate and germinate (Doughty et al. 1993). In *Arabidopsis*, pollen mutants that are defective in the synthesis of lipids fail to hydrate on the stigma (Edlund et al. 2004, Hulskamp et al. 1995b, Preuss et al. 1993, see also Zinkl et al. 1999). In maize, flavonoids which form a part of the pollen ECM have been shown to play a role in pollen germination. Flavonoid-deficient mutants produce flavonoid-deficient pollen which fail to function on mutant stigmas (van der Meer et al. 1992). Exogenous application of kaempferol, a flavonoid aglycone, either to the stigma or to the pollen grains restored the function of mutant pollen (Mo et al. 1992). As the dry stigma is covered by a layer of cuticle, pollen tubes have to digest the cuticle by activation of suitable enzymes for entering the tissues of the stigma (*see* Shivanna 2010). Several enzymes including cutinases have been shown to be present in pollen components of some species (*see* Lord 2003).

Pollen adhesion is not critical in wet stigma as the pollen grains readily adhere to the exudates present on the surface of the stigma. A number of experiments conducted on tobacco and *Petunia* (wet stigma, solid style) have highlighted the role of stigmatic exudates, particularly its lipid components in pollen germination and pollen tube entry into the stigma (Goldman et al. 1995, Wolters-Arts et al. 1998, 2002).Transgenic tobacco plants in which the secretory zone of the stigma was ablated through genetic transformation fail to produce stigmatic exudates and their stigma resembles that of the dry type (Goldman et al 1995). Such stigmas do not support pollen hydration and germination. Application of a suitable germination mediumon to the stigma improved pollen germination, but failed to promote pollen tube entry into the stigma. Interestingly, application of stigmatic exudates from the wild type *Nicotiana* stigma or *Petunia* stigma (both of which are largely lipidic) restored pollen function on transgenic tobacco. However, carbohydrate-rich aqueous exudates of *Lilium* stigma, although induces pollen germination on ablated stigma, did not permit pollen tube entry into the stigma indicating the role of lipids present in the exudates in pollen tube entry. Application of olive oil, which has a fatty acid composition similar to that of the stigma exudates, to the ablated stigma also restored the function of the stigma (Wolters-Arts et al. 1998, 2002). In wet stigma types with solid style, cuticle does not act as a barrier for pollen tube entry as the cuticle is disrupted during the secretion of stigmatic exudates. In wet stigma with hollow style, no tissue penetration is required as the pollen tubes grow over the cells of the stigma surface into the stylar cavity.

10.5.2 Pollen Tube Growth through the Style

Following the entry of pollen tubes into the stigma, they grow through the tissues of the stigma and style, and enter the ovary. In species with solid styles pollen tubes grow through the intercellular spaces filled with ECM (see Figure 10.4) and in species with hollow styles, pollen tubes grow on the surface of the canal cells and come in contact with the ECM present on the surface of the canal cells or in the exudates that fills the stylar canal. The pollen tube tip also releases proteins and other components into the stylar ECM as they grow through the tissues of the style. The stylar ECM contains free sugars, polysaccharides, amino acids, proteins, glycoproteins, proteoglycans and phenolic compounds. As pointed out earlier, a number of arabidogalactan proteins (AGPs), which are heavily glycosylated, have also been shown to be present along the pathway of pollen tube growth starting from the stigma and they are likely to play an important role in pollen-pistil interaction (Losada et al. 2014, Periera et al. 2014). These components present in the ECM and those released by the growing pollen tubes interact during pollen-pistil interaction and initiate the processes required for promoting/inhibiting pollen tube growth. In recent years the details of this interaction are being unrevealed through the application of the techniques of molecular biology and molecular genetics (Cheung and Wu 2008, Guan et al. 2013, Huang et al. 2014, Rounds et al. 2011, *see* Shivanna 2003).

Pollen tube growth involves the synthesis of an enormous amount of wall components required for pollen tube growth and respiration plays a major role in generating and regulating the energy flow to sustain this growth (Rounds et al. 2011). The length of the style in different species vary greatly from a few mm to several centimeters. For example, in some species such as *Nicotiana, Datura* and *Lilium* the style may reach up to 15 cm. In the silk of maize, pollen tubes may have to grow up to 25 cm when the pollen grain lands at the tip of the silk. Pollen grains, being very small, contain only a limited amount of nutrients which cannot support the growth of pollen tubes through the entire length of the style. Many evidences indicated that the stylar ECM provides the required nutrients. For example, by using labeled sugars it has been shown that sugars are taken up by the growing pollen tubes from the style (*see* Shivanna 2003). Also, depletion in the amount of nutrient reservesin the transmitting tissue following pollen tube growth indicates that tube growth occurs at the expense of the stylar reserves (Herrero and Hormaza 1996).

A specific glycoprotein rich in proline and galactose, termed transmitting tissue specific (TTS) protein, has been isolated and characterized in the stylar transmitting tissue of *Nicotiana* (Cheung et al. 1995). TTS protein forms a gradient of increasing glycosylation from the stigma to the ovary. Its molecular weight increases from 50 kD in the stigma region to 100 kD in the ovary region (Cheung et al. 1995; Wu et al. 1995, Cheung and Wu 2008); chemical deglycosylation of TTS protein isolated at any level of the pistil results in a 30 kD protein, indicating increasing molecular weight of TTS protein from the stigma to the ovary is due to increasing level of glycosylation. TTS protein is deglycosylated by the growing pollen tubes. TTS proteins bind to pollen tube surface and the tube tip, and are incorporated into the pollen tube wall indicating that the resulting carbohydrates may provide nutrients for pollen tube growth. TTS glycoproteins markedly stimulate pollen tube growth in vitro. Transgenic plants with reduced level of TTS proteins in the pistil showed reduced pollen tube growth rate and such plants yielded smaller fruits with reduced seed number.

A cysteine-rich,12-kD protein termed stigma-specific protein1 (STIG1) was isolated from the stigmatic exudates of tobacco and *Petunia* (Goldman et al. 1994, Verhoeven et al. 2005). Thus, STIG1 expression was sustained throughout pistil development and was confined to the stigma and upper part of the style. In tomato also, a 7-kD cystine-rich STIG1 protein was reported to be abundantly present in the stigma exudates and shown to promote pollen tube growth (Huang et al. 2014, Tang et al. 2004). Significant reduction in STIG1 expression (through generation of transgenic plants carrying an inverted repeat sequence against STIG1) did not affect the growth of the plants as well as pollen germination on the stigma but significantly reduced pollen tube growth as well as seed set. Pollen tube growth promoting signals of STIG1 have been shown to act through its binding with pollen specific receptor kinases (Huang et al. 2014). The pollen receptor kinases seem to interact with the plasma membrane of the pollen tube tip to regulate the activity of RAC/ROPs, which are the key regulators of polarized tip growth in pollen tubes (*see* Zou et al. 2011).

10.5.3 Pollen Tube Guidance in the Pistil

Pollen tube growth from the stigma to the ovary invariably follows a predetermined path irrespective of the length of the style and orientation of the flower. It was believed that pollen tubes are guided in the pistil by the presence of a chemotropic substance which forms an increasing gradient from the stigma to the ovary; pollen tubes were thought to follow this gradient. For some time in the 1960s calcium was considered to be the chemotropic substance (*see* Shivanna 2003). However, subsequent studies did not support the role of calcium as the chemotropic substance (*see* Heslop-Harrison 1987, Heslop-Harrison and Heslop-Harrison 1986). For example, deposition of pollen grains in the transmitting tissue of *Nicotiana* (Mulcahy and Mulcahy 1987) and *Torenia* (Higashiyama and Hamamura 2008) resulted in the growth of pollen tubes toward the stigma as well as the ovary at similar frequencies. Also, when the ovarian end of the excised style of *Lilium* was pollinated, pollen tubes grew normally through the style towards the stigma (Ascher 1977). These evidences clearly ruled out the presence of a chemotropic gradient in the style.

Subsequently it was suggested that the architecture of the transmitting tissue of the style guides the pollen tubes from the stigma towards the ovary. It was thought that the cells of the transmitting tissue forming files of elongated narrow cells surrounded by massive intercellular spaces filled with ECM provide a continuous pathway of least resistance for the growing pollen tubes. When once the pollen tubes enter the transmitting track, they follow this path unidirectionally from the stigma to the ovary without the need for a chemotropic gradient (*see* Shivanna, 2003). However, there is still some confusion about the presence of chemotropic substance at least in some species. In tobacco for example, pollen tubes have been reported to show positive chemotropism toward TTS glycoproteins, which form a gradient in the style; deglycosylated TTS proteins fail to attract pollen tubes there by suggesting their possible role in guiding the pollen tubes in the style (Cheung and Wu 2008). In *Lilium* also, studies of Lord and her associates (*see* Lord 2003) have provided evidences for the involvement small, heat-stable cystine-rich peptide molecule (SCA) and a pectic polypeptide in pollen tube guidance. The SCA/pectin together binds to the surface of the canal cells and induce adhesion of pollen tubes onto the canal cells (Jauh et al.1997). It was suggested that pollen tube growth is analogous to migrating cells of animal systems and that the stylar ECM actively facilitates pollen tube extension by way of biochemical recognition-adhesion system. In *Arabidopsis thaliana* pollen tubes seem to be guided in the pistil by the formation of a gradient in γ-amino butyric acid (GABA) formed as a result of the activity of a transaminase (Edlund et al. 2004, Palanivelu et al. 2003).

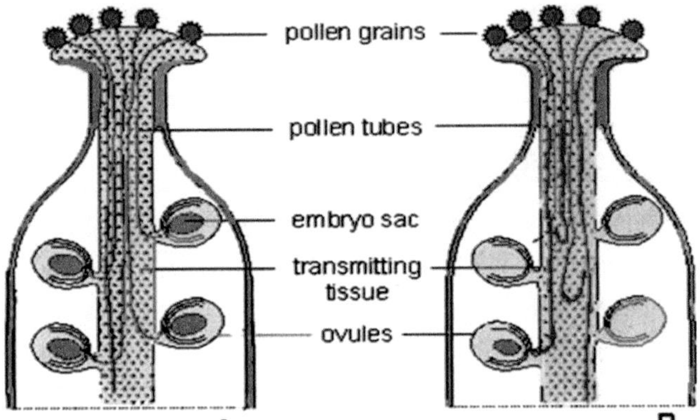

Fig. 10.7. Diagrammatic representation of pollinated pistils of *Arabidopsis thaliana* to show the role of embryo sac in pollen tube guidance. A. Wild type. Embryo sac is present in all the ovules and pollen tubes growing along the placenta turn by about 90º and enter all the ovules without any difficulty. B. A mutant in which three of the ovules lack embryo sac. The ovule with the embryo sac has received the pollen tube and the other pollen tubes grow randomly and fail to enter the ovules without the embryo sac (based on Hulskamp et al. 1995)

Irrespective of the presence or absence of pollen tube guidance in the style, there is a need for the guidance of pollen tubes on the surface of the stigma and in the ovary. On the stigma surface pollen tubes have to be guided to grow into the stigma instead of growing on its surface or along the outer wall around the stigma margin. In the ovary, pollen tubes have to be guided to turn about 90° from the placenta towards the micropyle of the ovule (Figure 10.7A). Considerable progress has been made in recent years on these aspects, although the studies so far are confined to a limited number of systems.

10.5.4 Pollen Tube Guidance on the Stigma Surface

The components involved in pollen tube guidance on the stigma surface seem to vary between species. A number of experiments carried out on tobacco and *Petunia* have highlighted the role of several components of the stigmatic exudates not only in pollen germination but also in pollen tubes guidance into the stigma (Goldman et al. 1995, Wolters-Arts et al. 1998, 2002). The stigma surface of tobacco, Petunia and several other members of Solanaceae is covered with a lipidic exudates. A thin layer of water is present on the stigma surface just below the exudates and forms a water gradient from the surface of the stigma to the surface of the exudates. Thus the concentration of water gradually increases from the surface of the exudates to the surface of the stigma where the water layer is located. This gradient seems to be necessary for guiding pollen tubes into the stigma (Wolters-Arts et al. 2002). Any disturbance in the water gradient affects the entry of pollen tubes into the stigma.

The stigma of *Lilium* is also of the wet type. However, the exudates is aqueous; therefore, lpids are unlikely to play a role in guiding the pollen tubes into the stigma. In this species pollen tube guidance seems to be mediated by two substances (see Higashiyama 2010, Kim et al.2003); one of them is asmall peptide with a molecular mass of 9.9 kDa, named chemocyanin and the other is a small, heat-stable cystine-rich peptide molecule (Park et al. 2000). Both of them have to combine for chemotropic activity.

The condition of the dry stigma is very different from that of the wet stigma. Studies carried out on *Arabidopsis* and *Brassica* have shown that the lipids present in the pollen coat in dry stigma types have taken over the function of the lipids present in the stigmatic exudates (Elleman and Dickinson 1996, Preuss et al. 1993, Wolters-Arts et al. 1998). As pointed out earlier, the pollen coat components from the pollen surface move out of the pollen and form a contact zone on the stigma surface which provides a medium to establish directional water gradient for pollen hydration, germination and pollen tube guidance into the stigma. In *Brassica*, aquaporin-like proteins in the stigma seem to act as water channels in facilitating pollen hydration (Dixitetal. 2001).

10.5.5 Pollen Tube Guidance in the Ovary

Pollen tube guidance in the ovary from the placenta toward the micropyle of the ovule and its subsequent entry into the embryo sac is the final and critical step of pollen-pistil interaction (Dresselhaus and Franklin-Tong, 2013). Since long this was considered to be in response to a chemotropic substance coming from the micropyle of the ovules (*see* Shivanna 2003). Accumulation of high level of calcium in the synergids was reported in some species and calcium released from the synergids was suggested to form the chemotropic gradient (Chaubal and Reger 1992). The micropyle of some species such as *Gasteria, Lilium* and *Arabidopsis* is reported to be covered with a drop of exudates (Franssen-Verheijen and Willemse 1993) which may play a role in pollen tube guidance. Recent report on the presence of carbohydrates and proteins in the micropylar region of several species also support such a concept (Sengupta and Tandon 2010).

Many experimental studies in recent years have confirmed that female gametophyte, particularly the synergids, play a key role in micropylar guidance of pollen tubes (Kessler and Grossniklaus 2011, Okuda et al. 2009, Takeuchi and Higashiyama 2011). A series of experiments carried out, particularly on *Arabidopsis* and *Torenia*, since late 1990s has given more definitive information on pollen tube guidance in the ovary. In an embryo-sac defective mutant of *Arabidopsis* (Hulskamp et al. 1995a, b) ovules either lacked an embryo sac or contained a partial or normal embryo sac. In such mutant over 90% of those with normal embryo sac and only 28% of ovules with defective embryo sac received pollen tubes; none of the ovules that lacked embryo sac received pollen tubes (Figure 10.7 A, B). These studies clearly showed the importance of the embryo sac in pollen tube attraction. Further studies using in vitro assay in *Torenia fournieri* have highlighted the specific cells of the embryo sac involved in chemotropic guidance. In this species, the micropylar part of the embryo sac protrudes beyond the ovule. Higashiyama et al. (1997) used a semi-vivo method of pistil culture to allow the pollen tubes to emerge into the culture medium (*see* Shivanna and Rangaswamy 1992) in which isolated ovules were also embedded. The pollen tubes that emerged from the cut end of the style grew towards the ovule and arrived at the filiform apparatus (wall ingrowths into the cytoplasm of the synergids at their tips) (Figure 10.8). Ovules with immature or damaged embryo sac and also ovules with fertilized embryo sac failed to attract pollen tubes. These studies clearly showed that pollen tubes are especially attracted to filiform apparatus of the synergids.

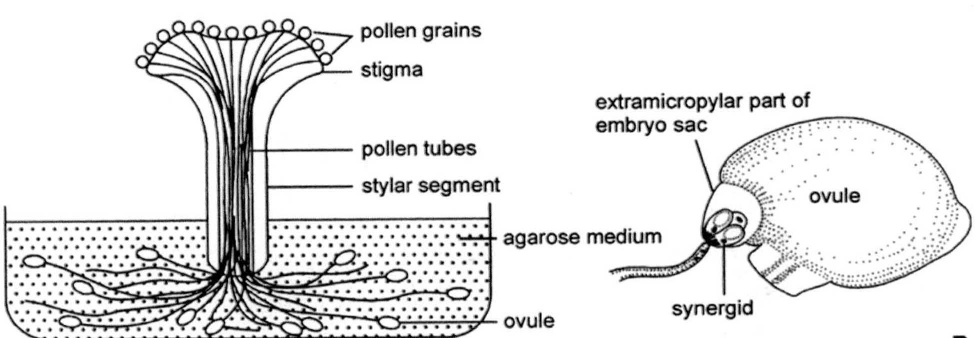

Fig. 10.8. A diagrammatic representation of semi-vivo culture set up to show pollen tube guidance to partially naked micropylar region of the embryo sac of *Torenia fournieri* (based on Hagashiyama et al. 1998). A. Pollen tubes following the emergence from the cut-end of the style grow towards the ovules. B. One of the ovule enlarged to show extramicropylar region of the embryo sac and the arrival of the pollen tube near the filiform apparatus of the embryo sac.

Subsequently, Hagashiyama et al. (2001) were able to ablate each cell of the embryo sac by a UV laser beam and study the arrival of the pollen tubes into the filiform apparatus following the semi-vivo method used in their earlier studies. Pollen tube attraction was not affected when central cell or egg cell was ablated. Even when one of the synergids was ablated, the frequency of pollen tube arrival was reduced but did not cease. However, when both the synergids were ablated, pollen tube attraction ceased completely. These studies clearly confirmed the role of synergid(s) in attracting pollen tubes. When the ovule with receptive embryo sac is moved farther or above or below the approaching pollen tube using a micromanipulator, the pollentube precisely chased the micropylar end of the embryo sac (Higashiyama and Hamamura 2008) suggesting that some diffusible signal derived from the micropylar end of the embryo sac is involved in pollen tube attraction. Basically similar results were obtained in the semi vivo assay (similar to that in *Torenia)* in *Arabidopsis* also (Palanivelu and Preuss 2006).

Interestingly, studies using the semi-vivo assay in *Torenia* and *Arabidopsis* have also highlighted the requirement for pollen tubes to grow through the stigma and style to become competent to respond to signals released from the ovules. In *Torenia*, pollen tubes emerging directly from the pollen grains cultured along with the ovules failed to grow towards the embryo sac (Higashiyama et al. 1998, 2001). In *Arabidopsis* also growth of the pollen tubes through the stigma and style greatly enhanced (from about 3 to 50%) the capacity of pollen tubes to target the ovules (Palanivelu and Preuss 2006).

Attempts have been made to identify active components involved in pollen tube attraction (see Dai et al. 2014, Jones-Rhoades et al. 2007, Marton et al. 2005, and Okuda et al. 2009). In *Torenia* two cyctein-rich polypeptides (CRPs), termed LURE1 and LURE2 have been shown to be secreted by the synergids (Okuda et al. 2009). These CRPs are expressed specifically in the synergid cells and are detected in the filiform apparatus.Down-regulation of each LURE decreased the frequency of pollen tube attraction, suggesting that LUREs are involved in pollen tube attraction (see Lin et al. 2014).

In maize, an egg apparatus-specific gene, *ZmEA1*, seems to be involved in pollen tube guidance (Marton et al. 2005). When *ZmEA1* (EGG APPARATUS 1) is down regulated, pollen tubes grew in close proximity to the micropyle, but failed to enter the embryo sac (Marton et al. 2005). In a mutant of maize lacking functional embryo sac but fully differentiated sporophytic tissues of the ovule, pollen tubes grew on the surfaceof the ovule but failed to enter the micropyle (Lausser et al. 2010). ZmEA1 protein attracts maize pollen tubes in vitro also (Dresselhaus and Marton 2009, Marton and Dresselhaus 2010) further supporting ZmEA1 protein as the attractant for micropylar guidance.

For effective functioning of pollen tube guidance, pollen tubes need to perceive the female guidance signals at the tip to initiate directional growth toward the micropyle. Many investigators have attempted to understand the components involved in pollen tube perception of the synergid signal (von Besser et al. 2006, Li et al. 2011, Liu et al. 2013). In *Arabidopsis* two receptor-like kinases, LIP1 and LIP2 in the plasma membrane at the tip of the growing pollen tubes have been shown to perceive the LURE1 signal from the synergids (Liu et al. 2013).

10.6 Double Fertilization

The details of fusion events involved in double fertilization were known through classical microtome studies since the end of the 19th century (*see* Maheshwari 1950). Application of electron microscopy since 1960s added another dimension to the structural detail of fertilization (*see* Russell 1992). One of the highlights of recent studies has been in understanding the role of synergid in fertilization events. Pollen tube invariably enters one of the synergids through the filiform apparatus. Even in species such as *Plumbago* in which synergids are absent, the egg itself develops filiform apparatus at the micropylar region through which pollen tube enters the egg cell. In some species, one of the synergids degenerates just before the arrival of the pollen tube and the pollen tube invariably enters the degenerated synergid. In others the synergid that receives the pollen tube degenerates soon after pollen tube entry (*see* Russell 1992, Raghavan 1999, 2003, Shivanna 2003). Pollen tubes cease growth in the synergid and discharge the two male gametes into the synergid. The signal responsible for cessation of pollen tube growth in the synergid is not known.

In recent years, application of the techniques of in vitro culture, molecular biology and high resolution imaging have added additional information on double fertilization (Berger 2011, *see* Raghavan 2003, Shivanna 2003, Kawashima and Berger 2011). Effective communication between pollen tube and the synergid has been shown to be an essential feature of these fusion events (see Kawashima and Berger 2011). In the absence of such communication, pollen tubes continue to grow in the embryo sac without releasing the gametes. The movement of sperm cells in the synergid (one of them towards the egg and the other towards the central cell) for double fertilization was thought to be an active process mediated through the cytoskeleton elements, particularly the actin filaments (Russell 1992). However, recent evidences indicate that the cytoplasmic flow induced by pollen tube rupture is sufficient for the sperm cells to reach the site of gamete fusion (Hamamura et al. 2011).

The discovery of sperm dimorphism in some species, particularly in *Plumbago* (Russell 1984) and *Brassica* (Knox and Singh 1987) showing distinct differences in the morphology and contents of the two sperm cells, raised the possibility of preferential fertilization involving the fusion of specific sperm cell with the egg cell and the central cell. In *Plumbago* for example, one of the sperm cells is plastid rich and the other is mitochondria rich; the two sperms also express different sets of transcripts (Gou et al. 2009). Subsequent studies showed that in *Plumbago* the plastid-rich sperm cell fuses with the egg cell in 94% of fusions, indicating that at least in this species fusion of the sperm cells with the egg and the central cell are not random but specific; this would then involve gamete recognition in the embryo sac (Russell 1985). The details of this recognition event, if any, are yet to be elucidated.

However in many other species such as *Arabidopsis*, the two sperm cells are isomorphic and do not show any differences (Ge et al. 2011). Both the sperm cells are able to fertilize the egg cell and the central cell with similar probability, indicating that in *Arabidopsis* the two sperm cells cannot be distinguished functionally (Hamamura et al. 2011).

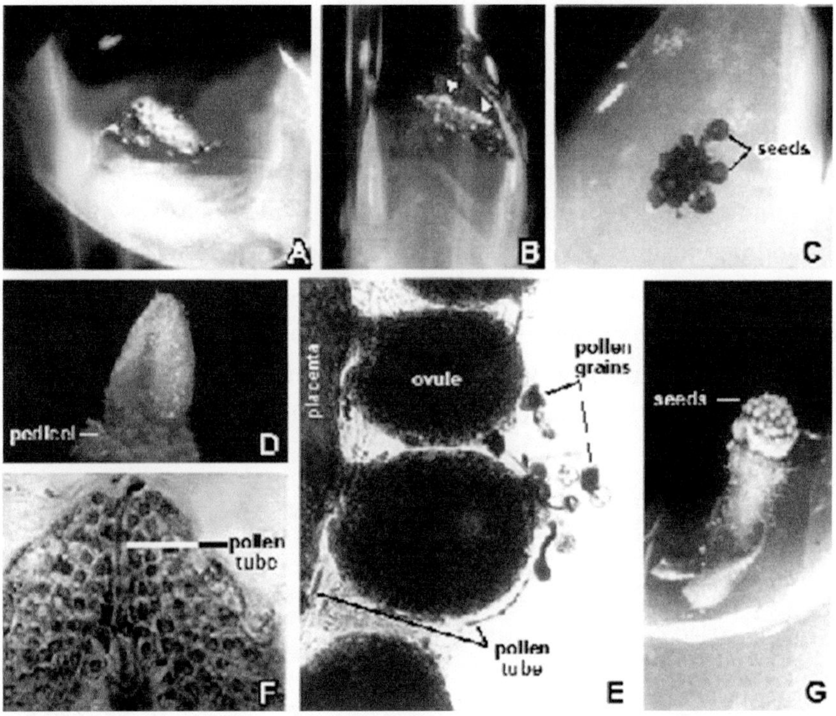

Fig. 10.9. In vitro pollination of ovules. A-C. Pollination of cultured ovules in *Papaver somniferum* (A,B) and *Argemone mexicana* (C). Arrowheads in B show developing seeds. D-G. Placental pollination in *Petunia axillaris*. D. A placental explant along with the pedicel made ready for culture. E. A part of a section from pollinated placental culture to show pollen germination and pollen tube growth. F. Longisection of the micropylar part of the ovule to show pollen tube entry into the ovule. G. Mature seeds developed on cultured placenta.

10.6.1 In Vitro Pollination and Fertilization

In flowering plants, fertilization takes place deep inside the embryo sac covered by the tissues of the ovule and the ovary. Inaccessibility of the site of fertilization has been a major difficulty in conducting experimental studies on fertilization. Thus, information on fertilization in flowering plants has lagged far behind when compared to that in animals and some of the lower groups of plants. One of the dreams of experimental embryologists has been to develop techniques to achieve in vitro fertilization similar to that in animals so that the details can be studied more effectively. A beginning was made in this direction in the 1960s at the University of Delhi by Professor Maheshwari and his associates. They were able to achieve fertilization in isolated cultured ovules (Kanta et al. 1962, Rangaswamy and Shivanna 1967, *see* Rangaswamy 1977, Shivanna 2003). A group of ovules of *Papaver somniferum* were excised from the ovary and cultured on a nutrient medium, and pollen grains were dusted around the cultured ovules. Following ovules pollination, pollen grains germinated and the resulting pollen tubes entered the ovules and effected double fertilization (Kanta et al. 1962). Fertilized ovules developed into seeds (Figure 10.9 A-C). Subsequently a modified technique of in vitro pollination of ovules, termed placental pollination was developed (Rangaswamy and Shivanna 1967). The modified technique involved retaining the entire ovule mass intact on the placentae (after removing the ovary wall) along with a short length of pedicel (Figure 10.9 D) and culturing the ovule mass by inserting only the pedicel into the medium. This eliminated the injury to the ovules and also prevented the ovules from coming in contact with the medium. The ovule mass was dusted with pollen grains. This modified technique has been shown to be effective in achieving fertilization and seed set in vitro (Figure 10.9 E-G) in a number of species and was used to overcome self- as well as interspecific-incompatibility barriers (see Zenkteler 1990, Shivanna 2003). Further progress to accomplish in vitro fertilization during the 1960s was hampered for lack of effective techniques for isolating male and female gametes.

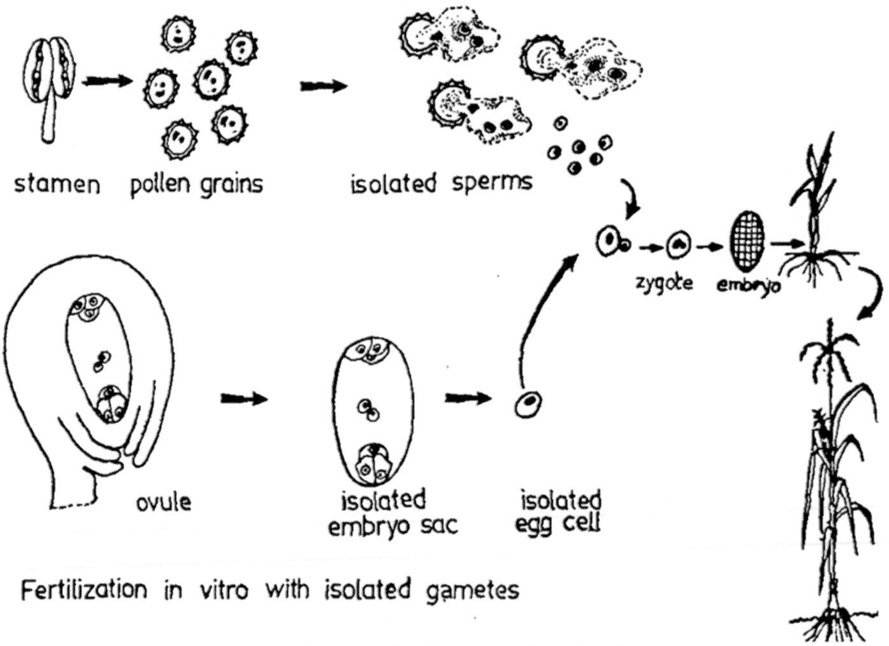

Fig. 10.10. Diagrammatic representation of in vitro fertilization with isolated sperm and egg cells in maize (based on Kranz and Lorz 1993).

Following the progress in protoplast technology and somatic hybridization in the 1970s and 1980s, many investigators started working on isolation of sperms and egg cells in flowering plants and soon success was achieved in isolating male and female gametes in viable condition in several species (*see* Matthys-Rochon et al. 1997). By 1990 all the basic requirements to attempt in vitro fertilization were available. The pioneering success in achieving in vitro fertilization by using isolated egg and sperm cells was achieved in maize by Kranz and his associates (see Kranz 1997, Kranz and Lorz 1993). In vitro fertilized zygotes were able to divide and develop into embryos under nurse culture conditions. The embryos developed into plantlets; they were transplanted and grown into fertile plants (Figure 10.10). Subsequently, the technique has been extended to several other species and also it has been possible to achieve interspecific fusion. Japanese group has established protocol for isolating rice gametes and achieving in vitro fertilization and subsequent development of in vitro fertilized zygotes into fertile plants (Uchiumi et al. 2006, 2007). Using this in vitro system, they have been able to analyze gene expression profiles in gametes and zygotes, the changes associated with the polarity during zygote development and the dynamics of karyogamy (*see* Ohnishi et al. 2014 and references there in). These studies indicated that the migration of sperm nucleus adjacent to the egg nucleus is mediated through actin cytoskeleton and the egg chromatin appears to move into the sperm nucleus during karyogamy. With the completion of karyogamy, the sperm chromatin begins to decondense followed by de novo zygotic gene expression. Apart from its application in achieving interspecific and intergeneric hybrids and genetic transformation, in vitro pollination and fertilization have great promise in generating new knowledge on a range of fundamental problems of double fertilization and embryogenesis in flowering plants.

10.7 Concluding Remarks

Pollen-pistil interaction, a prerequisite for fertilization in flowering plants, remained neglected by earlier embryologists due to lack of effective techniques. Following the realization of its significance in reproductive biology of flowering plants and availability of techniques, studies were initiated in the 1960s and since then enormous data has accumulated on both the structural and functional aspects of pollen-pistil interaction. Application of the techniques of molecular biology and molecular genetics has opened up new vistas to unravel further knowledge on pollen-pistil interaction. The molecular aspects of interaction between the ECM of the pollen grains and pollen tubes, and of the pistil are becoming apparent in recent years.

Studies on fertilization also remained confined for long to microscopic studies of fixed materials. Applications of the techniques of electron microscopy and in vitro fertilization, combined with those of molecular biology have started unrevealing the mysteries of double fertilization. There has been a significant progress during the last decade in understanding the molecular aspects of fertilization, although it is so far confined to a few systems. This is the result of application of the advanced techniques including the analyses of gamete transcriptomes and high resolution imaging of double fertilization in vivo in living cells (see Kawashima and Burger 2011).With these integrated techniques, the progress in our understanding of pollen-pistil interaction and fertilization is expected to be more rapid in the coming years. Such advances would enable not only in gaining new knowledge on the fundamental aspects of these important events of reproduction in flowering plants but also in manipulating more effectively pollen-pistil interaction and fertilization in a desirable way for practical benefits.

Acknowledgements

I thank Indian National Science Academy for the award of a position of INSA Honorary Scientist.

References

Ascher P D 1977. Localization of the self– and inter–specific incompatibility reactions in style sections of *Lilium longiflorum*. Plant Sci Lett 10: 199–203.

Berger F 2011. Imaging fertilization in flowering plants, not so abominable after all. J Exp Bot 62: 1651–1658.

Berger F, Hamamura Y, Ingouff M, Higashiyama T 2008. Double fertilization—caught in the act. Trends in Plant Sci 13: 437–443.

Brooks J, Shaw G 1971. Recent developments in the chemistry, biochemistry and geochemistry and post–tetrad ontogeny of sporopollenin derived from pollen and spore exines. In Heslop–Harrison J (Ed) Pollen development and physiology, Butterworths, London, pp 99–114.

Chapman L A, Goring D R 2010. Pollen–pistil interactions regulating successful fertilization in the Brassicaceae. J Exptl Bot 61:1987–1999.

Chaubal R, Rejer BJ 1992. The dynamic of calcium distribution in the synergid cells of wheat after pollination. Sex Plant Repro 5: 206–213.

Chen G, Zhang B, Zhao Z, et al. 2010. 'A life or death decision' for pollen tubes in S–Rnase–based self–incompatibility. J Exptl Bot 61: 2027–2037.

Cheung AY, Wu H–M. 2008. Structural and signaling networks for the polar cell growth machinery in pollen tubes. Ann Rev Plant Biol 59: 547–572.

Cheung A Y, Wang H, Wu H–M 1995. A floral transmitting tissue–specific glycoprotein attracts pollen tubes and stimulates their growth. Cell 82: 383–393.

Clarke A E, Considine J A, Ward R, Knox R B 1977. Mechanism of pollination in *Gladiolus*: roles of the stigma and pollen tube guide. Ann Bot 41: 15–20.

Dafni A, Kevan P G, Husband B C (Eds) 2005. Practical pollination biology. Enviroquest, Cambridge, Ontario.

Dai XR, Gao X–Q, Chen GH, et al. 2014. ABNORMAL POLLEN TUBE GUIDANCE1, an endoplasmic reticulum–localized mannosyltransferase homolog of GLYCOSYLPHOSPHATIDYLINOSITOL10 in yeast and PHOSPHATIDYLINOSITOL GLYCAN ANCHOR BIOSYNTHESIS B in human, is required for *Arabidopsis* pollen tube micropylar guidance and embryo development. Plant Physiol. 165: 1544–1556.

Dardella F, Lehner A, Ramdani Y, et al. 2010. Biochemical and immunocytological characterizations of *Arabidopsis* pollen tube cell wall. Plant Physiol. 153: 1563–1576.

De Nettancourt D 2001. Incompatibility and incongruity in wild and cultivated plants. Springer–Verlag, Berlin.

Dixit R, Rizzo C, Nasrallah M E, Nasrallah J B 2001. The Brassica MIP–MOD gene encodes a functional water channel that is expressed in the stigma epidermis. Plant Mol Biol 45: 51–62.

Doughty J, Hedderson F, McCubbin A, Dickinson H G 1993. Interaction between coating–born peptides of the *Brassica* pollen grain and stigmatic S (self–incompatibility)–locus–specific glycoproteins. Proceed Nat Acad Sci USA, 90: 467–471.

Dresselhaus T, Franklin–Tong N 2013. Male–female crosstalk during pollen germination, tube growth and guidance, and double fertilization. Mol Plant 6: 1018–1036.

Dresselhaus T, Marton M L 2009. Micropylar pollen tube guidance and burst: adapted from defense mechanisms? Current Opinion Plant Biol 12: 773–780.

Edlund A F, Swanson R and Preuss D 2004. Pollen and stigma structure and function: The role of diversity in pollination. Plant Cell 16 (supplement): S84–S97.

Elleman C J, Dickinson H G 1996. Identification of pollen components regulating pollination–specific responses in the stigmatic papillae of *Brassica oleracea*. New Phytol, 133:197–205.

Endress P K 2011. Evolutionary diversification of flowers in angiosperms. Am J Bot 98:370–396.

Franssen–Verheijen AAM, Willemse MTW 1993. Micropylar exudates in *Gasteria* (Aloaceae) and its possible function in pollen tube growth. Am J Bot 80: 253–262.

Ge L, Gou X, Yuan T 2011. Migration of sperm cells during pollen tube elongation in *Arabidopsis thaliana*: behavior during transport, maturation and upon dissociation of male germ unit associations. Planta 233: 325–332.

Goldman MHS, Goldberg RB, Miriani C 1994. Female sterile tobacco plants are produced by stigma–specific cell ablation. EMBO J 13: 2976–2984.

Gou X, Yuan T, Wei X, Russell SD 2009. Gene expression in the dimorphic sperm cells of *Plumbago zeylanica*: transcript profiling, diversity, and relationship to cell type. Plant J 60: 33–47.

Guan Y, Guo J, Li H, Yang Z 2013. This article appears in:Pollen tube and reproduction signaling in pollen tube growth: Crosstalk, feedback, and missing links. Mol Plant 6: 1053–1064.

Hamamura Y, Saito C, Awai C, et al. 2011. Live–cell imaging reveals the dynamics of two sperm cells during double fertilization in *Arabidopsis thaliana*. Curr Biol 21: 497–502.

Hepler P K, Rounds CM, Winship LJ 2013. Control of cell wall extensibility during pollen tube growth. Mol Plant 6: 998–1017.

Herrero M, Hormaza J I 1996. Pistil strategies controlling pollen tube growth. Sex Plant Repro 9: 343–347.

Heslop–Harrison J 1987. Pollen germination and pollen–tube growth. Inter Rev Cytol 107:1–78.

Heslop–Harrison J, Heslop–Harrison Y 1970. Evaluation of pollen viability by enzymatically induced fluorescence; intracellular hydrolysis of fluorescein diacetate. Stain Tech 45: 115–120.

Heslop–Harrison J, Heslop–Harrison Y 1986. Pollen tube chemotropism; fact or delusion? In Cresti M, Dallai R (Eds) Biology of reproduction and cell motility in plants and animals. Univ Siena, Italy, pp 169–174.

Heslop–Harrison Y, Shivanna K R 1977. The receptive surface of the angiosperm stigma. Ann Bot 41: 1233–1258.

Higashiyama T 2010. Peptide signaling in pollen–pistil interactions. Plant Cell Physiol 51:177–189.

Higashiyama T, Hamamura Y 2008. Gametophytic pollen tube guidance. Sex Plant Repro 21:17–26.

Higashiyama T, Kuroiwa H, Kawano S, Kuroiwa T 1998. Guidance in vitro of the pollen tube to the naked embryo sac of *Torenia fournieri*. Plant Cell 10: 2019–2031.

Higashiyama T, Yabe S, Sasaki N, et al. 2001. Pollen tube attraction by the synergid cell. Science 293:1480–1483.

Huang W–J, Liu H–K, McCormick S, Tang W–H 2014. Tomato pistil factor STIG1 promotes in vivo pollen tube growth by binding to phosphatidylinositol 3–phosphate and the extracellular domain of the pollen receptor kinase LePRK2. Plant Cell 26: 2505–2523.

Hülskamp M, Schneitz K, Pruitt R E 1995a. Genetic evidence for a long–range activity that directs pollen tube guidance in *Arabidopsis*. Plant Cell 7: 57–64.

Hülskamp M, Kopozak S D, Horejsi T M, et al. 1995b. Identification of genes required for pollen–stigma recognition in *Arabidopsis thaliana*. Plant J 8: 703–714.

JauhG Y, Eckard K J, Nothnagel E A, Lord EM 1997. Adhesion of lily pollen tubes on an artificial matrix. Sex Plant Repro 10: 173–180.

Jones–Rhoades M W, Borevitz J O, Preuss D 2007. Genome–wide expression profiling of the *Arabidopsis* female gametophyte identifies families of small, secreted proteins. PloS Genet 3: 1848–1861.

Kanta K, Rangaswamy N S, Maheshwari P 1962. Test tube fertilization in a flowering plant. Nature 194:1214–1217.

Kawashima T, Berger F 2011. Green love talks; cell–cell communication during double fertilization in flowering plants. AoB Plants plr015, doi:10.1093/aobpla/plr015

Kessler S A, Grossniklaus U 2011. She's the boss: signaling in pollen tube reception. Curr Opin Plant Biol 14: 622–627.

Kim S, Mollet J C, Dong J, et al. 2003. Chemocyanin, a small basic protein from the lily stigma, induces pollen tube chemotropism. Proc Natl Acad Sci USA 100: 16125–16130.

Knox R B, Singh MB 1987. New perspectives in pollen biology and pollen–pistil interaction. Ann Bot Suppl 4: 15–37.

Knox R B, Willimas E G, Dumas D 1986. Pollen, pistil and reproductive function in crop plants. Plant Breed Rev 4: 9–79.

Knox R B, Clarke A E, Harrison S, et al.1976. Cell recognition in plants: determinants of the stigma surface and their pollen interactions. Proc Natl Acad Sci USA 73: 2788–2792.

Kormutak A 1999. Self–incompatibility in silver fir (*Abies alba*). Biologia 54: 101–105.

Kranz E 1997. In vitro fertilization with single isolated gametes. In Shivanna K R, Sawhney V K (Eds) Pollen biotechnology for crop production and improvement. Cambridge Univ Press, New York. NY, pp 377–391.

Kranz E, Lorz H 1993. In vitro fertilization with single isolated gametes results in zygotic embryogenesis. Plant Cell 5: 739–746.

Kranz E, Hoshino Y, Okamoto T 2008. In vitro fertilization with isolated higher plant gametes. Methods Mol Biol 427: 51–69.

Lausser A, Kliwer I, Srilunchang K, Dresselhaus T 2010. Sporophytic control of pollen tube growth and guidance in maize. J Expl Bot 61: 673–682.

Li HJ, Xue Y, Jia DJ, et al. 2011. POD1 regulates pollen tube guidance in response to micropylar female signaling and acts in early embryo patterning in *Arabidopsis*. Plant Cell 23: 3288–3302.

Lin S–Y, Chen P–W, Chuang M–H, et al. 2014. Profiling of translatomes of in vivo–grown pollen tubes reveals genes with roles in micropylar guidance during pollination in *Arabidopsis*. Plant Cell 26: 602–618.

Little S, Prior N, Pirone C, Aderkas Pv 2014. Pollen–ovule interactions in Gymnosperms. In: Ramawat K G, Merillon J M, Shivanna K R (Eds) Reproductive biology of plants. CRC Press, Boca Raton, FL, pp 97–117.

Liu J, Zhong S, Guo X et al 2013. Membrane–bound RLCKs LIP1 and LIP2 are essential male factors controlling male–female attraction in *Arabidopsis*. Curr Biol 23: 993–998.

Lord E M 2003. Adhesion and guidance in compatible pollination. J Expl Bot 54: 47–54.

Lord E M, Russell R D 2002. The mechanisms of pollination and fertilization in plants. Annu Rev Cell Develop Biol 18: 81–105.

Losada J M, Herrero M, Hormaza J I, Friedman W E 2014. Arabinogalactan proteins mark stigmatic receptivity in the portogynous flowers of *Magnolia virginiana*. Am J Bot 101: 1963 – 1975.

Ma H 2005. Molecular genetic analyses of microsporogenesis and microgametogenesis in flowering plants. Ann Rev Plant Biol 56:393–434.

Maheshwari P 1950. An introduction to the embryology of angiosperms. McGraw–Hill, New York, NK.

Maheshwari P (Ed) 1963. Recent advances in the embryology of angiosperms. Intl Soc Plant Morphl, Delhi.

Marton M, Dresselhaus T 2010. Female gametophyte–controlled pollen tube guidance. Bioche Soc Trans 38: 627–630.

Marton M, Cordts S, Broadhvest J, Dresselhaus T 2005. Micropylar pollen tube guidance by egg apparatus of maize. Science 307: 573–576.

Mathys–Rochon S, Mol R, Faure C, et al. 1997. Isolation and micromanipulation of the embryo sac and egg cell in maize. In Shivanna KR, Sawhney VK (Eds) Pollen biotechnology for crop production and improvement. Cambridge Univ Press, New York, NY, pp 363–376.

Mo Y, Nagel C, Taylor L P 1992. Biochemical complementation of calchone synthase mutants define a role for flavonols in functional pollen. Proc Natl Acad Sci USA, 89: 1713–1717.

Mulcahy D L 1979. Rise of the angiosperms: a genecological factor. Science 206: 20–23.

Mulcahy D L 1984. Manipulation of gametophytic populations. In Lange W, Zeven AC, Hogenboom D (Eds) Efficacy in plant breeding. PUDOC, Wageningen, The Netherlands. pp 114–119.

Mulcahy D L, Mulcahy GB 1975. The influence of gametophytic competition on sporophytic quality of *Dianthus chinensis*. Theore Appl Genet 46: 277–280.

Mulcahy D L, Mulcahy G B 1987. Induced pollen tube directionality. Am J Bot 74: 1458–1459.

Ohnishi Y, Hoshino R, Okamoto T 2014. Dynamics of male and female chromatin during karyogamy in rice zygotes. Plant Physiol 165:1533–1543.

Okuda S, Tsutsui H, Shiina K, et al. 2009. Defensin–like polypeptide LUREs are pollen tube attractants secreted from synergid cells. Nature 458: 357–361.

Owens J N, Colangeli A M, Morris S J 1990. The effect of self–, cross–, and no pollination on ovule, embryo, seed, and cone development in western red–seeder (*Thuja plicata*). Can J Forest Res 20: 66–75.

Palanivelu R, Preuss D 2006. Distinct short–range ovule signals attract or repel *Arabidopsis thaliana* pollen tubes *in vitro*. BMC Plant Biol6: 7.

Palanivelu R, Brass L, Edlund A F, Preuss D 2003. Pollen tube growth and guidance is regulated by *POP2*, an *Arabidopsis* gene that controls GABA levels. Cell 114: 47–59.

Pannell J R, Labouche A M 2013. The incidence and selection of multiple mating in plants. Phil Trans R Soc B 368: Article No. 20120051.

Park S Y, Jauh G Y, Mollet J C, et al. 2000. A lipid transfer–like protein is necessary for lily pollen tube adhesion to an in vitro stylar matrix. Plant Cell 12:151–163.

Pereira A M, Pereira L G, Coimbra S 2015. Arabinogalactan proteins: rising attention from plant biologists. Plant Reprod 28: 1–15.

Pereira A M, Masiero S, Nobre M S, Costa M L, et al. 2014. Differential expression patterns of arabinogalactan proteins in *Arabidopsis thaliana* reproductive tissues. J. Exp Bot 65: 5459–5471.

Preuss D, Lamieux B, Yen G, Davis RW 1993. A conditional sterile mutation eliminates surface components from *Arabidopsis* pollen and disrupts cell signaling during fertilization. Genes Develop 7: 974–985.

Punwani J A, Drews GN 2008. Development and function of the synergid cell. Sex Plant Repro 21: 17–26.

Qin X, Liu B, Soulard J et al. 2006. Style–by–style analysis of two sporadic self–compatible *Solanum chacoense* lines supports a primary role for S–Rnases in determining pollen rejection thresholds. J Expl Bot 57: 2001–2013.

Raghavan V 1999. Molecular Embryology of Flowering Plants. Cambridge University Press, New York, NY.

Raghavan V 2003. Some reflections on double fertilization, from its discovery to the present. New Phytol 159: 565–583.

Rangaswamy N S 1977. Applications of in vitro pollination and in vitro fertilization. In Reinert J, Bajaj YPS (Eds) Applied and fundamental aspects of plant cell tissue and organ culture. Springer–Verlag, Berlin, pp 412–425.

Rangaswamy N S, Shivanna K R 1967. Induction of gamete compatibility and seed formation in axenic cultures of a diploid self–incompatible species of Petunia. Nature 216: 937–939.

Rea A C, Liu P, Nasrallah J B 2010. A transgenic self–incompatible *Arabidopsis thaliana* model for evolutionary and mechanistic studies of crucifer self–incompatibility. J Exptl Bot 61: 1897–1906.

Rounds C M, Winship LJ, Hepler PK 2011. Pollen tube exudates: respiration, fermentation and the race to the ovule. AoB Plants 2011: plr019.

Russell S D 1984. Ultrastructure of the sperms of *Plumbago zeylanica*: 2. Quantitative cytology and three–dimentional reconstruction. Planta 162: 385–391.

Russell S D 1985. Preferential fertilization in *Plumbago*: ultrastructural evidence for gamete–level recognition in an angiosperm. Proc Natl Acad Sci USA 82: 6129–6132.

Russell S D 1992. Double fertilization. Intl Rev Cytol 140: 357–388.

Sanchez A M, Bosch M, Bots M, et al. 2004. Pistil factors controlling pollination. Plant Cell 16: S 98–S 106.

Sari–Gorla M, Frova C 1997. Pollen tube growth and pollen selection. In Shivanna KR, Sawhney V K (Eds), Pollen biotechnology for crop production and improvement. Cambridge Univ Press, New York, NY, pp 333–351.

Sengupta S, Tandon R 2010. Assessment of ovule receptivity as a function of expected brood size in flowering plants. Intl J Plant Reprod Biol 2: 51–63.

Shivanna K R 2003. Pollen Biology and Biotechnology. Science Publishers, Enfield, USA (Special Indian Edition: Oxford–IBH Publishers, New Delhi).

Shivanna K R 2010. Pollen–pistil interaction: A complex mating game required for fertilization in flowering plants. In Nair P K K (Ed) Vistas of palynological science (Prof M S Swaminathan Festschrift Vol), J Palynology 46: 97–120.

Shivanna K R 2014. Biotic pollination: How plants achieve conflicting demands of attraction and restriction of potential pollinators. In Ramawat KG, Merillon JM, Shivanna K R (Eds) Reproductive biology of plants. CRC Press, Boca Raton, FL, pp 218–267.

Shivanna K R, Bahadur Bir 2014. Efficacy of biotechnological approaches to raise wide sexual hybrids. In Bahadur Bir, et al. (Eds) Plant biology and biotechnology, Springer India, New Delhi, pp 347-362.

Shivanna K R, Johri BM 1985. The Angiosperm pollen: Structure and function. Wiley Eastern, New Delhi. Reprinted 1989 by John Wiley, USA.

Shivanna K R, Rangaswamy N S 1992. Pollen biology: A laboratory manual. Springer–Verlag, Berlin, Heidelberg, New York. (Indian Edition: Narosa Publishers, New Delhi).

Shivanna K R, Tandon Rajesh 2014. Reproductive biology of flowering plants: A manual. Springer India, New Delhi.

Shivanna K R, Heslop–Harrison Y, Heslop–Harrison J 1978. Pollen–stigma interaction: bud pollination in the Cruciferae. Acta Bot Neerl 27: 107–119.

Skogsmyr I, Lankinen A 2002. Sexual selection: an evolutionary force in plants. Biol Rev 77: 537–562.

Stanley R G, Linskens H F 1974. Pollen biology, biochemistry and management. Springer–Verlag, Berlin, Heidelberg, New York.

Suzuki G 2009. Recent progress in plant reproduction research: The story of the male gametophyte through to successful fertilization. Plant Cell Physiol 50: 1857–1864.

Takayama S, Isogai A 2005. Self–incompatibility in plants. Ann Rev Plant Biol 56:467–489.

Takeuchi H, Higashiyama T 2011. Attraction of tip–growing pollen tubes by the female gametophyte. Curr Opin Plant Biol 14: 614–621.

TangW, Kelley D, Ezcurra I, et al. 2004. LeSTIG1, an extracellular binding partner for the pollen receptor kinases LePRK1 and LePRK2, promotes pollen tube growth in vitro. Plant J. 39: 343–353.

Uchiumi T, Komatsu S, Koshiba T, Okamoto T 2006. Isolation of gametes and central cells from *Oryza sativa* L. Sex Plant Reprod 19: 37–45.

Uchiumi T, Uemura I, Okamoto T 2007. Establishment of an in vitro fertilization system in rice (*Oryza sativa* L.). Planta 226: 581–589.

Van der Meer I M, Stam M E, van Tunen J, et al. 1992. Antisense inhibition of flavonoid biosynthesis in *Petunia* anthers results in male sterility. Plant Cell 4: 253–262.

Verhoeven T, Feron R, Wolters–Arts M, et al. 2005. STIG1 controls exudates secretion in the pistil of petunia and tobacco. Plant Physiol. 138: 153–160.

Von Besser K, Frank A C, Johnson M A, Preuss D 2006. *Arabidopsis* HAP2 (GCS1) is a sperm–specific gene required for pollen tube guidance and fertilization. Development 133: 4761–4769.

Wheeler M J, Vatovec S, Franklin–Tong VE 2010. The pollen *S*–determinant in *Papaver*: comparisons with known plant receptors and protein ligand partners. J Exptl Bot 61: 2015–2025.

Williams J H, Taylor M L, O'Meara BC 2014. Repeated evolution of tricellular (and bicellular) pollen. Am J Bot 101: 559–571.

Willmer P 2011. Pollination and floral ecology. Princeton Univ. Press, New Jersey.

Wolters–Arts M, Lush WM, Mariani C 1998. Lipids are required for directional pollen–tube growth. Nature 392: 818–821.

Wolters–Arts M, van Der Weerd L, van Aelst A C, et al. 2002. Water–conducting properties of lipids during pollen hydration. Plant Cell Environ 25: 513–519.

Wu H M, Wang H, Cheung A Y 1995. A pollen–tube growth–stimulatory glycoprotein is deglycosylated by pollen tubes and displays a glycosylation gradient in the flower. Cell 82: 395–403.

Zinkl G M, Zwiebel B I, Grier D G, Preuss D 1999. Pollen–stigma adhesion in *Arabidopsis*: a species–specific interaction mediated by lipophilic molecules in the pollen exine. Development 126: 5431–5440.

Zenkteler M 1990. In vitro fertilization and wide hybridization in higher plants. Cri Rev Plant Sci 9:267–279.

Zou Y, Aggarwal M, Zheng WG 2011. Receptor–like kinases as surface regulators for RAC/ROP mediated pollen tube growth and interaction with the pistil. AoB Plants 2011: plr017.

Chapter 11

Recent Advances on Biotic Seed Dispersal by Vertebrates

Palatty Allesh Sinu, VK Shibil, K Manoj, TP Rajesh*
sinupa@gmail.com

Abstract

Seed dispersal is a fundamental process that helps the seeds to naturally get distributed over space and time. The mode of seed dispersal is attributed to the fruit and seed traits. Some seeds are adapted for mechanical dispersal which is usually facilitated by wind, water and gravity. Some seeds are adapted for dispersal externally through attaching on the body of the animals and human beings. Yet, a large proportion of seeds are dispersed by frugivorous animals: birds, reptiles, bats, primates, small and large mammals, canids, rodents and fishes. This chapter brings out the recent advances in biotic mode of seed dispersal particularly by vertebrates. After giving the theoretical concept of seed dispersal, it explains the characteristics of the vertebrate-dispersed fruits, the characteristics of some common and rarely studied biotic seed dispersal agents and the evolutionary significance of poisonous fruits in the dispersal process. It highlights the concerns on the efficacy of biotic dispersal, particularly posed by the defaunation leading to the loss of medium and large frugivores and its probable consequences on the tropical forest recruitment structure and patterns.

11.1 Introduction

The plant distribution and community composition is fundamentally governed by the efficacy of seed dispersal (Condit et al. 2002, Wang and Smith 2002). Seed dispersal is the movement of viable seeds away from the parent plant either by means of abiotic (wind and water) or biotic (animals) agents to the sites favourable for the germination and survival of the seedlings. In pioneer species (early-successional species) the seed dispersal is primarily through wind or other mechanical means and in climax species (late-successional and emergent species) it is primarily facilitated by animals.

DOI: 10.1201/9781003505471-11

Globally, over 30% of the angiosperm families have fruits and seeds dispersed exclusively by animals; 22% have fruits and seeds dispersed by both biotic and abiotic agents (Tiffny 2004). In Gymnosperms, about 64% of the families have fruits and seeds dispersed by animals (Herrera 1989). About 50-90% of tropical tree and shrub species depend on frugivorous vertebrates to disperse their seeds (Howe and Smallwood 1982). In humid tropics, over 70% of the woody species are dispersed by vertebrate frugivores. It accounts to nearly 80% of woody species in African tropical rainforests, 88% in Australian rainforests and 70-94% in Neotropical rainforests (Jordano 2000). Frugivore vertebrate seed dispersal accounts to 78% for Indian tropical forests in north-eastern India (Datta and Rawat 2008) and 70% for northern Western Ghats (Tadwalkar et al. 2012).

Biotic seed dispersal by frugivorous animals has two aspects connected to each other –frugivory (consumption of fruits) and dispersal of viable seeds. Biotic seed dispersal in which the seeds/fruits are attached to the body of animals, do not have the former part, i.e., fruit consumption. The behaviour of dispersers is an important factor determining the seed shadow of the plants generated by the vertebrates, and this is rarely accounted in seed dispersal studies (but see, Russo et al. 2006, Wenny 2000, Westcott et al. 2005). This chapter deals both frugivory and seed dispersal, and highlights the foraging and associated behaviour of the key dispersal agents.

A seed under its parent tree is subjected to competition for natural resources and physical requirements for growth with the parent tree and siblings (Harper et al. 1970). The dispersal of seeds reduces the competition within the individuals of a species and increases the survival of individuals and populations. On the other hand, the concentration of seeds under the parent plant attracts the seed predators and pathogens, and is considered an important biotic factor limiting the recruitment of the plants under the parent tree (Janzen 1970, Peters 2003) (see Box 11.1).

Box 11.1

Janzen-Connell Hypothesis

This simple hypothesis explains why plants have evolved to invest their energy to produce fruits and seeds adapted for the biotic and abiotic long-distance seed dispersal. It predicts that *seed mortality is density- and distance-dependent with reference to the parental plant* (Janzen 1970; Connell 1971). It has direct implications to herbivory of seeds (seed predation) and seedlings, and the pattern of seed shadow rather than having a direct implication on the seed dispersal process. But, this gives sufficient explanation why seeds have to be dispersed. The first part of this hypothesis predicts that the seed and seedling mortality increases with the density of the seeds. Had the seeds not been dispersed, the seed density will be high under the parent tree, and according to this hypothesis, the high density of seeds attract more natural seed mortality agents (seed predators and fungi). The second part of this hypothesis, which is non-exclusive to the former part, predicts that the seed mortality decreases with the distance from the parent tree.

Therefore, through the seed dispersal process, the seeds get the advantages of: a) escape from high seed and seedling mortality under the parent tree (escape hypothesis), b) colonize an unpredictable, transient, or newly created microsite, such as new canopy gaps due to tree fall (colonization hypothesis), and c) directed dispersal to certain favourable microclimate conditions for germination, such as perching sites and roosting sites (directed dispersal hypothesis) (Wenny 2001).

11.2 Quality of Seed Dispersal

Dispersal quality predicts the area of plant recruitment, spatial pattern of plant recruitment or seed-shadow, coexistence of species and community formation, gene flow and genetic heterogeneity within and between the populations. The seed shadow is the combined effect of all primary and secondary dispersal agents.

The dispersal quality is generally assessed by two factors: the distance the seeds are moved and the quality of the habitat receiving the dispersed seeds (Cain et al. 2000). A distance over 100 m is generally regarded as long-distance dispersal (Cain et al. 2000, Nathan 2006). To this effect, the behaviour of the frugivorous animals, particularly the foraging and post-foraging behaviour is a prognostic factor determining the quality of the biotic seed dispersal (Sinu et al. 2012, Wenny 2000). The foraging behaviour of animals includes fruit handling, such as pulp-removal, gut retention of seeds, defecation site (Traveset et al. 2008), and the post-foraging flight pattern, the distance and the habitat the animal moved to after the frugivory (Wenny 2001). The seed in the gut of the animal is always heavy for the frugivores and therefore they defecate the seeds as early as possible. For the case of frugivorous birds, both the seed and bird size determines the gut retention time and the dispersal distance (Sun et al. 1997). Since most frugivorous birds move to perching sites or roosting sites or breeding sites after consuming the fruits, the seeds are likely to get dispersed to those sites (Sinu et al. 2012, Wenny and Levey 2000). Hence, the seed shadows generated by most frugivores are rather unpredictable, but often quite close to the fruiting tree. However, a frugivore's suitability as the seed dispersal agent can be predicted only through our knowledge on the behavioural and ecological traits of the frugivores.

The quality of the dispersed site is determined on the basis of the ability of seeds to germinate and the extent of post-dispersal seed predation (Anderson et al. 2009, Gómez-Aparicio et al. 2005, Wenny 2000). The characteristics of the microsites receiving the dispersed seeds that predict the quality of the seed dispersal include prevailing canopy gaps, light conditions, moisture content of the soil and chances of subsequent seed predation by animals and microbes (post-dispersal seed predation). Therefore, for gaining an optimum benefit out of this interaction, a frugivorous animal should disperse seeds to most favourable conditions required for its germination and growth. For the wind-dispersed seeds, the movement of the diaspores is determined by the morphology of the diaspore, the behaviour of the wind (wind speed and direction) and the prevailing weather conditions (Nathan and Muller-Landau 2000). Since the activity of the wind is predominant in the emergent layer (top-most layer of a forest), the most wind-dispersed plants are emergents. For instance, most dipterocarpaceaen trees are emergent and wind dispersed (Osada et al. 2001). However, the wind dispersal is not considered as a quality type of dispersal as the seeds are likely to fall in unfavourable microsites for germination (Nathan et al. 2002). Solomon Raju (this vol.) elaborates on the abiotic seed dispersal processes in the flowering plants.

The seeds at the primary dispersal sites might sometimes undergo a secondary seed dispersal through some entirely unrelated biotic dispersal agents, such as ants, crabs, rodents and dung beetles (Clifford and Monteith 1989, Estrada et al. 1993, Levey and Byrne 1993, Nogales et al. 1996, Roberts and Heithaus 1986, Roberts and Milleron 1991, Wenny 1999) or by an abiotic process, such as the wind or water (Vander Wall and Joyner 1998). Most of the secondary biotic dispersal agents are either invertebrates, such as dung beetles and ants or scatter-hoarding rodents. The roller and the tunneller classes of dung beetles remove the dung ball and fix the dung in the soil, the former at a distance from the dung pat and the latter under the dung pat. In both the cases, there is a great chance of seeds getting trapped in the dung ball. Ants, on the other hand, selectively remove the seeds from the dung pat. However, ants give preference to small seeds or those with oil appendages, called elaiosomes. This further increases the dispersal distance and escapes the seeds from post-dispersal agents to certain extent.

11.3 Frugivory and Seed Dispersal

11.3.1 Characteristics of Vertebrate-dispersed Fruits

According to Fleming (1991) the families of woody plants produce fleshy fruits and those of herbaceous plants produce capsular or other types of dry fruits. This indicates that vertebrates are important seed dispersal agents of the woody plant families, and the dispersal syndrome in herbaceous plants is by other agents. The frugivore-dispersed plants decrease with the increasing latitude. The woody plant families are richer in tropical regions around the world and therefore, their vertebrate associates are also diverse in tropics. About 80% of all the bird and mammal species that live in the tropical regions are frugivorous (Terborgh 1986). Among the frugivorous birds and mammals, greater endemism and greater richness at the family and species levels is observed in the neotropics (Fleming and Kress 2011).

Recent phylogenetic evidences suggest that the frugivorous diet in the birds and mammals have evolved independently on several occasions in the course of evolution (Fleming and Kress 2011). Majority of the vertebrate-dispersed fruits are brightly coloured and succulent with edible pulp. The fleshy fruits that are consumed by vertebrates belong to either one or more of the following clades of fruits: a) seeds with fleshy seed coats (e.g., Arecaceae, Euphorbiaceae, Meliaceae etc.); b) seeds with aril (arilloid fruits; e.g., Annonaceae, Fabaceae, Myristicaceae, Sapindaceae etc.); c) fruits with a fleshy endocarp protected by a hard pericarp (pulpy fruits; e.g., Clusiaceae, Fabaceae, Rutaceae etc.); d) fruits with fleshy pericarp (drupes and berries; e.g., Rubiaceae, Malvaceae, Moraceae, Sapotaceae etc. (Van der Pijl 1982).

The nutritional characteristics of the fruits vary between the species and the geographical locations. Corlett (1996) based on his study in 153 species of fruiting trees in Hong Kong reports that the vertebrate-dispersed fruits have an average (median) of 69.2% of pulp (range: 10-99.2%), 78% water (11.1-94%), 2% lipid (0-84%), 53% sugar (4-88%), ~1% nitrogen and 14.3% fibre (1-44%). Kitamura et al. (2002) report an average of 12% sugar (2-26%) in 123 vertebrate-dispersed fruits in the Southeastern Asia. The Median fruit and the seed weight of vertebrate-dispersed plants (N=259 species) are 1 g (range: 0.02-275 g) and 0.1 g (range: 0.01-21 g), respectively. The median fruit and the seed diameter of the vertebrate dispersed plants are 10 mm (2-84 mm) and 6 mm (0.1-28 mm), respectively. The mammal-dispersed seeds are generally lipid-poor (Corlett 1996, Debussche and Isenmann 1989).

Since the fruit pulp attracts fungal growth, the seeds in fleshy fruits are rapidly damaged by their attack (Heer et al. 2010). The fruit pulp of some plant species also has secondary metabolites that limit the germination of the seeds (Robertson et al. 2006). Sometimes, the high solute concentration of the fruit pulp might damage the seeds through osmotic pressure (Samuels and Levey 2005). Seed dispersal by animals often help the seeds to overcome those inhibitory factors and facilitate germination as the gut retention depulps the fruits and deposits the seeds often in the dung that acts as an organic fertilizer (Robertson et al. 2006, Traveset 1998).

11.3.2 Quality of Frugivorous Seed Dispersal Agents

The animals having fruit as a major component in their regular or seasonal diet are broadly classified as frugivores. However, to qualify for the dispersal agent, frugivores must move potentially-viable seeds away from the parent plant either through defecating, regurgitating, spitting or dropping (Corlett 1998). The following steps are involved in the process of a successful internal seed dispersal process: a) seed uptake (seeds taken into the mouth), b) seed ingestion, c) seed retention time in the stomach, d) seed survival (the probability of seeds surviving after gut passage), e) seed germination (the probability of seeds germinating after gut passage), f) germination rate after gut passage, g) time required for seed germination (germination facilitation, germination inhibition or no effect on germination). Although the seed dispersal effectiveness should have been assessed as "the number of new adults produced by the dispersal activities of a disperser" (Schupp 1993), it is very difficult to investigate this aspect for many plant species.

Over 5000 species of the vertebrates have been predominant frugivores and likely to act as dispersal agents of seeds worldwide. Birds (hornbills, bulbuls, parrots, pigeons, thrushes, starlings, mynas, barbets and the neotropical trogons (the paleotropical species including Malabar trogon are strictly insectivores), mammals (bats, bears, civets, martens, foxes, raccoons, coyotes, monkeys, porcupine, scatter-hoarding rodents, ungulate animals and elephants) and fishes are the predominant frugivores in tropical region. Fishes, reptiles and amphibians although contribute in quality seed dispersal in the tropics (Corlett 1998, Anderson et al. 2011; Hansen et al. 2008), it has not received much attention in paleotropics. Insects also contribute to the primary dispersal of seeds, not through the process of endozoochory (see Solomon Raju in this volume for the terminologies used to explain different dispersal types), but externally through attaching the sticky seeds to the body (epizoochory). Epizoochory is quite often a passive mechanism of seed dispersal. Manohara (this volume) demonstrates an unusual type of seed dispersal by a wasp species which intentionally move the seeds from fruiting trees. The following part covers the major types of vertebrate-mediated seed dispersal mechanisms.

11.4 Major Types of Biotic Seed Dispersals

11.4.1 Ichthyochory

Ichthyochory refers to the seed dispersal by fishes and is considered a primitive type of seed dispersal. Early fishes might have been the first vertebrate frugivores and dispersal agents of plants (ca. 300 million years ago; Correa et al. 2015, Tiffny 2004). This hypothesis gets stronger with the recent evidences that a) early angiosperms had lived in freshwater systems and wetlands (Eriksson et al. 2000) which were producing fleshy fruits with small seeds (Friis et al. 2010) and b) small vertebrates had an evolutionary radiation only 80 mya after the first angiosperms evolved (Eriksson 2008). It is interesting that some of the present day fish dispersal agents are in existence from 15 mya (Anderson et al. 2011, Correa et al. 2015, Lundberg et al. 1986).

Generally, the swamp and riparian plants and those living in seasonally flooded forests are the direct beneficiaries of the primary fish seed dispersal (Anderson et al. 2011, Corlett 1998, Horn 1997). Plant species growing in several riparian and swampy habitats, such as *Aglaia*, *Artabotrys*, *Dysoxylum*, *Gonystylus*, *Hydnocarpus*, *Pandanus*, *Quassia*, *Sandoricum* and *Syzygium* are likely to be dispersed by the fishes. Euryphagous fishes (gulps anything that fall on water) are the major frugivores and dispersal agents of riverine plants. Cyprinid fishes (eg. *Puntius* spp, *Leptobarbus* spp, *Tor* spp) and Clariid fishes (catfishes belong to *Clarias* spp and *Pangasius* spp) are the major frugivorous species in paleotropics (Corlett 1998); Characid fishes (eg. *Colossoma* spp, *Brycon* spp) are the major frugivorous species in neotropics (Anderson et al. 2011). A recent study (Horn et al 2011) suggests that 275 species of the fishes globally consume fruits and may contribute to the dispersal of the seeds of numerous plant species. The Indomalayan (55 species, 25 genera and 9 families of plants) and Australasian (6 species, 5 genera and 4 families of plants) regions have documented less cases of Ichthyochory when compared to those from Neotropical region (149-151 species, 75 genera and 17 families of plants) (Horn et al 2011). Correa et al (2015) gave a fine-scale analysis of the fish-fruit interaction in the neotropics and found that 69 Neotropical fish species have 566 species of fruits and seeds from 82 plant families in their diet.

The fish-dispersed fruits are buoyant and pulpy; the seeds vary in size from small to large; the defecated viable seeds are deposited in river/ flood plain beds, which are not buoyant. There are evidences that the fishes in flood plains move seeds more towards favourable microsites, such as other patches of flood plain forests rather than to permanent deep waters of rivers, which are unfavorable microsites for seed germination (Anderson et al. 2009, 2011). Sometimes the seed dispersal efficiency of fishes (in terms of the distance and quality of the habitats) is comparable to that predicted for other large-bodied terrestrial dispersal agents, such as hornbills, elephants and deers (Anderson et al. 2011). The size of the fish-dispersed seeds varies between 2 and 4 cm in diameter. The gut retention time of seeds, an important attribute of long-distance dispersal agent, ranges between 34 h and 147 h in fishes, depending on the seed size (Anderson et al. 2011). Despite growing evidences, only a few studies have experimentally assessed the seed dispersal quality of the fishes (Pollux 2011).

11.4.2 Saurochory

Saurochory refers to seed dispersal by reptiles and is another primitive type of seed dispersal. Presently the reptiles are considered to be lesser important vectors of seed dispersal. This is mainly because of the lesser abundance of herbivorous reptiles in the present day ecosystems (Cobo and Andreu 1988). However, in the historic ecosystems of Mesozoic period reptiles and amphibians were likely to be more important dispersal agents of seeds, because that was an era of large herbivorous reptiles, such as archosaurs (Marron and Moore 2013). In the modern world Saurochory – seed dispersal by the reptiles through ingestion of fruits (Moll and Jansen 1995; Valido and Olesen 2007) – is predominant in ecosystems where large number of herbivorous reptiles live, such as the islands of Galapagos and Aldabaran, where the giant tortoises are free-rangers (Marlow 1989). The predominant Saurochorous animals are giant tortoises (Hnatiuk 1978) and turtles (Moll and Jansen 1995, Ridley 1930) that are attracted by strong smell, color and pulpiness of the fruits (Moll and Jansen 1995); they are predominantly gulpers than chewers (Strong and Fragoso 2006). Therefore both turtles and tortoises are good dispersal agents as they discharge good number of intact seeds (Hansen et al. 2008, Hansen et al. 2010, Hnatiuk 1978, Jerozolimski et al 2009, Strong and Fragoso 2006, Traveset 1998). Hansen et al (2010) through an experimental study showed that the gut passage time of *Syzygium mammillatum* seeds by the modern Aldabran tortoise is 2-3 weeks. The long gut-passage time along with the mobility pattern of the tortoises shows that they contribute to long distance seed dispersal (Hamilton and Coe 1982, Hansen et al. 2008, Rick and Bowman 1961). It is believed that the early reptiles were important selective forces in the evolution of basicaulicarpy (fruits develop on stems and stalks near the ground for easy access to ground-dwelling potential seed dispersers) (Tiffny 1986, van der Pijl 1982). Although we do not have much direct reports on the seed dispersal by reptiles in India, indirect observations (diet analysis of tortoises and turtles by the herpetologists) indicate that they are likely to contribute to the dispersal of several large-seeded fruits such as *Dillenia pentagyna*, *Cordia obliqua* and the jackfruit, *Artocarpus heterophyllus* (see Vijaya 1982).

Crocodiles, among saurochorous animals, have received less attention from the seed dispersal ecologists. But, a recent review (Platt et al. 2013) that studied the dietary composition of crocodiles suggests that fruit is a part of the diet of 72% of crocodilian species. Fruits and seeds of 46 genera and 34 families of plants have been identified in the gut analyses of crocodiles (Platt et al. 2013). The crocodile fruits compose of many types (nut, drupe, berry, fleshy/pulpy, legume, nutlets, capsule, cone, grain) and many colours (although brown predominates). There is no clear evidence for the fruit traits used by the crocodiles to consume fruits. However, it is believed that fruit falling on water surface elicits the feeding response of the crocodiles as euryphagous fishes do. The dispersal is either through regurgitation or defecation.

11.4.3 Chiropterochory and Ornithochory

Chiropterochory and ornithochory refer to seed dispersal by bats and birds, respectively. They are the examples of long distance seed dispersal. Bats (Chiroptera) are the major frugivores of the tropical regions. Bats' adaptability to the disturbed and degraded habitats makes them important seed dispersal agents in fragmented forests (Griscom et al. 2007) and other managed habitats including the restoration sites (Gorresen and Willig 2004, Silveria et al. 2011). The Bat seed dispersal is particularly important for small-seeded pioneer tree species (Lobova and Mori 2004, Medellin and Gaona 1999, Muscarella and Fleming 2007, Silveria et al. 2011) although some adult climax forest trees are also benefited (Melo et al. 2009). However, their feeding piecemeal habit of large fruits on fruiting tree itself (bites of the flesh/pulp of the fruit and leaving the seed attached to the plant) might reduce the quality of dispersal of large seeds (but see Melo et al. 2009). For example, in mango, the large bats may take the entire small mango fruit to feeding roosts, but most others process the large fruit through piecemeal on tree itself, leaving the seeds attached to the tree. Bat fruits are predominantly green or dull in colour, medium to large, smell strong, held away from the foliage and composed of single large seed (dispersed through epizoochory) or hundreds of small seeds (dispersed through endozoochory) (Lobova et al. 2003, Lobova and Mori 2004, Marshall 1983). Fruits of Anacardiaceae, Guttiferae, Myrtaceae, Moraceae, Palmae and Sapotaceae are frequently fed and dispersed by the frugivorous bats. *Ficus* (Moraceae) fruits are the commonest wild fruit sources of frugivorous bats in the tropics (Bleher et al. 2003, Herre 1996, Janzen 1979). Medium and large fruits are also taken to the feeding roosts within some meters (20-100 m) from the fruiting tree and dispersed not far from the parent tree. Only some small seeds are likely to be transported to the roosting sites or carried for long-distances (Corlett 2011).

Bats also have the habit of defecating seeds during flight increasing their long-distance seed-dispersal ability (Phua and Corlett 1989). The bat-ingestion influences the germination parameters of the seeds, particularly the rate and the time required for germination. With reference to *Ficus* seeds, some studies report that the bat ingestion of figs accelerated the germination rate and reduced the time required for germination (Heer et al. 2010, Lopez and Vaughan 2004). But, some studies report that the germination parameters are not significantly affected by the bat-ingestion when compared to that of the other mammals (Izhaki et al. 1995, Teixeira et al. 2009, Traveset 1998). Relatively, short gut-passage time in bats over other mammals has been explained as a reason for this difference (Laska 1990). The gut retention time of seeds in birds is also short (Murray 1988, Spiegel and Nathan 2007).

Bats and birds, the two predominant flying long-distance seed dispersal agents of trees including the canopy trees, differ on the fruit trait preferences and the quality of seed dispersal. This difference is partly explained by the adaptations of these animals for frugivory. Birds have flatter bills, wider gapes and smaller gizzards. This permits the birds to gulp the entire fruit, even large fruits in case of larger birds with wider gape (e.g., Hornbill, wood pigeon). The bats on the other hand, have shorter and wider palates. Bird-dispersed fruits are often small, purple-coloured, with thin skin and succulent. Mammal-dispersed fruits (including bat-fruits) are often large, dull-coloured, with tough rind or husk. The bird dispersed fruits remain attached to the tree at maturity, while the mammal-dispersed fruits fall at maturity (Herrera 1989, van der Pijl 1982, Vander Wall and Beck 2012). Mammal dispersed fruits use odour as a major advertisement tool; therefore the bats have good sense of odour and olfaction which permits them to distinguish toxic fruits from edible fruits. The bird dispersed fruits produce no odour as the birds are poor at sensing the odour. For the same reason some bird dispersed fruits are toxic that are either not detected by the birds or if consumed the physiological adaptations of birds do not permit the uptake of toxic elements into the digestive tract and sometimes the digestive enzymes detoxify the toxic elements of the fruits (Levey et al. 2006, Struempf et al. 1999, Witmer and Van Soest 1998).

The fruit sources of the birds and the bats do not show much overlap (Fleming and Kress 2011, Galindo-González et al. 2000); but, in some common fruit species such as *Ficus* and some pioneer plant community, both the bats and the birds consume fruits concomitantly (Coates-Eastrada and Estrada 1986). Jacomassa and Pizo (2010) in their studies on the effectiveness of birds and bats in dispersing the seeds of a pioneer species found that the bats are efficient in transporting considerably more number of seeds when compared to the birds. However, the germination rate and the time taken for germination were noticeably shorter for bird-dispersed seeds. So, the bats contribute to quantitative aspect of the seed dispersal and the birds contribute to the qualitative aspect of the seed dispersal (Jacomassa and Pizo 2010). Since the bats do defecate seeds during flight or in the feeding roosts which are usually isolated trees inside or outside the forest, there is a greater probability that the seeds (particularly small seeds) are getting deposited in forest clearings (crucial for pioneer species) (Galindo-González et al. 2000, Muscarella and Fleming 2007, Phua and Corlett 2011) when compared to the birds, in which defecation of seeds is always while perched under closed canopy cover, and mostly inside the forest itself. Therefore, the bats also play a key role in quality dispersal of seeds. Above all, most frugivorous, specialist birds are large canopy birds, such as hornbills and wood pigeons, which are often affected by the disturbance.

Unlike fishes and reptiles, which are non-obligate frugivores, some bird and mammal families have independently evolved as obligate frugivores; the mutualistic interaction between the frugivorous birds and mammals and the flowering plants dates back to only 90 mya (Fleming and Kress 2011). Some examples for the oldest interaction between the fleshy-fruited plants and the frugivorous birds and mammals are Myrtaceae-Casuariidae (Cassowary group of birds; 68 mya), Elaeocarpaceae-Casuariidae (66 mya), Anacardiaceae-Pteropodidae (flying foxes (bats); 56 mya), Loranthaceae-Meliphagidae (honeyeater group of birds; 53 mya), Burseraceae-Cotingidae (cotinga group of birds; 51 mya), Meliaceae-Bucerotidae (hornbill group of birds; 40 mya) and Cactaceae-Phyllostomidae (leaf-nosed bats; 30 mya). However, the major radiations (but not coevolution) in the fleshy-fruited plants and the frugivorous birds and mammals that we see today have been taken place in the last 30 mya (Fleming and Kress 2011).

Fig. 11.1. A palm civet fecal matter having the seeds of *Mimusops elengi*.

11.4.4 Seed Dispersal by Carnivores: Role of Civets

Many carnivores eat fruits at least a part of the year and defecate relatively good quality seeds in the tropical, the mediterranean and the temperate forest communities (Corlett 1998, Herrera 1989, Koike el al. 2008, Traveset 1998). Among all the small terrestrial vertebrate carnivores, civets including the largest Asian civet, binturong (Viverridae: Carnivora) are the most frequently-encountered frugivores in paleotropics (Patou et al. 2010). Among them the palm civets belong to the genera *Paguma* and *Paradoxurus* are the widespread frugivores (Cranbrook 1991, Joshi et al. 1995). Unlike other canids, fruit is a major dietary component of civets (84% of 193 scats examined had fruits in them; Joshi et al. 1995) although small vertebrates and invertebrates constitute a minor part of their diet (Wang and Fuller 2003). A well-demarcated shift in the diet from frugivory to carnivory from the month of April through June is reported in palm civets in Nepal (Joshi et al. 1995). Palm civets are strictly nocturnal and active during 18:00-04:00 h. Civets have large home ranges (64-451 ha) (Zhou et al. 2014) which overlap each other (Joshi et al. 1995). They move longer distances daily (Zhou et al. 2014) qualifying themselves as long-distance seed dispersal agents of fleshy fruits (Nakashima and Sukor 2010). Civets are adapted to live in highly disturbed habitats and commonly seen in rural and urban climates. Civets use forests in the fringes, close to forest gravel roads than interior part of the forests (Nakabayashi et al. 2014). Thus civets are equivalent terrestrial counterparts of bats, another important long-distance quality flying dispersal agent of disturbed habitats. Although civets live solitarily, occasionally they share resources on fruiting trees (Nakabayashi et al. 2012). The physical appearance of fecal matter also indirectly helps one to distinguish the civet species in the field. For example, the feces of Palm Civets are strung lengthwise and gelatinous when fresh (Figure 11.1). The feces of Large Indian Civet are strung coiled in heaps or piles (Joshi et al. 1995).

The wide gapes in the civets permit them to swallow entire fruits. Civets disperse seeds primarily through defecation and direct seed dispersal nonrandomly to open sites, such as abandoned trails, tree fall gaps, banks of small rivers, dried-up rainflow paths (Nakashima et al. 2010, Su and Sale 2007), on fallen trees, primary branches of large trees and narrow bridges (Joshi et al. 1995). Direct observations of feces and mitochondrial haplotype studies give enough evidences that civets repeatedly use the same spot for defecation (Corlett 1998, Nakashima et al. 2010). This has direct implications for the seed shadow generated by the civets. Since the seeds are moved to relatively dry open places by the canids, including the civets, the survival of seeds and seedlings in the dispersed habitat is dependent to a greater extent on the traits of the seeds. Zhou et al. (2008) and Nakashima et al. (2010) suggest that the seeds in certain nonrandom unique dispersed sites are susceptible to greater mortality at the seed and seedling levels due to the physical climatic conditions. Civet passes 90 to 99% of intact seeds between 9 hours and three days after fruit consumption, and the gut passage greatly improves the time and rate of seed germination (Colon and Campos-Arceiz 2013). In India Brown-Palm Civets consume fruits of 57 wild species (Mudappa 2010) and consume several domesticated fruits, such as Papaya (Caricaceae) and Sapota (Sapotaceae) in the rural and semi-urban areas (Jothish 2011). The fruits consumed by the carnivores overlap with that of the birds and the bats. In the Mediterranean region the 'carnivore fruits' are large, multi-seeded, brown, scented, often falling to the ground after ripening, pulp and fibre-rich, but protein and mineral-poor (Herrera 1989).

Apart from civets, other carnivores that routinely consume bulk quantities of fruits in oriental and adjoining regions are Raccoon dogs (Canidae; Palaearctic region), Asiatic jackal (Canidae; Indian subcontinent), red foxes (Canidae; Indian subcontinent and Mediterranean region), the Sloth Bear and Asiatic Black Bear (all Ursidae: Carnivora) (Satyakumar and Viswanath 2003, Sreekumar and Balakrishnan 2002), Marten (Mustelidae: Carnivora; widespread), Badger (Mustelidae: Carnivora; widespread) and the Indian Grey Mongose (Herpestidae: Carnivora; Indian subcontinent) (Corlett 1998, Herrera 1989).

Fig. 11.2. a) a fruit of *Chilocarpus* sp (a liana species of the Western Ghats); b) fruits of *Gnetum ula* (a liana species belong to Gymnosperm) in the Western Ghats; the arils of these fruits are important food sources of macaques in the Western Ghats.

11.4.5 Seed Dispersal by Primates

Macaques may be the largest vertebrate species seen in small forest fragments including those in the urban and the rural areas. Therefore, they are important frugivores of all types and sizes of forests. The primate-fruits are mostly large often with arillate-seeds, orange or brown coloured and with a protective rind (Leighton 1993, Lucas and Corlett 1991, Mabberley et al. 1995) (Figure 11.2). The macaques generally spit out or drop the large and very large seeds under the fruiting tree itself, unless they carry the fruits in cheek pouches (Corlett and Lucas 1990). Only the small seeds are swallowed and defecated, to bring out effective dispersal. Unlike macaques, monkeys (langurs) prefer foliages rather than fruits. Gibbons also consume foliages more and selectively consume some preferred fruits, such as figs. The Gibbon-fruits are mostly small and unprotected (rind free).

Fig. 11.3. Seed-predated shells of *Terminalia bellirica*; teeth marks are important identification marks of seed predators; a) seeds preyed upon by giant squirrel; b) seeds preyed upon by dormouse; c) seeds preyed upon by a rat species in caches; d) seeds preyed upon by porcupine

11.4.6 Scatter-hoarding Animals and Seed Dispersal

Rodents are generally considered seed predators as most of them depulp the fruits and consume only the seed kernels, eventually kill the endosperm part of the seed. They either consume the seed kernels directly (Sinu 2012) or cache (store) them in nests (burrows in the ground or in tree holes) for later consumption during the lean-fruiting period or overwintering period (Figure 11.3). The seeds stored in the soil burrows alone are likely to germinate and establish. Those scatter-hoarders that cache seeds above ground in tree-holes and nests in leaf litter are poor seed dispersal agents, because the associated microhabitats are not suitable for germination and survival of the seedlings. The seeds stored deep inside the underground burrows by the larder-hoarding animals are also less likely to germinate, unless the seeds are capable of deep site germination. Only very few frugivores are adapted with this caching behaviour and play a significant role in the seed dispersal (Vander Wall and Beck 2012). Tree squirrels including giant squirrels, chipmunks, spiny dormouse, and some birds, such as jays, tits, nuthatches and nutcrackers are the major scatter-hoarding animals. The primary adaptation of scatter-hoarding animal is the presence of cheek pouch which is used to transport the seeds to nests or holes. The second major adaptation is their exceptional spatial memory to locate the cache sites when required to access them. Quite often the seeds of the scatter-hoarding animals are hard-shelled. So the third adaptation of the scatter-hoarding animals is the presence of strong bills (jays, nuthatches, nutcrackers) or strong incisor teeth (rodents) (Vanderwall and Beck 2012) to break hard shells.

The frugivorous animals stop consuming fruits once they are satiated with a fruit. The scatter-hoarders do not eat the fruits or seeds while picking, but transport them to cache sites, therefore, their 'hunger' for seeds will not be stopped quickly. The scatter-hoarders cache more seeds than they could eat during the lean fruiting period or overwintering period. Sometimes the animal may die due to a number of causes before the stored seeds are completely consumed. Such seeds are likely to germinate and survive. Since the scatter-hoarders store the seeds inside the soil (an optimum micro-site that favours germination), scatter-hoarding is often considered to be a directed seed dispersal mechanism (Briggs et al. 2009). The second advantage of caching is that the seeds when buried under the soil are not easily detected by the seed predators (Forget and Milleron 1991, Vander Wall et al. 2006).

The seed dispersal through scatter-hoarding animals is critical for arid and alpine species, because, the seeds buried under the soil maintain the seed moisture content and escape from the extreme climatic conditions (Chambers and MacMohan 1994). According to Vander Wall and Beck (2012) the spatial pattern of seeds dispersed by the scatter-hoarders is at par with that dispersed by the birds and bats, and certainly superior to that dispersed by the mammals. Birds and bats disperse seeds either through defecation or regurgitation because their gut can retain only a limited number of seeds, and scatter the seeds individually. The scatter-hoarders also cache seeds either singly (for larger nuts) or in small clutches. The large mammalian frugivores, however, deposit hundreds to thousands of seeds in a single fecal matter (e.g. dung). Because of the large number of seeds in dung, they are likely to compete amongst themselves for resources or attract the seed predators (LoGiudice and Ostfeld 2002) which eventually kill a significant number of seeds and seedlings. The probability of seeds survival in scatter-hoarding places is also determined by the seed traits; dormant seeds may survive better than the non-dormant seeds. There is also the possibility that the seeds brought for caching contain insect larvae (seed predator) which cause seed mortality. This is a loss for both the animal and the seed (Steele et al.

1996). Some rodents have exceptional ability to distinguish between the insect-infested and non-infested fruits (Dixon et al. 1997, Sinu 2012, Xiao et al 2003). The scatter-hoarders incapable of detecting the infection at the spot remove the pulp of the fruits and cache only the intact seed (Yi et al. 2012; Yi et al. 2014).

Fig. 11.4. Fruits of *Holegarna arnotteana*, an example of toxic fruit from the Western Ghats, India

11.5 Vertebrate Frugivores and Toxic Fruits

The fruit pulp and seed tissues of most of the fruits are not only the 'rewards' to the dispersal agents, but also contain secondary metabolites (condensed and hydrolysable tannins and saponins, alkaloids and phenolics, cynogenic glycosides etc., Mali and Borges 2003) that may not be palatable for vertebrate frugivores. The major source of secondary compound in a pulpy fruit is the seed. However, flavour compounds (e.g. favonones), scent compounds (e.g., aglycones, sesquiterpene, laxatives and resins (e.g., sorbitol and glucosides)), pulp pigments (e.g. anthocyanins, Carotinoids) are also present as the secondary compounds in the fruits. Almost all fruits have some level of secondary metabolites that function as deterrents to non-mutualistic frugivores (antagonists), such as insects, pathogens and seed-predatory vertebrates (Figure 11.4). The animals that swallow entire fruits are likely to get exposed to the secondary metabolites in the gut system. However, there is less chance that the legitimate vertebrate frugivores are exposed to the toxins, because, the fruit pulp and the seed tissues have different profile of the secondary metabolites in ripe and unripe fruits (Cipollini and Stiles 1993). Most of the vertebrate frugivores consume ripe fruits. However, if there is high amount of toxins present in the seeds of ripe fruits, it is hypothesized that it is used to deter the vertebrate granivores (seed predators) and the insects and microbes (directed toxicity and defense trade-off hypotheses) (Cipollini and Levey 1997). The legitimate vertebrate frugivores, especially the birds have physiological adaptations which prevent the intestinal uptake of toxins. As mentioned earlier, frugivorous birds have no sense of odour, so they are likely to consume many toxic fruits as well. The second adaptation to escape from the toxicity of the secondary metabolites of the ingested seeds is the quick gut passage of the seeds. Mixing fruits and foliages of different plant species in the diet of some vertebrate frugivores is also considered as a measure to combat the toxins from their primary fruits (Struempf et al. 1999, Witmer 1998, Witmer 2001, Witmer and Van Soest 1998).

11.6 Implications of Defaunation on Seed Dispersal

Defaunation is defined as the loss or decline of populations of medium or large native vertebrates from a given ecosystem (Galetti and Dirzo 2013). Natural processes, such as ice age, hurricanes and floods, were the major reasons for the heavy loss of megafauna in the past (Corlett 2013). However, in the anthropocene period, the large vertebrates are being lost directly to poaching, hunting, illegal trading of animals and animal parts and commercial bush meat trade (Corlett 2007, Vanthomme et al. 2010). The hunters generally target the medium or large sized animals for bush meat trade (Poulsen et al. 2000). The bush meat consumption in the tropics is estimated to reach about two million tons per year (Nasi et al. 2008). The population density of the large vertebrates has also declined drastically due to the ecosystem processes, such as land-use change, fragmentation, introduction of invasive species and wildlife diseases (Galetti and Dirzo 2013). Most of the vertebrates of this order function as keystone species in all forms of ecological networks due to their large body size (Beune et al. 2013). In fruit-frugivore interaction network, the large fruits are consumed and effectively dispersed by the large vertebrates whose removal therefore can not be replaced by the small frugivores. The large frugivore species that act as the seed dispersers are the large birds (wood pigeon, hornbills, and toucans), macaques, civets, apes, gibbons, and the large fruit bats. Their number in a given community is relatively small when compared to that of the small mammals and the small passerines (perching) birds. Defaunation of the large vertebrate frugivores affect the specialized type of seed-dispersal or long-distance seed dispersal, and it is not compensated by the small vertebrates. Possible consequence of the loss of the large frugivores is that the small-seeded plants occupy all available spaces and gaps facilitated by the small vertebrate dispersal agents including the birds and the bats. Selective logging of emergent and canopy trees also lead to population decline of the large birds as they use such plants as perches during their flight or as nest sites. Sometimes, the emergent trees also offer food for the large birds and monkeys. So the direct loss of the large trees and the large frugivores expedite the decline of the population density of both the taxa (Vidal et al. 2013). Galetti and Dirzo (2013) edited and compiled thirteen excellent research articles on "defaunation and seed dispersal" that were presented in a symposium in the 2011 conference of the Association for Tropical Biology and Conservation in the

Journal, Biological Conservation. The readers are urged to go through those papers to have more information on the effect of defaunation on the seed dispersal process.

Acknowledgements

VKS would like to thank UGC for the RGNF award; PAS, KM and TPR thank SERB-DST for a research grant.

References

Anderson JT, Saldana Rojas J, Flecker AS 2009. High–quality seed dispersal by fruit–eating fishes in Amazonian floodplain habitats. Oecologia 161: 279–290.

Anderson JT, Nuttle T, Saldana Rojas J, Pendergast TH, Flecker AS 2011. Extremely long–distance seed dispersal by an overfished Amazonian frugivore. Proc R Soc B (2011) 278: 3329–3335.

Beaune D, Bretagnoll F, Bollache L, Hohmann G, Surbeck M, Fruth B 2013. Seed dispersal strategies and the threat of defaunation in a Congo forest. Biodivers Conserv 22: 225–238.

Bleher B, Potgieter CJ, Johnson DN, Böhning–Gaese K 2003. The importance of figs for frugivores in a South African coastal forest. J Trop Ecol 19: 375–386.

Briggs JS, Vander Wall SB, Jenkins SH 2009. Forest rodents provide directed dispersal of Jeffrey pine seeds. Ecology 90: 675–687

Cain M, Milligan B, Strand A 2000. Long–distance seed dispersal in plant populations. Am J Bot 87: 1217– 1227.

Chambers JC, MacMahon JA 1994. A day in the life of a seed: movements and fates of seeds and their implications for natural and managed systems. Annu Rev Ecol Evol S 25: 263–292.

Cipollini ML, Levey DJ 1997. Secondary metabolites of fleshy vertebate–dispersed fruits: adaptive hypotheses and implications for seed dispersal. Amer Nat 150: 346–372.

Cipollini ML, Stiles EW 1993. Fruit rot, antifungal defense, and palatability of fleshy fruits for frugivorous birds. Ecology 74: 751–762.

Connell J 1971. On the role of natural enemies in preveting competitive exclusion in some marine animals and in rain forest trees. In: Boer PJD, Gradwell GR (Eds.), Popuation Dynamics. Centre for Agricultural Publishing and Documentation, Wageningen, pp. 298–310.

Coates–Estrada R, Estrada A 1986. Fruiting and frugivores at a strangler fig in the tropical rain forest of Los Tuxtlas, Mexico. J Trop Ecol 2: 349–357.

Condit R, Pitman N, Leigh EG, Chave J, Terborgh J, Foster RB, Núñez P, Aguilar S, Valencia R, Villa G, Muller–Landau HC, Losos E, Hubbell SP 2002. β– diversity in tropical forest trees. Science 295: 666–669.

Correa SB, Costa–Pereira R, Fleming T, Goulding M, Anderson JT 2015. Neotropical fish–fruit interactions: eco–evolutionary dynamics and conservation. Biol Rev 90: 1263-1278.

Colon CP, Campos–Arceiz A 2013. The impact of gut passage by Binturongs (*Arctictis binturong*) on seed germination. Raffl Bull Zool 61: 417–421.

Corlett RT 1996. Characteristics of Vertebrate–Dispersed Fruits in Hong Kong. J Trop Ecol 12: 819–833

Corlett RT 2007. The impact of hunting on the mammalian fauna of tropical Asian forests. Biotropica 39: 292–303.

Corlett R 2013. The shifted baseline: prehistoric defaunation in the tropics and its consequences for biodiversity conservation. Biol Conserv 163: 13–21.

Datta A, Rawatt GS 2008. Dispersal modes and spatial patterns of tree species in a tropical forest in Arunachal Pradesh, north–east India. Trop Conserv Sci 1: 163–185.

Debussche M, Isenmann P 1989. Fleshy fruit characters and the choices of bird and mammal seed dispersers in a Mediterranean region. Oikos 56: 327–338.

Dixon MD, Johnson WC, Adkisson CS 1997. Effects of weevil larvae on acorn use by jays. Oecologia 111: 201–208

Eriksson O 2008. Evolution of seed size and biotic seed dispersal in angiosperms: paleoecological and neoecological evidence. Int J Plant Sci 169: 863–870.

Eriksson O, Friis EM, Löfgren P 2000. Seed size, fruit size, and dispersal systems in angiosperms from the early Cretaceous to the late Tertiary. Am Nat 156: 47–58.

Fleming TH, Kress WH 2011. A brief history of fruits and frugivores. Acta Oecol 37: 521–530

Fleming TH 1991. Fruiting plant–frugivore mutualism: the evolutionary theater and the ecological play. In: Price PW, Lewinsohn TM, Fernandes GW, Benson WW (Eds.), Plant–Animal Interactions: Evolutionary Ecology in Tropical and Temperate Regions. J Wiley and Sons, New York, pp.119–144

Forget P–M, Milleron T 1991. Evidence for secondary seed dispersal by rodents in Panama. Oecologia 87:596–599.

Friis EM, Pedersen KR, Crane PR 2010. Diversity in obscurity: fossil flowers and the early history of angiosperms. Phil Trans R Soc B 365: 369–382.

Galindo–González J, Guevara S, Sosa VJ 2000. Bat– and Bird–Generated Seed Rains at Isolated Trees in Pastures in a Tropical Rainforest. Conserv Biol 14: 1693–1703.

Gómez–Aparicio L, Gómez JM, Zamora R 2005. Microhabitats shift rank in suitability for seedling establishment depending on habitat type and climate. J Ecol 93: 1194–1202.

Gorresen PM, Willig MR 2004. Landscape responses of bats to habitat fragmentation in Atlantic forest of Paraguay. J Mammal 85: 688–697.

Griscom HP, Kalko EKV, Ashton MS 2007. Frugivory by small vertebrates within a deforested, dry tropical region of Central America. Biotropica 39: 278–282.

Gruezo W, Soligam AC 1990. Identity and germination of seeds from feces of the Philippine Palm Civet (*Paradoxurus philippinensis* Jourdan). Nat Hist Bull Siam Soc 38: 69–82.

Hamilton J, Coe M 1982. Feeding, digestion and assimilation of a population of giant tortoises (*Geochelone gigantea* (Schweigger)) on Aldabra atoll. J Arid Env 5: 127–144.

Hansen DM, Donlan CJ, Griffiths CJ, Campbell KJ 2010. Ecological history and latent conservation potential: large and giant tortoises as a model for taxon substitutions. *Ecography* 33: 272–84.

Hansen DM, Kaiser CN, Müller CB 2008. Seed Dispersal and Establishment of Endangered Plants on Oceanic Islands: The Janzen–Connell Model, and the Use of Ecological Analogues. PLoS ONE 3: e2111

Harper JL, Lovell PH, Moore KG 1970. The shapes and sizes of seeds. Ann Rev Ecol Evol S 1: 327–356.

Heer K, Albrecht L, Kalko EKV 2010. Effects of ingestion by Neotropical bats on germination parameters of native free–standing and strangler figs (*Ficus* sp., Moraceae). Oecologia 163: 425–435.

Herre EA 1996. An overview of studies on a community of Panamanian Figs. J Biogeogr 23:593–607.

Herrera CM 1989a. Seed dispersal by animals: a role in angiosperm diversification? Am Nat 133: 309–322.

Herrera CM 1989b. Frugivory and seed dispersal by carnivorous mammals, and associated fruit characteristics, in undisturbed Mediterranean habitats. Oikos 55: 250–262.

Hnatiuk SH 1978. Plant dispersal by the Aldabran giant tortoise, *Geochelone gigantea* (Schweigger). Oecologia 36: 345–350.

Horn MH, Correa SB, Parolin P, Pollux BJA, Anderson JT, Lucas C, Widmann P, Tjiu A, Galetti M, Goulding M 2011. Seed dispersal by fishes in tropical and temperate fresh waters: the growing evidence. Acta Oecol 37: 561–577.

Horn MH 1997. Evidence for dispersal of fig seeds by the fruit–eating characid fish Brycon guatemalensis Regan in a Costa Rican tropical rain forest. Oecologia 109: 259–264.

Howe HF, Smallwood J 1982. Ecology of seed dispersal. Annu Rev Ecol Evol S 13: 201–228.

Horn MH 1997. Evidence for dispersal of fig seeds by the fruit–eating characid fish Brycon guatemalensis Regan in a Costa Rican tropical rain forest. Oecologia 109: 259–264.

Izhaki I, Korine C, Arad Z 1995. The effect of bat (*Rousettus aegyptiacus*) dispersal on seed germination in eastern Mediterranean habitats. Oecologia 101: 335–342.

Jacomassa FA, Pizo, MA 2010. Birds and bats diverge in the qualitative and quantitative components of seed dispersal of a pioneer tree. Acta Oecol 36: 493–496.

Janzen DH 1970. Herbivores and the number of tree species in tropical forests. Am Nat 104: 501–528.

Janzen DH 1979. How to be a fig. Annu Rev Ecol Evol S 10: 13–51.

Jerozolimski A, Ribeiro MBN, Martins M 2009. Are tortoises important seed dispersers in Amazonian forests? *Oecologia* 161: 517–528.

Jordano P, Garcia C, Godoy JA, Garcia–Castanõ JL 2007. Differential contribution of frugivores to complex seed dispersal patterns. Proc Natl Acad Sci USA 104: 3278–3282.

Jordano P 2000. Fruits and frugivory. Pages 125–165 in M Fenner (Ed) Seeds: The ecology of regeneration in plant communities. CAB International, Wallingford.

Joshi AR, Smith JLD, Cuthbert FJ 1995. Influence of food distribution and predation pressure on spacing behavior in palm civets. J Mammal 76: 1205–1212.

Jothish P 2011. Diet of the common palm civet *Paradoxurus hermaphroditus* in a rural habitat in Kerala, India, and its possible role in seed dispersal. Small Carniv Conserv 45: 14–17.

Kitamura S, Yumoto T, Poonswad P, Chuailua P, Plongmai K, Maruhashi T, Noma N 2002. Interactions between Fleshy Fruits and Frugivores in a Tropical Seasonal Forest in Thailand. Oecologia 133: 559–572.

Kitamura S, Yumoto T, Poonswad P, Suzuki S, Wohandee P 2008. Rare seed–predating mammals determine seed fate of *Canarium euphyllum*, a large–seeded tree species in a moist evergreen forest, Thailand. Ecol Res 23: 169–177.

Koike S, Morimoto H, Goto Y, Kozakai C, Yamazaki K 2008. Frugivory of carnivores and seed dispersal of fleshy fruits in cool–temperate deciduous forests. J For Res 13: 215–222.

Laska M 1990. Food transit times and carbohydrate use in three phyllostomid bat species. Z Säugetierk 55: 49–54.

Levey DJ, Tewksbury JJ, Cipollini ML, Carlo TA 2006. A field test of the directed deterrence hypothesis in two species of wild chili. Oecologia 150: 61–68.

Lobova TA, Mori SA 2004. Epizoochorous dispersal by bats in French Guiana. J Trop Ecol 20: 581–582.

Lobova TA, Mori SA, Blanchard F, Peckham H, Charles– Dominique P 2003. Cecropia as a food resource for bats in French Guiana and the significance of fruit structure in seed dispersal and longevity. Am J Bot 90: 388–403.

LoGiudice K, Ostfeld RS 2002. Interactions between mammals and trees: predation on mammal dispersed seeds and the effect of ambient food. Oecologia 130: 420–425.

Lopez JE, Vaughan C 2004. Observations on the role of frugivorous bats as seed dispersers in Costa Rican secondary humid forests. Acta Chiropt 6:111–119.

Lundberg JG, Machado–Allison A, Kay RF 1986. Miocene characid fishes from Colombia: evolutionary stasis and extirpation. Science 234: 208–209.

Mali S, Borges RM 2003. Phenolics, fibre, alkaloids, saponins, and cyanogenic glycosides in a seasonal cloud forest in India. Biochem Syst Ecol 31: 1221–1246.

Marlow RW 1989. Co–evolution of a Galapagos Tortoise– *Psidium* Interaction. Am Zool 29: 145a.

Marron AO, Moore JR 2013. Evidence of frugivory and seed dispersal in Oligocene tortoises from South Dakota. Geol Mag 150: 1143–1149.

Marshall AG 1983. Bats, flowers and fruit: evolutionary relationships in the Old World. Biol J Linn Soc 20: 115–35.

Medellin RA, Gaona O 1999. Seed dispersal by bats and birds in forest and disturbed habitats of Chiapas, Mexico. Biotropica 31: 478–485.

Melo PL, Rodriguez–Herrera B, Chazdon RL, Medellin RA, Ceballos GG 2009. Small Tent–Roosting Bats Promote Dispersal of Large–Seeded Plants in a Neotropical Forest. Biotropica 41: 737–743.

Moll D, Jansen KP 1995. Evidence for a role in seed dispersal by two tropical herbivorous turtles. *Biotropica* 27: 121–127.

Mudappa D, Kumar A, Chellam R 2010. Diet and fruit choice of the brown palm civet *Paradoxurus jerdoni,* a viverrid endemic to the Western Ghats rainforest, India. Trop Conserv Sci 3: 282–300.

Murray KG 1988. Avian seed dispersal of three Neotropical gap–dependent plants. Ecol Monogr 58: 271–298.

Muscarella R, Fleming TH 2007. The role of frugivorous bats in tropical forest succession. Biol Rev 82: 573–590.

Nakabayashi M, Nakashima Y, Bernard Y, Kohshima S 2014. Utilisation of gravel roads and roadside forests by the common palm civet (*Paradoxurus hermaphroditus*) in Sabah, Malaysia. Raffl Bull Zool 62: 379–388.

NakashimaY, Sukor JA 2010. Importance of common palm civets (Paradoxurus hermaphroditus) as a long–distance disperser for large–seeded plants in degraded forests. Tropics 18: 221–229.

Nakashima Y, Inoue E, Inoue–Murayama M, Sukor JRA 2010. Functional uniqueness of a small carnivore as seed dispersal agents: a case study of the common palm civets in the Tabin Wildlife Reserve, Sabah, Malaysia. Oecologia 164: 721–730.

Nakabayashi M, Bernard H, Nakashima Y 2012. An observation of several Common Palm Civets *Paradoxurus hermaphroditus* at a fruiting tree of *Endospermum diadenum* in TabinWildlife Reserve, Sabah, Malaysia: comparing feeding patterns of frugivorous carnivorans. Small Carniv Conserv 47: 42–45.

Nasi R, Brown D, Wilkie D, Bennett E, Tutin C, van Tol G, Christophersen T 2008. Conservation and use of wildlife–based resources: The bushmeat crisis. Technical series 33. pp 50.

Nathan R, Katul GG, Horn HS, Thomas SM, Oren R, Avissar R, Pacala S, Levin SA 2002. Mechanisms of long–distance dispersal of seeds by wind. Nature 418: 409–413.

Nathan R (2006) Long–distance dispersal of plants. Science 313: 786–788.

Patou ML, Wilting A, Gaubert P, Esselstyn JA, Cruaud C, Jennings AP, Fickel J, Veron G 2010. Evolutionary history of the *Paradoxurus* palm civets – a new model for Asian biogeography. J Biogeogr 30: 2077–2097.

Peters HA 2003. Neighbour–regulated mortality: the influence of positive and negative density dependence on tree populations in species–rich tropical forests. Ecol Lett 6:757–765.

Platt SG, Elsey RM, Liu H, Rainwater TR, Nifong JC, Rosenblatt AE, Heithaus MR, Mazzotti FJ 2013. Frugivory and seed dispersal by crocodilians: an overlooked form of saurochory? J Zool 291: 87–99.

Pollux BJA 2011. The experimental study of seed dispersal by fish (Ichthyochory). Freshwater Biol 56: 197–212.

Poulsen J, Clark C, Mavah G, Elkan P 2009. Bushmeat supply and consumption in a tropical logging concession in northern Congo. Conserv Biol 23: 1597–1608.

Rick CM, Bowman RI 1961. Galápagos tomatoes and tortoises. Evolution 15: 407–417.

Ridley HN 1930. The dispersal of plants throughout the world. Reeve, Ashford, U.K.

Russ SE, Portnoy S, Augspurger CK 2006. Incorporating Animal Behaviour into Seed Dispersal Models: Implications for Seed Shadows. Ecology: 3160–3174.

Robertson AW, Trass A, Ladley JJ, Kelly D 2006. Assessing the benefits of frugivory for seed germination: the importance of the deinhibition effect. Funct Ecol 20: 58–66.

Samuels IA, Levey DJ 2005. Effects of gut passage on seed germination: do experiments answer the questions they ask? Funct Ecol 19: 365–368.

Sathyakumar S, Viswanath S 2003. Observations on food habits of Asiatic black bear in Kedarnath Wildlife Sanctuary, India: preliminary evidence on their role in seed germination and dispersal. Ursus 14: 99–103.

Schupp EW, Jordano P, Gómez JM 2010. Seed dispersal effectiveness revisited: a conceptual review. New Phytol 188: 333–353.

Silveria M, Trevelin L, Port–Carvalho M, Godoi S, Mandetta EN, Cruz–Neto A 2011. Frugivory by phyllostomid bats (Mammalia: Chiroptera) in a restored area in Southeast Brazil. Acta Oecol 37: 31–36.

Sinu PA, Shivanna KR, Kuriakose G 2012. Frugivorous bird diversity and their post–feeding behaviour in fruiting *Syzygium cumini* (Myrtaceae) in fragmented forests of central Western Ghats, India. Curr Sci 103: 1146–1148.

Sinu PA 2012. Seed predators of an old–world tropical deciduous tree (*Terminalia bellirica*: Combretaceae) in wet habitats of the Western Ghats, India. Curr Sci 103: 309–315.

Spiegel O, Nathan R 2007. Incorporating dispersal distance into the disperser effectiveness framework: frugivorous birds provide complementary dispersal to plants in a patchy environment. Ecol Lett 10: 718–728.

Sreekumar PG, Balakrishnan M 2002. Seed Dispersal by the Sloth Bear (*Melursus ursinus*) in South India. Biotropica 34: 474–477.

Steele MA, Hadj–Chikh LZ, Hazeltine J 1996. Caching and feeding decisions by *Sciurus carolinensis*: responses to weevil–infested acorns. J Mammal 77: 305–314.

Strong JN, Fragoso JMV 2006. Seed dispersal by *Geochelone carbonaria* and *Geochelone denticulata* in northwestern Brazil. Biotropica 38: 683–686.

Struempf HM, Schondube JE, Martinez del Rio C 1999. The cyanogenic glycoside amygdalin does not deter consumption of ripe fruit by cedar waxwings. Auk 116: 749–758.

Su S, Sale J 2007. Niche differentiation between Common Palm Civet *Paradoxurus hermaphroditus* and Small Indian Civet *Viverricula indica* in regeneration degraded forest, Myanmar. Small Carniv Conserv 36: 30–34.

Teixeira RC, Correa CE, Fischer E 2009. Frugivory by Artibeus jamaicensis (Phyllostomidae) bats in the Pantanal, Brazil. Stud Neotrop Fauna Environ 44:7–15.

Terborgh J 1986. Community aspects of frugivory in tropical forests. In: Estrada, A,Fleming, T (Eds), Frugivores and Seed Dispersal. Dr W Junk, Publishers, The Hague, The Netherlands, pp. 371–384.

Tiffny BH 2004. Vertebrate dispersal of seed plants. Annu Rev Ecol Evol S 35: 1–29.

Tiffny BH 1986. Evolution of seed dispersal syndromes according to the fossil record. In DR Murray (Ed.) Seed dispersal, Academic Press, Sydney, pp. 273–305.

Traveset A 1998. Effect of seed passage through vertebrate frugivores' guts on germination: a review. Persp Plant Syst Evol1/2: 151–190.

Traveset A, Rodriguez–Perez J, Pias B 2008. Seed trait changes in dispersers' guts and consequences for germination and seedling growth. Ecology 89: 95–106.

Valido A, Olesen JM 2007. The importance of lizards. In Dennis J (Ed) Seed Dispersal: Theory and its Applications in a Changing World, CABI Publishing, Wallingford, pp. 124–47.

Van der Pijl L (Ed) 1982. Principles of Dispersal in Higher Plants. Springer–Verlag, Berlin.

Vander Wall SB, Briggs JS, Jenkins SH, Kuhn KM, Thayer TC, Beck MJ 2006. Do food–hoarding animals have a cache recovery advantage? Determining recovery of stored food. Anim Behav 72: 189–197.

Vander Wall SB, Beck MJ 2012. A Comparison of Frugivory and Scatter–Hoarding Seed–Dispersal Syndromes. Bot Rev 78: 10–31.

Vanthomme H, Bellé B, Forget P–M 2010. Bushmeat hunting alters recruitment of large–seeded plant species in Central Africa. Biotropica 42: 672–679.

Vidal M, Pires MM, Guimarães PR 2013. Large vertebrates as the missing components of seed dispersal networks. Biol Conserv 163: 42–48.

Vijaya J 1982. Rediscovery of the Forest Cane Turtle (*Heosemys silvatica*) of Kerala. Hamadryad 7: 2 – 3.

Wang BC, Smith TB 2002. Closing the seed dispersal loop. Trends in Ecol Evol 17: 379–386.

Wang BC, Sork VL, Leong MT, Smith TB 2007. Hunting of mammals reduces seed removal and dispersal of the Afrotropical tree *Antrocaryon klaineanum* (Anacardiaceae). Biotropica 39: 340–347.

Wang H, Fuller TK 2003. Food habits of four sympatric carnivores in southeastern China. Mammalia 67: 513–519.

Wenny DG 2000. Seed dispersal, seed predation, and seedling recruitment of a Neotropical montane tree. Ecol Monogr 70: 331–351.

Wenny DG, Levey DJ 1998. Directed seed dispersal by bellbirds in a tropical cloud forest. Proc Natl Acad Sci USA 95: 6204–6207.

Westcott DA, Bentrupperbäumer J, Bradford MG, McKeown A 2005. Incorporating patterns of disperser behaviour into models of seed dispersal and its effects on estimated dispersal curves. Oecologia 146:57–67.

Wheelwright NT, Orians GH 1982. Seed Dispersal by Animals: Contrasts with Pollen Dispersal, Problems of Terminology, and Constraints on Coevolution. Am Natural 119: 402–413.

Witmer MC 2001. Nutritional interactions and fruit removal: cedar waxwing consumption of *Viburnum opulus* fruits in spring. Ecology 82: 3120–3130.

Witmer MC 1998. Do seeds hinder digestive processing of fruit pulp? Implications for plant/frugivore mutualisms. Auk 115: 319–326.

Witmer MC, Van Soest PJ 1998. Contrasting digestive strategies of fruit–eating birds. Funct Ecol 12: 728–741.

Xiao ZS, Zhang ZB, Wang YS 2003. Rodent's ability to discriminate weevilinfested acorns: potential effects on regeneration of nut–bearing plants. Acta Theriol Sin 23: 312–320.

Yi X, Zhang M, Bartlow AW, Dong Z 2014. Incorporating Cache Management Behavior into Seed Dispersal: The Effect of Pericarp Removal on Acorn Germination. PLoS ONE 9: e92544.

Yi XF, Steele MA, Zhang ZB 2012. Acorn pericarp removal as a cache management strategy of the Siberian chipmunk, *Tamias sibiricus*. Ethology 118: 87–94.

Zhou Y, Newman C, Palomares F, Zhang S, Xie Z, MacDonald DW 2014. Spatial organization and activity patterns of the masked palm civet (*Paguma larvata*) in central–south China. J Mammal 95: 534–542.

Chapter 12

Seed Dispersal and Seed Predation: Case Studies from Eastern Ghats

A J Solomon Raju
ajsraju@yahoo.com

Abstract

Seed dispersal syndromes are important for the survival and expansion of the distribution ranges of the plant species. Broadly, two types of seed dispersal syndromes exist, autochory and allochory; each syndrome is, again, classified based on the mode and agent of the seed dispersal. Seed predation is an inevitable event in the reproductive cycle of most of the plant species. It is divided into two distinctive temporal categories, pre-dispersal and post-dispersal predation, which may involve different strategies and requirements and have different implications at the individual and population level. The seed dispersal syndromes and the seed predation have been explained in this paper based on some case studies made in the Eastern Ghats forest of Andhra Pradesh, India. They include both the flowering plants and the gymnosperms; some of them are rare, endangered and endemic species. Flowering plants: *Hildegardia populifolia, Eriolaena lushingtonii, Syzygium alternifolium, Shorea roxburghii, Decalepis hamiltonii, Boswellia ovalifoliolata, Terminalia pallida, Wendlandia glabrata* and *Wendlandia tinctoria*; Gymnosperms: *Cycas beddomei* and *Cycas sphaerica*. The studies indicate that there are certain evolved characters that assist in the efficient functioning of the seed dispersal syndromes concerned. The fruit or seed predation in these species has a negative effect and thus curtail the success of the sexual reproduction and subsequently, impacting the restoration and expansion of populations of these plant species.

12.1 Introduction

Seed dispersal is a critical event in plant life history for the survival of populations, and the survival is linked to various biotic and abiotic factors (Vanschaik et al. 1993). The diaspores of many plant species have characteristic morphological structures and traits that enhance

DOI: 10.1201/9781003505471-12

their probability of being dispersed away from the mother plant (Griz and Machado 2001). Among these traits are fruit color, seed size and shape, and time of fruit ripening (Willson and Whelan 1990). The principal agents of dispersal are either abiotic or biotic, and the dispersal syndromes are termed, autochory and allochory respectively. The proportion of the dispersal modes in a particular vegetation type is defined as the dispersal spectrum influenced by community attributes, environmental circumstances and floristic composition (Hughes et al. 1994). The seed dispersal by animals predominates in the tropical forest plant species (Willson et al. 1989), and involves a tremendous diversity of animal species and their behaviour. Animals may consume fruit and drop, spit or discharge the seeds, carry seeds in their coats or scatter-hoard seeds for later consumption. Abiotic strategies such as wind, water and ballistic dispersal form the main mode of seed movement for a small percentage of the tropical tree species (Willson et al. 1989). Yanjun et al. (2009) reported that zoochory dominates the dry season in the sub-tropical broad-leaved forests and these authors contradicted the previous study by Griz and Machado (2001) that zoochory is the most common mode during the rainy season. Further, Yanjun et al. (2009) stated that dry seasons favour seed dispersal by animals and wind in the subtropical understorey forests.

Dispersal by wind has been considered to be inefficient relative to dispersal by animals. The wind dispersal occurs in several members of Leguminosae and Asteraceae (Renner 1987). The role of water in seed dispersal in Amazonia has been stressed by Kubitzki (1986). There have been no recent studies of the implications of water dispersal. The seed dispersal by animals is linked to co-evolution, fruit and seed characters and fruiting phenology. Plants need not direct the dispersal of their seeds to particular locations with a degree of exactness comparable to pollen dispersal; constancy and specificity of the disperser is not necessary for the interaction to work and be of mutual benefit. This fundamental difference between the pollination systems and the seed dispersal partly explains the virtual absence of species-specific interactions in seed dispersal (Renner 1987).

Many tropical birds, bats and monkeys depend on a constant supply of fruits and may have adapted their movements and reproductive schedules to the seasonal abundance of this supply. Wheelwright (1985) made a detailed study on frugivory in Costa Rican forests. He showed that the phenologies and habitat distributions of the plants may dictate the birds' seasonal movements. Herrera (1984a, b) stated that the fruit-bearing plants influence the kinds and numbers of birds attracted to a certain habitat. Simultaneously, fruiting species sometimes compete for the dispersal and the fruiting environment affects the probability that a plant will have its own seed disperser. The colourful, most often black or red, displays of ripe fruits are attractive to birds (Morden-Moore and Willson 1982, Willson and Thompson 1982). On the contrary, Wheelwright (1983) reported that there are many bird fruits that are not brightly coloured. Denslow and Moermond (1982) found that the birds prefer fruits near horizontal petioles to those along the main stem and out of reach from convenient perches. Moermond and Denslow (1983) showed that the terminal placement of infrutescences may maximize the probability for removal by birds and minimize the risk of seed destruction by rodents. Snow (1971) and Wheelwright (1985) reported that the plants producing the small fruits attract more species of birds than the large-fruited plants because most fruit eating birds typically swallow whole fruits. Levey et al. (1984) and Levey (1987) stated that the seed size also influences the physical treatment of seeds in bird guts. Pratt and Stiles (1983) mentioned that the large-gaped birds eat even very small fruits; large fruits exclude narrow-gaped birds. Further, large fruits are more likely to have their seeds dropped under the parent trees because the large frugivorous birds frequently make long visits to fruiting plants during which they defecate seeds. The least known factor that might affect avian fruit choice and subsequent seed dispersal is the chemical composition of fruits (Herrera 1982). Herrera (1985) described the mutualistic dispersal of seeds by vertebrates as a determinant of plant-animal co-evolution. Piper (1986) explained the relationship between the seasonality of fruit characters and the seed removal by birds. Gorchov (1987, 1990) suggested that synchronous flowering may be a constraint in fruiting synchrony. Karasov and Levey (1990) presented on the digestive system trade-offs and the adaptations of frugivorous passerine birds.

Bats take large, dry seeds. As the bats are large in size they may accidentally lose some of the seeds during flight and they thus escape being eaten (Kevan and Gaskell 1986). Removal of fruit pulp by bats may increase seed survival (Janzen 1982). The fruit diet of Old World monkeys overlaps to a large extent with that of the birds while New World monkeys differ substantially from the birds in their fruit preferences. The monkeys move long distances daily and respond to the temporal and spatial dispersion of fruiting trees and search out individual ones which are consistently used year after year. They have also been considered to be important in seed dispersal (see Renner 1987).

Janzen (1981) provided an elaborate account on the relationships between the fruits/seeds and their predators. Viewing a fruit and its contained seeds as a habitat, the distribution of coleopterous seed predators within that habitat is certainly not a random selection of the total behaviour shown by the large array of beetle species that attack the seeds. The female beetles attach eggs directly to the surface of the full-sized and nearly mature fruit, and the larvae hatch within one or two weeks. The larvae then mine through the fruit wall into a seed, the coat of which is thick and firm but still soft at this time. One or a few larvae grow to full size feeding on the contents of one seed. The last instar larva cuts an exit hole nearly through the seed coat and pupates in the seed. The adult emerges one to three weeks later. If the fruit has not dehisced by this time, the adult then cuts another exit hole through the fruit wall or uses one cut by a previous emergent. The chances of survival of adults are much greater as free-living individuals than when encapsulated in a seed or fruit that may be eaten by fungi, vertebrates, or insects, or that would be either severely heated and desiccated or water-soaked. There are several more likely causes for delayed oviposition. First, the younger a seed, the more likely that a plant will perceive the damage and cut off nutrient flow to a doomed offspring. Second, the larger the attacked seed, the larger are the unattacked seeds in the same fruit and the greater the return to the parent for each further resource unit invested in them. Thus, the more developed the attacked seed the less likely is the plant to abort the entire fruit. Fruit abortion dumps the larva into a hostile environment with fewer defenses and nutrients than it were the fruit to remain attached to the plant. Finally, the beetle larvae feed exclusively on the seed contents. When they also feed on the fruit tissue, it is only as they are moving from seed to seed or moving from the outside of the fruit to the seed. Obviously, the seed contains the highest nutrient value per bite and to the beetle larvae with the ability to detoxify the seed's defenses, the seed is defenseless. Samuni-Blank et al. (2013) reported that fruits produce toxins to control and show physiological and behavioural effects on the seed predators. But, certain predators are tolerant to these toxins which enable them to avoid their predators. Keeping this in view, the authors stated that the extreme ecological/evolutionary lability of this plant-animal relationship resulted in a shift from predation to mutualism and vice-versa. With this backdrop, the paper is contemplated to provide the types of seed dispersal syndromes and the occurrence of seed predation (Caswell et al. 2003, Corlett 1998, Fedriani and Manzoneda 2005, Figueroa et al. 2002, Gurevitch et al. 2006, Howe and Smallwood 1982, Janzen 1970, Lengyel 2010, Maron and Simms 1997, Vander Wall et al. 2005, Wichmann et al. 2009) and then the inputs supplementing the above stated reports based on the work carried out by the author.

12.2 Seed Dispersal: Types

A seed dispersal syndrome is a mutualistic plant-animal interaction. The seed dispersal syndromes are the morphological characters of seeds correlated to particular seed dispersal agents. Dispersal is the event by which the individuals move from the site of their parents to establish in a new area. A seed disperser is the vector by which a seed moves from its parent to the resting place where the individual is likely to establish. The characteristics for the seed dispersal syndromes are commonly fruit colour, mass, and persistence. These syndrome characteristics are often associated with the fruit that carries the seeds. Fruits are the packages of seeds, composed of nutritious tissues to feed the animals. However, fruit pulp is not commonly used as a seed dispersal syndrome because the pulp nutritional value does not enhance the seed dispersal success. Animals interact with these fruits because they are a common food source for them. Although, all seed dispersal syndromes do not have fruits all seeds are not dispersed by animals. Suitable biological and environmental conditions of the dispersal syndromes are needed for the seed dispersal and invasion success such as temperature and moisture. The seed dispersal syndromes have several benefits for the plant species. First, seed survival is often better when the seed is away from the parent plant. This improved survival may result from the actions of the density-dependent seed and seedling predators and pathogens, which often target the high concentrations of seeds beneath adults. Competition with adult plants may also be lower when the seeds are transported away from their parent.

The seed dispersal syndromes are classified broadly into two types – autochory and allochory. The autochory refers to self-dispersal while the allochory to the dispersal via vector.

12.2.1 Autochory: It is classified into different types based on the mode of self-dispersal.

1. *Barochory* – the seed transport via gravity in which a plant's seeds fall beneath the parent plant. The effect of gravity on the heavier fruits causes them to fall from the plant when ripe. Fruits exhibiting this type of dispersal include apples, coconuts, etc. and those with harder shells (which often roll away from the plant to gain more distance). The gravity dispersal also allows for later transmission by water or animal.
2. Blastochory – dispersal via runners
3. *Ballochory* - self-seeder - the seed is forcefully ejected by dehiscence and squeezing.
   Zooballochory – impetus provided by animals
   Anemoballochory – impetus provided by wind
   Hydroballochory – impetus provided by water
   Autoballochory – propulsion mechanisms based on sap pressure or drying
4. *Herpochory* - dispersal via active creeping – the seeds move themselves short distances - ex. Awns (hair or bristle-like appendage) with hygroscopic characteristics twist with changes in humidity dispersing the seeds.

12.2.2 Allochory is also classified into different types of seed dispersal based on the vector type.

1. Anemochory – transport via wind. The characteristics that make up this syndrome include wing structures and brown or dull coloured seeds without rewards. Seed flyers, rollers or throwers are typically categorized as dust diaspores, balloons, plumed or winged, the wings evolved to increase dispersal distance to promote gene flow. This syndrome is commonly found in open habitats, canopy trees and dry season forests. Anemochorous seeds mature in the dry season for optimum long-distance dispersal to increase the possibility of germination. The anemochory is one of the most primitive means of dispersal. The wind dispersal can take on one of the two primary forms: seeds can float on the breeze or alternatively, they can flutter to the ground. An important constraint of the wind dispersal is the need for abundant seed production to maximize the likelihood of seed landing in a site suitable for germination. There are also strong evolutionary constraints on this dispersal mechanism.
   Chamaechory - diaspore is rolled along the ground surface by wind.

Meteochory – transport in air for small seeds.

Boleochory – transport started by wind and later by other mechanisms.

2. Hydrochory – transport in water. The plants that exhibit this mode of seed dispersal show certain characteristics. The seed disperse by rain or ice or submerged in water. The seeds dispersed by water need to have the ability to float and resist water damage. They often have hair to assist with enlargement and floating. Seeds have air space, light weight tissues and corky tissues. This mode of dispersal is most common in aquatic plants.

   Nautochory – transport by movement in the sea.

   Bythisochory – drifting in flowing water.

   Ombrochory – transport via rain drops.

3. Zoochory – dispersal via animals. This is further divided based on the animals involved in the dispersal. This syndrome is characterized by certain features such as coloured fruits, scented fruits, presence of rewards for animals and different textures for different animals.

   Ornithochory – transport by birds. The characteristics this syndrome are the small fleshy fruits with bright colours and without husks. This mode of seed dispersal is common in temperate zones and oceanic islands because of the absence of native mammals. It can increase the reproductive success of a plant because a bird's digestive tract increases the seed germination after it has been bypassed and dispersed by the bird.

   Mammalochory – transport by mammals. The characteristics are large fleshy, green or dull coloured fruits with or without husks. The seeds tend to have more protection to prevent mechanical destruction. Mammals rely on the smell more than the vision for feeding, causing the seeds they disperse to be more scented compared to the bird dispersed seeds. The animal-dispersed seeds ripen in rainy season when the feeding activity is high, resulting in fleshy diaspores.

   Chiropterochory – transport of seeds by bats.

   Malacochory – transport of seeds by molluscs, mainly terrestrial snails.

   Anthropochory – seed dispersal by humans. Human dispersers differ from the animal dispersers by a much higher mobility based on the technical means of the human transport. Dispersal by humans may act on large geographical scales and lead to invasive species. This form of dispersal acts on smaller, regional scales and drives the dynamics of existing biological populations. Dispersal on human clothes is up to 250 m, on shoes up to 5 km, or by cars up to 250 m. Deliberate seed dispersal also occurs as seed bombing. This has risks as unsuitable provenance may introduce genetically unsuitable plants to new environments.

Myrmecochory – transport by ants. The common syndrome traits are the presence of elaisomes and hard seed coat. Elaisomes are the structures that attract ants because they are high in lipid content, providing important nutrients for the ants. The ants carry the seeds into their colonies, feed the elaiosome to their larvae and discard the otherwise intact seed in an underground chamber. They also bury the seeds underground and thus avoid the adverse environment effects such as fire or drought and reach nutrient-rich microsites and survive longer than other seeds. Ants do clearly interact with the seeds and transport them but they do not travel very long distance.

Epizoochory – transport by attachment of propagules to animals. This syndrome has burrs or spines to transport the seeds on the outside of animals and highly associated with the animals that have fur, while burrs would be lacking on the seeds that are dispersed by reptiles because of their smooth skin. Epizoochorous transport can be highly effective if the seeds attach to the wide-ranging animals. This form of seed dispersal has been implicated in rapid plant migration and the spread of invasive species. Herbaceous Apiaceae and Asteraceae members are the best examples.

Endozoochory – transport of propagules during passage through the gut. The characteristics of this syndrome would develop based on the palatability of the fruit by a vector type. This syndrome has evolved in seeds of certain plants which are to be ingested by animals and later bypassed in a new environment so the seed can germinate.

Stomatochory – the seeds are carried in the mouth of the animal and dropped after the fleshy reward has been eaten from around the seed.

Dysochory – transport of accidentally ingested propagules.

4. Hemerochory – dispersal by humans and human activities.
   Ethelochory – deliberate dispersal.
   Speirochory – dispersal is due to the accidental dispersal of seeds.
   Agochory – an unintentional dispersal.

12.3 Seed Predation

Seed predation, often referred to as granivory, is a type of plant-animal interaction in which granivores feed on the seeds of plants as a main or exclusive food source, in many cases leaving the seeds damaged and not viable. Granivores are found across many families of vertebrates (especially mammals and birds) as well as invertebrates (mainly insects); thus, the seed predation occurs virtually in all terrestrial ecosystems. The seed predation is commonly divided into two distinctive temporal categories, pre-dispersal and post-dispersal predation, which may involve different strategies and requirements and have different implications at the individual and population level. Nevertheless, regardless of timing, the seed predation has great implications on plant population dynamics. To counterbalance the effects of predation, plants have evolved defenses such as the seed morphology (size, shape, toughness) and the chemical defenses (secondary compounds such as tannins and alkaloids) to defend against their seed predators. However, as the plants have adapted defenses to seed predation using the chemical compounds, so have the seed predators adapted to the plant defenses withthe ability to detoxify the chemical compounds. The plant seeds are important sources of nutrition for animals across most ecosystems. The seeds contain food storage organs such as endosperm and/or cotyledons that provide nutrients to the developing seedling. This makes the seeds an attractive food source for animals because they are a highly concentrated and localized nutrient source in relation to the other plant parts.the seeds of many plants have evolved a variety of defenses to deter the predation from the seed predators. Plants may allocate resources towards physical or chemical defenses. The seeds are often contained inside the protective structures or fruit pulp that encapsulate the seeds until they are ripe. Other physical defenses include spines, hairs, fibrous seed coats and hard endosperm. The seeds, especially in arid areas, may have a mucilaginous seed coat that can glue the soil to seed hiding it from granivores. Some seeds have evolved strong anti-herbivore chemical compounds. In contrast to the physical defenses, the chemical defenses in seeds to deter consumption by the seed predators by using the chemicals that are toxic to granivores or inhibit the digestibility of the seed. These chemicals include non-protein amino acids, cyanogenic-glycosides, protease and amylase inhibitors and phytohaemaglutinins. Seeds face trade-offs between the allocation toward defenses and the size and number of the seeds produced. Additionally, plants have trade-offs in seed defense allocation between deterring the seed predators and attracting the seed dispersers.Plants may reduce the severity of the seed predation by making the seeds spatially or temporally scarce to granivores. The seed dispersal away from the parent plant is considered to reduce the severity of the seed predation. Seed masting is an example of how the plant populations are able to temporally regulate the severity of the seed predation. The masting refers to a concerted abundance of the seed production followed by a period of paucity. This strategy has the potential to regulate the size of the population of seed predators.

The seed predation occurs during pre- and post-dispersal stages. The pre-dispersal seed predation takes place when the seeds are removed from the parent plant before the dispersal, and it has been most often reported in invertebrates, birds, and granivorous rodents that clip fruits directly from trees and herbaceous plants. The post-dispersal seed predation arises once the seeds have been released from the parent plant. Birds, rodents, and ants are known to be among the most pervasive post-dispersal seed predators. Furthermore, the post-dispersal seed predation can take place in two contrasting stages: predation on the 'seed rain' and predation on the 'seed bank' . Whereas the predation on the seed rain occurs when the animals prey on the released seeds usually flush with the ground surface, the predation on the seed bank takes place after the seeds have been incorporated deeply into the soil. Both pre- and post-dispersal seed predations are extremely frequent in nature. However, the fundamental differences occur between the two. Pre-dispersal predators are most often specialists, as they are more likely to adapt to the clustered resources. The predator specialization requires that these animals utilize specific cues like plant chemistry (volatile compounds), colour and size to attack the seeds. In addition, specialization requires that the life cycle of the predator matches the production of seeds by the host plant. Therefore, the animals with shorter life span, like insects, are more common pre-dispersal predators. Common insect groups of pre-dispersal predators are Coleoptera, Hemiptera, Hymenoptera and Lepidoptera. For this reason, a greater diversity of predators feed on the seeds prior to dispersal. Nevertheless, there are important vertebrate pre-dispersal predators, especially birds and small mammals. The post-dispersal seed predation is extremely common in virtually all ecosystems. Given the heterogeneity in both resource types (seeds from different species are available), quality (seeds of different ages and/or different status of integrity or decomposition) and location (seeds are scattered and hidden in the environment), most post-dispersal predators have generalist habits. These predators usually belong to a diverse array of animals. In addition, not only is the assemblage of the post-dispersal seed predators diverse, but its composition varies considerably among the ecosystems.

The complex relationship between the seed predation and the plant demography is an important topic of plant-animal interactive studies. The plant population structure and size over time is closely associated with the effectiveness at which the seed predators locate, consume, and disperse the seeds. In many cases this relationship is dependent upon the type of seed predator (specialist vs generalist) or the particular habitat in which the interaction is taking place. The role of seed predation on plant demography may be either detrimental or in particular cases actually beneficial to the plant populations. The seed predation is predicted to be highest where the seeds are most abundant under the parent tree. As the distance from the parent tree increases, the seed abundance and thus the seed predation are predicted to decrease as the seed survival increases. In many cases the seed predators support the plant populations by dispersing the seeds away from the parent plant. By dispersing the seeds away from the parent plant the seed predator is in effect supporting gene flow among the populations. Other seed predators collect the seeds and then store or cache them for later consumption. In the case that the seed predator is unable to locate the buried or hidden seed, there is a chance that it will later germinate and grow, supporting the species dispersal.

12.4 Case Studies

12.4.1 *Hildegardia populifolia* (Sterculiaceae)

In *Hildegardia* genus, *H. populifolia* is the only species which occurs in India. Its conservation status has been reported differently by different workers such as critically endangered, endangered and vulnerable. Its distribution is restricted to certain places of Eastern Ghats in Andhra Pradesh and Tamil Nadu in South India. It occurs on the Nigidi Hills of Anantapur and Ankalammakonda of Chittoor in Andhra Pradesh, and Kalyaran Hills and Pakkamalai Hills in Villupuram and also in Salem.

In fertilized flowers, the ovary gradually bulges and the carpels get separated forming follicles in two weeks. The follicles take about 10 months to mature and fall to the ground. The fruits are initially greyish, turn rusty brown and then green and later greyish brown as they mature. The follicled fruits are thinly woody, falcately ovate-reniform, winged, inflated and on their outside covered with pubescent hair. A fruit produces 1-5 follicles. Likewise, a follicle produces 1 or 2 seeds affixed at the base. The seeds are ovate-oblong, 13-14 cm long, dented, green initially and pale brown when dry. The fruits with the seeds inside begin to fall to the ground from February and the complete fruit shedding by the end of March. The light weight fruits or follicles are dispersed effectively by wind. The fallen follicles find their way into rock openings/crevices and settle there. The rocky nature of the area greatly reduces the dispersal distance of the fruits and in effect, the seeds remain mostly within the plant area. The follicle sheath gradually disintegrates and exposes the seeds by the end of May/early June. As soon as the first monsoonal rains fall in June/July, the healthy and viable seeds germinate and produce new plants.

The wing-like membranous follicles of *H. populifolia* with 1 or 2 seeds being very light in weight are adapted for dissemination by wind and referred to as anemochory. The dry follicles disseminate by the prevailing wind speeds. There is no possibility for the fallen follicles to disperse further on the ground due to the rocky terrain; they finally settle within the vicinity of the parental plants. It is effective only under the ideal wind speeds and even then the rocky terrain limits the dispersal distance of follicles. The membranous follicle sheath disintegrates by the end of the dry season exposing the seeds to the soil environment. As the seed coat is very hard, it takes time for the emergence of radical and subsequent steps leading to the seedling growth. The disintegration of seed coat and subsequent germination depends on the soil chemical and nutrient environment; the rocky habitat of the plant does not favour the regeneration from seeds. The seeds located in the litter rich areas of rock crevices/openings germinate and produce seedlings immediately during rainy season and hence, they are non-dormant.

H. populifolia appears to have evolved hard seed coat for the protection against regular seasonal fire. Fire influences seed germination in a variety of ways. It reduces inhibitory substances in soil and litter, provides chemical stimulation from charred wood, fractures hard seed coats, stimulates seed embryos, causes desiccation of seed coats and enhances nutrient supply due to nutrient mobilization from the burnt plant biomass and soil microbes. The seeds of *H. populifolia* are not buried in the soil but stored on the soil or rocky surface; here the fires are mostly lethal for seeds especially when the fuel load is high. Partly burned seeds germinate and produce seedlings but their continuous growth is dependent on water and nutrient status of the soil.

12.4.2 *Eriolaena lushingtonii* (Sterculiaceae)

It is an endemic and endangered deciduous tree species of southern Peninsular, India where it is restricted to open slopes of moist deciduous forests at 350-900m. In fertilized flowers, the ovary gradually bulges, transforms into fruit and produces seeds. The total duration from fruit initiation to maturation and dispersal is six - seven months. A bruchid beetle (unidentified) was found to breed in the fruits and the fruits used by it do not produce viable seeds. Each infested fruit shows a single larva and the mature and adult beetle exits through a hole made by it. These beetles emerge out during February-March when the fruits are almost ready to dehisce. The fruit is a 3.5-3.7 cm long woody 4-5 lobed capsule with 13-22 ascending seeds and wool filling the spaces between them. The seeds possess apical membranous wings which are about as long as the body of the seed. The mature, dry capsules dehisce loculicidally a bit explosively liberating the winged seeds into the air. The seeds thus released are driven away by the prevailing wind during February-March when the ambient conditions are dry and hence the plant is anemochorous.

12.4.3 *Syzygium alternifolium* (Myrtaceae)

It is an endemic and globally endangered species as per the criteria of IUCN but not yet included in IUCN list. It occurs in the tropical dry deciduous forests of Kurnool, Cuddapah and Chittoor Districts of Andhra Pradesh, Chengalpattu and North Arcot Districts of Tamil Nadu and Bangalore District in Karnataka in India. Its distribution is related to the geology and rock structure along with elevation. The fertilized flowers grow, mature and ripen within 2 months. The fruit is a globose berry, luscious, fleshy, 25-30 mm in diameter and edible. It has a combination of sweet, mildly sour and astringent flavour and colours the tongue purple when eaten. The green and light purple fruits are very tasty and sweet while the dark purple and violet ones are sweet and bitter. Each fruit produces a single large seed. The fruits fall off during late July-August. The locals collect the ripen fruits from the trees and fallen fruits from the ground since they are edible and have commercial value. Since the fruits are pulpy with soar and sweet taste, they are naturally adapted for feeding by birds and mammals as well. But, the seed dispersal is almost non-functional in this species due to the collection of the fruits by locals as and when they are ripe.

12.4.4 *Shorea roxburghii* (Dipterocarpaceae)

It is a semi-evergreen endangered and red listed tree species of Peninsular India included in the list of medicinal plants of conservation areas of Eastern Ghats of Andhra Pradesh. The fruits mature quickly within the dry season. They are winged, light weight and characteristically produce a single seed. The winged character of fruits is seen in most dipterocarps and it is an important adaptation for dissemination by wind. The winged structure of the sepals allows 1-seeded fruits to gyrate toward the ground and hence the seed dispersal is anemochorous. The seeds disperse by wind to a small distance only due to semi-closed nature of the canopy cover of the forest. The dispersal of the winged fruits takes place much more efficiently by wind if the forest is open seasonal dry deciduous type. The seeds fallen on the ground have no possibility for further dispersal by sweeping action of the wind due to litter accumulation and grass growth in the study sites during the rainy season. A bruchid beetle attacks at the early stage of development of the fruit. Its larva leaves through the exit hole created by it when the mature fruits fall to the ground. It pupates in the soil and perhaps, produces adult only when the favourable conditions return in the forest floor to repeat its life cycle. The seed predation is 87% suggesting that this tree is threatened by bruchid beetle in terms of reproductive success.

12.4.5 *Shorea tumbuggaia* (Dipterocarpaceae)

It is an endemic and globally endangered semi-evergreen tree species restricted to Southern Eastern Ghats up to 1000 m distributed in Seshachalam and Veligonda Hills in Cuddapah, Tirupati hills in Chittoor district in Andhra Pradesh to North Arcot and Chengalpattu districts of Tamil Nadu. The fruits are winged, mature quickly within a month and each fruit invariably produces a single seed with a large chlorophyllous embryo. The winged character of fruits is seen in most dipterocarps and is an important adaptation for dissemination by wind. The winged structure of the sepals allows the single seeded fruits to gyrate toward the ground and hence the seed dispersal is anemochorous. The distance of seed dispersal by wing is up to 10 m only due to semi-closed nature of the canopy cover of the forest. An unidentified bruchid beetle species attacks the flower buds and growing fruits/seeds for the production of its offspring. It produces offspring within a span of three weeks. It causes 27% of flower bud infestation and 70% of seed infestation. Some batches of larvae pupate in the soil due to the fall of fruits/seed to the ground; such larvae transform into adults in the soil. The mass fruiting appears to favour the seed predators, but it can also be a strategy to escape the complete seed destruction. The seed predation by bruchid beetle is the principle limiting factor for its regeneration.

12.4.6 *Decalepis hamiltonii* (Periplocaceae)

It is endemic to Southern States (Karnataka, Kerala, Tamil Nadu and Andhra Pradesh) of India. It has also been included as an endangered species in the red list by the International Union for Conservation of Nature. The ripe follicles are ellipsoid, cylindric and wrinkled; each follicle is 60 mm long and 17 mm wide. Each follicle contains 70 + 11 ovate 6 mm long, 4 mm wide and 10 mg weight reddish brown coloured seeds crowned with a tuft of 28 mm long silky hairs, and are well seated on the denticles of thick cylindric placenta of the ovary. The fruits take a long period to mature and dehisce to disseminate seeds. The dry season provides ideal conditions for fruit dehiscence and dispersal of seeds. The timing of fruit maturation and ripening is so perfect that it occurs during the dry season when the plant is completely leafless. The light weight of the seed crowned with a tuft of silky hair is an important aerodynamic trait characterizing 'anemochory'. The placenta with seeds well seated on its denticles collectively comes out of the dehisced follicle, settles on the ground and then the individual seeds gradually disseminate by the wind at low height. This mode of seed dispersal is more efficient than the individual seed dispersal directly from the fruit.

12.4.7 *Boswellia ovalifoliolata* (Burseraceae)

B. ovalifoliolata occurs on the foothills of the Seshachalam hill ranges of Eastern Ghats in Chittoor, Cuddapah and Kurnool districts of Andhra Pradesh up to an altitude of about 600-900 m. It is a globally endangered, strict endemic medium-sized deciduous medicinally important tree species and listed in CITES Red Data Book under the medicinal plants list. The fruits mature in a short period and dehisce along the septa to disperse the seeds for which the dry conditions are essential. Their dispersal takes place in the month of May when the temperature is at its maximum and provides the ideal conditions for the seed dispersal by wind. The seed characteristics such as small size, light weight, papery and winged nature are adapted for anemochory. As the study site is windy, the anemochory is very effective, dispersing seeds up to 400 m away from the parental site. Therefore, the dry season seems to be the pre-requisite for flowering, fruiting and seed dispersal in this plant species.

12.4.8 *Terminalia pallida* (Combretaceae)

It is an endemic medicinal semi-evergreen tree species of India. It has been given endangered status at global level based on IUCN criteria by the Conservation Assessment and Management Planning Workshop help in 2001 in Hyderabad. It is now included in the First Red List of Medicinal Plants of Andhra Pradesh. It occurs on rocky hilly areas of dry deciduous forests of Chittoor, Cuddapah and Kurnool districts at 700-800 m elevation in the Eastern Ghats but it is mainly centered at Tirumala Hills of Chittoor. The Fruit predation during pre- and post-dispersal stage is very heavy due to its palatability by beetles and rodents. Out of the small percentage of natural fruit set, a major chunk at maturing stage forms the food for a rodent species (unidentified) leaving a small quantity of fruits for maturation. Further, the same rodent consumes some more quantity of dry fruits on the ground. The rodent fruit feeding activity at the fruit maturation and fallen stage is highly detrimental for the success of sexual reproduction since such fruits and seeds are destroyed. Dry fruits fall to the ground due to abscission at the point of fruit stalk. The wind also causes fruit fall. The fallen fruits serve as food for the rodent and are also susceptible to a fungal species. The fallen fruits are carried away by rain water up to a distance of 1 km suggesting that the rain water is effective in fruit dispersal. The fruits thus dispersed and settled eventually decompose exposing the seed for subsequent germination.

12.4.9 *Cycas beddomei* (Cycadaceae)

Cycas beddomei is a critically endangered species with a population size less than 1000 trees extended in an area of 50 km^2 at an altitude of 650-1500 m. It is restricted to dry, open hill slopes, in open grassy woodland or grassland of Seshachalam hills of Chittoor and Cuddapah districts of Andhra Pradesh in eastern India. This species was listed in Appendix I (it prohibits trade in wild-collected plants, except for scientific or conservation purposes) of CITES in 1985. Further, it was included in the list of specified plants as per Schedule 6 of Indian Wildlife (Protection) Act 1972. This plant attracts *Alphitobius* beetles at the time of maturation of male and female cones. In male cones, the beetle feeds on the parenchymatous tissue of microsporophylls; the parenchyma has been reported to be rich in starch. After feeding, it mates and produces a single generation of offspring within two weeks. The adults thus emerged carry pollen and in search of other maturing male cones they also visit female cones by mistake. The beetle is a short distance flier and its flight activity takes place during dusk and early evening hours due to its sensitivity to bright sunlight. In cycads, almost all parts of the plant produce toxins, the concentration of which varies with each part. The seeds also contain toxins, usually have brightly coloured sarcotesta, and attract a variety of birds and mammals. These animals feed on the fleshy sarcotesta and disperse the seeds which are protected by a hard sclerotesta. The sarcotesta is barely toxic or if it does contain toxins, the animals feed on this layer somehow cope with them, like many animals do with toxic angiosperm fruits. The squirrels feed on the sarcotesta of *C. beddomei* and disperse the seeds with sclerotesta intact. Other animals such as rodents, bears and birds are also present in the study areas but we never found them feeding on the seeds of *C. beddomei.* Such defensive chemicals might be present in *C.beddomei* also. Rain water also disperses the seeds but the rugged terrain of the area does not permit dispersal far away from the parent sites. An unidentified beetle raises a single generation of offspring in the interior of the seed and the emerging adults diapause in the soil until the next seed season. This beetle is responsible for a larger percentage of seed infestation or predation and hence is regulating the reproductive success and population size.

12.4.10 *Cycas sphaerica* (Cycadaceae)

C. sphaerica although prominent in the literature, its distributional areas and taxonomic characters are poorly described. This species is listed under Data Deficient category of IUCN Redlist and included in CITES Appendix II that allows trade in wild-cultivated seeds or plants as long as there is no detrimental impact on the survival of wild populations. It occurs in the eastern ghats of Srikakulam District, Andhra Pradesh, but it was wrongly documented as *C. circinalis* in the Flora of Srikakulam, Andhra Pradesh. This species attracts

Derelomus weevils at the time of maturation of male and female cones. The male cones provide a platform for the weevils for communal sexual display to choose their partners. The weevils feed on the cone axis tissue, possibly because it is starch-rich. They do not feed on pollen although it is protein-rich. After feeding, they mate and use sporophyll tissue for feeding and breeding; they produce a single generation of offspring within two weeks on a male cone and like-wise they may produce eight generations within the three-month period of the annual coning episode. The adults thus emerged get covered with pollen and then depart the male cone in search of other male cones during which they also visit female cones thus effecting pollination. The weevils are short distance fliers and fly during dusk and early evening hours due to their sensitivity to sunlight. The only reward that the weevils get from the female cones is that the latter provide them warmth during the night. Female cones attract only a small number of weevils since they do not offer feeding and breeding environment. The Spotted deer and Sloth Bear feed on the sarcotesta of *C. sphaerica* and may disperse the seeds with sclerotesta intact. Other animals such as rodents and birds are also present at the plant site but we never found them feeding on the seeds of *C. sphaerica*. The rain water also disperses the seeds but the rugged terrain of the area does not permit dispersal far away from the parent sites. The seed infestation or predation is totally absent suggesting that its seeds are highly toxic and hence no insect uses them for breeding.

12.4.11 *Wendlandia glabrata* (Rubiaceae)

It has scattered distribution in the Eastern Ghats of Andhra Pradesh. The distribution pattern indicates that there are a few individuals occurring here and there along the slopes of the hills in this region. The hill slopes are characteristically rocky and the plant grows only in the rock crevices and produces a network of strong root system. Since the population is very small, it is considered to be a rare semi-evergreen tree species that blooms during dry season when most of the plant species go leafless and even without flowering. The fruit growth and development begins immediately after pollination and fertilization. The fruit is small (2 mm long and 3 mm wide), globose capsule, light green initially and brown when mature; each fruit produces 37-48 minute seeds (1 mm long). The fruits split apart to disperse seeds into the air during the summer season. As the seeds are very tiny, light in weight and fly easily in the air, they are adapted for effective dispersal by wind and hence are anemochorous.

12.4.12 *Wendlandia tinctoria* (Rubiaceae)

It is a semi-evergreen tree species confined to the southern Eastern Ghats of Andhra Pradesh, India. A small population occurs in this region. The fruit growth and development begins immediately after pollination and fertilization. The fruit set in open-pollinations is 7%. The fruit is a small globose dry capsule, light green initially and brown when mature. Each fruit produces 45-53 seeds out of which viable seeds range from 34 to 46 while the remaining ones are either half-filled or ill-formed. The fruits mature in about five months and dehisce to release the tiny seeds into the air. This occurs during late rainy season but it is unfavourable for the seed dispersal, germination and establishment.

12.5 Concluding Remarks

The case studies briefed above regarding the seed dispersal syndromes and the seed or fruit predation indicate that there are certain evolved characters that assist in the efficient functioning of the seed dispersal syndromes concerned. The fruit or seed predation noted in these species has a negative effect and hence curtailing the success of sexual reproduction. The detailed studies on the tree species in particular are required with reference to the seed dispersal and the seed predation since the tree cover is the backbone to the forest ecosystems, in the present case Eastern Ghats which is representative of deciduous forest ecosystem. Without this information, it is impossible to consider any conservation and management measures for the studied and to be studied plant species in the Eastern Ghats or elsewhere in India.

References

Caswell H, Lensink R, Neubert MG 2003. Demography and Dispersal: Life table response experiments for invasion speed. Ecology 84: 1968–1978.

Corlett RT 1998. Frugivory and seed dispersal by vertebrates in the Oriental (Indomalayan) Region. Biol Rev 73: 413–448.

Denslow JS, Moermond TC 1982. The effect of accessibility on removal rates of fruits from tropical shrubs: An experimental study. Oecologia 54: 170–176.

Fedriani JM, Manzoneda A 2005. Pre– and post–dispersal seed predation by rodents: balance of food and safety. Behav Ecol 16: 1018–1024.

Figueroa JA, Muñoz A A, Mella JE, Arroyo Mary TK 2002. Pre– and post–dispersal seed predation in a Mediterranean–type climate montane sclerophyllous forest in central Chile. Austr J Bot 50: 183–195.

Gorchov DL 1987. Sequence of fruit ripening in bird–dispersed plants: Consistency among years. Ecology 68: 223–225.

Gorchov DL 1990. Pattern, adaptation, and constraint in fruiting synchrony within vertebrate–dispersed woody plants. Oikos 58: 169–180.

Griz IMS, Machado ICS 2001. Fruiting phenology and seed dispersal syndromes in Caatina, a tropical dry forest in the northeast of Brazil. J Trop Ecol 17: 303–321.

Gurevitch J, Scheiner SM, Fox GA 2006. Plant Ecology, 2nd ed. Sinauer Associates, Inc., Massachusetts.

Herrera CM 1982. Seasonal variation in the quality of fruits and diffuse coevoluiton between plants and avian dispersers. Ecology 63: 773–785.

Herrera CM 1984a. Avian interference of insect frugivory: an exploration into the plant–bird–fruit pest evolutionary triad. Oikos 42: 203–210.

Herrera CM 1984b. A study on avian frugivores bird–dispersed plants, and their interaction in Mediterranean scrublands. Ecol Monogr 54: 1–23.

Herrera CM 1985. Determinants of plant–animal coevolution: the case of mutualistic dispersal of seeds by vertebrates. Oikos 44: 132–141.

Howe HF, Smallwood J 1982. Ecology of Seed Dispersal. Annu Rev Ecol Evol S 13: 201–228.

Hughes L, Dunlop M, French K, Leishman MR, Rice B, Rodgerson L, Westoby M 1994. Predicting dispersal spectra—a minimal set of hypotheses based on plant attributes. J Ecol 82: 933–950.

Janzen DH 1970. Herbivores and the number of tree species in tropical forests. Am Nat 102: 592–595.

Janzen DH 1981. Patterns of herbivory in a tropical deciduous forest. Biotropica 13: 271–282.

Janzen DH 1982. Differential seed survival and passage rates in cows and horses, surrogate Pleistocene dispersal agents. Oikos 38: 150–156.

Karasov WH, Levey DJ 1990 Digestive system trade–offs and adaptations of frugivorous passerine birds. Physiol Zool 63: 1248–1270.

Kevan PG and Gaskell BH 1986. Biotropica 18: 76–78. Cited from Renner SS 1987. Prog. Bot. 49: 413–432.

Kubitzki K 1986. An Acad Brasil Cienc 57: 513–516. Cited from Renner SS 1987. Prog Bot 49: 413–432.

Lengyel S 2010. Convergent evolution of seed dispersal by ants, and phylogeny and biogeography in flowering plants: a global survey. Perspect Plant EcolEvol S 12: 43–55.

Levey DJ 1987. Facultative ripening in *Hamelia petans* (Rubiaceae): effects of fruit removal and rotting. Oecologia 74: 203–208.

Levey DJ, Moermond TC, Denslow JS 1984. Fruit choice in Neotropical birds: the effect of distance between fruits on preference patterns. Ecology 65: 844–850.

Maron JL, Simms EL 1997. Effects of seed predation on seed bank size and seedling recruitment of bush lupine (*Lupinus arboreus*). Oecologia 111: 76–83.

Moermond TC, Denslow JS 1983. Fruit choice in Neotropical birds: effects of fruits type and accessibility on selectivity. J Anim Ecol 52: 407–420.

Morden–Moore AL, Willson MF 1982. On the ecological significance of fruit color in *Prunus serotina* and *Rubus occidentalis*: field experiments. Can J Bot 60: 1554–1560.

Piper JK 1986. Seasonality of fruit characters and seed removal by birds. Oikos 46: 303–310.

Pratt TK, Stiles EW 1983 Am Nat 122: 797–805. Cited from Renner, S S 1987. Prog Bot 49: 413–432.

Renner SS 1987 Seed dispersal. Prog Bot 49: 413–432.

Samuni–Blank M, Izhaki I, Dearing MD, Karasov WH, Gerchman Y, Kohl KD, Lymberakis P, Kurnath P, Arad Z 2013. Physiological and behavioural effects of fruit toxins on seed–predating versus seed–dispersing congeneric rodents. J Exp Biol 216: 3667–3673.

Snow D W 1971. Evolutionary aspects of fruit–eating by birds. Can J Bot 113: 194–202.

Vanchaik CP, Terborgh JW, Wright SJ 1993. The phenology of tropical forests–adaptive significance and consequences for primary consumers. Annu Rev Ecol Evol S 24: 353–377.

Vander Wall SB, Kuhn KM, Beck MJ 2005. Seed Removal, Seed Predation, and Secondary Dispersal. Ecology 86: 801–806.

Wheelwright NT 1983. Fruits and the ecology of Resplendent Quetzals. Auk 100: 286–301.

Wheelwright NT 1985. Competition for dispersers, and the timing of flowering and fruiting in a guild of tropical trees. Oikos 44: 465–477.

Wichmann MC, Alexander MJ, Soons MB, Galsworthy S, Dunne L, Gould R, Fairfax C, Niggemann M, Hails RS, Bullock JM 2009. Human mediated dispersal of seeds over long–distances. Proc Roy Soc B-Biol Sci 276: 523-532.

Willson MR, Thomson JN 1982. Phenology and ecology of colour in bird–dispersed fruits, or why some fruits are red when they are "green"? Can J Bot 60: 701–713.

Willson MF, Whelan CJ 1990. The evolution of fruit color in fleshy–fruited plants. Am Nat 136: 790–809.

Willson MF, Irvine AK, Walsh NC 1989. Vertebrate dispersal syndromes in some Australian and New–zealand plant communities, with geographic comparisons. Biotropica 21: 133–147.

Yanjun D, Xiangcheng M, Xiaojuan L, Lei C, Keping M 2009. Seed dispersal phenology and dispersal syndromes in a subtropical broad–leaved forest of China. Forest Ecol and Manag 258: 1147–1152.

Chapter 13

Wasp-mediated Seed Dispersal, a Rare Phenomenon

TN Manohara
manohara_tn@yahoo.com

Abstract

Effective seed dispersal is critical for reproductive success of species under natural habitat. So far studies on seed dispersal are vertebrate-centric. Amongst insects, ants are recognized as secondary dispersers in a large number of species. In recent years, wasps have been reported to be important seed dispersers in a few species in North America; this has been termed as 'vespicochory'. Recently, I have described one of the wasp species, *Vespa affinis* to be the only effective seed disperser in *Aquilaria malaccensis*, popularly known as 'agarwood plant', an economically important, critically endangered and overexploited species of northeast India. Seeds of *A. malaccensis* produce caruncle on which the wasp forages. The distance for which the wasp disperses the seeds was found to be over 500 m, much farther than ants. Wasp dispersal of seeds in *A. malaccensis* is the first report not only in a tree species but also in tropical forests. It would be profitable to look for seed dispersal by wasps, particularly in species producing elaiosomes, to understand the extent of the mutualism in seed dispersal between plants and wasps.

13.1 Introduction

Effective dispersal of seeds is one of the most critical aspects of reproductive ecology; it enables the progeny to escape from specialist predators and pathogens concentrated around the parent and results in deposition of at least some seeds in safe sites where they can successfully germinate and establish seedlings (Dennis et al. 2007, Fenner and Thompson 2005). Seed dispersal limitation/failure has been reported to be one of the main constraints that drive species to vulnerable category, particularly in tropical forests (Corlett 2007). Vertebrates are the main animal seed-dispersers. Extensive studies have

DOI: 10.1201/9781003505471-13

been carried out on their mode of dispersal and its role on seedling recruitment (Bronstein et al. 2006, Fenner and Thompson 2005, Howe and Smallwood 1982, Schaefer and Ruxton 2011, Tiffney 2004, Turner 2001). Amongst insects, ants are well recognized as secondary seed dispersers (myrmecochory) of species producing elaiosomes. According to an estimate, seeds of over 3000 plant species distributed in all the continents except Antarctica are myrmechochorus (Handel and Beattie 1990). Wasps are known to be effective pollinators but not seed dispersers (Faegri and van der Pijl 1971, Sühs et al. 2009). However, there are several reports, in recent years, on seed dispersal by wasps. This communication summarizes the details of wasp dispersal so far reported.

Fig. 13.1.a-e. Seed dispersal in *Aquilaria malaccensis*. ***a***. Habit ***b.*** Flowering Branch ***c.*** Dehisced fruit ***d.*** Wasp perching on to the seed caruncle ***e.*** Wasp feeding on the caruncle.

Pellmyr (1985) was the first to describe seed dispersal by wasp in *Vancouveria hexandra* (Berberidaceae) in Washington State, USA. Subsequently, Jules (1996) reported the involvement of both ants and wasps (the yellow jackets, *Vespula vulgaris)* in dispersing the seeds of *Trillium ovatum* (Liliaceae) from western North America.Yellow jacket perches on dehisced multiseeded fruit of *T. ovatum*, pull out a seed and flies away. The dispersal distance in the forest ranged from 30 m to 3 km, much farther than those dispersed by ants. Wasp dispersal was termed as 'vespicochory' (Jules 1996). Later, Zettler et al. (2001) reported the dispersal of seeds of three other *Trillium* species, *T. undulatum, T. catesbaei,* and *T. cuneatum* by *Vespula flavopilosa* and *Vespula maculifrons* from North America. The wasps foraged seeds from the open capsules and the dispersal distance ranged from 1.4 m to >20 m. Several conspecific wasps were found to forage fruits at a time and most of the seeds (94%) were removed within an hour. The wasps invariably feed on elaiosomes and discard the seeds.

13.2 Wasp dispersal in *Aquilaria malaccensis*

Recently, I described the details of wasp dispersal in a tree species, *Aquilaria malaccensis* (Malvales: Thymelaeaceae), popularly known as 'agarwood plant'. Agarwood plant is an economically important, critically endangered and highly exploited species of North East (NE) India (Manohara 2013). The genus has 15 species occurring mainly in the tropical forests of Asia and is native to South East Asia. In India, the genus is represented by two species, *A. malaccensis* and *A. khasiana*; the former exhibits wider range of distribution throughout North East (NE) India and the latter is endemic to Khasi Hills of Meghalaya. *Aquilaria malaccensis* is the major source of agarwood, a resinous heartwood, used in perfume and incense industry (Broad 1995, Ng et al. 1997). The resin is produced by the stem in response to infection by a parasitic ascomycetous fungus, *Phaeoacremonium parasiticum*, (formerly *Phialophora parasitica*) (Crous et al. 1996). The infected resinous heartwood called 'agar' or 'gaharu' has been traded internationally for centuries. As a result of overexploitation throughout its range, its natural population has reached its critical minimum (Barden 2000), and presently found only in few protected pockets in the NE India. It is considered to be globally threatened in the wild and is included in 'The World List of Threatened Trees' (Khan 2012, Oldfield et al. 1998). The species has been listed in Red List (IUCN Red List, Version 2009.2, Vulnerable A1cd, Year Assessed: 1998) and is included in Appendix II of CITES and all export of its products has been prohibited, although its trade is continuing illegally. Under the present rate of exploitation, it is likely to get extinct in the near future.

Effective conservation measures of the species are urgently needed. For any successful conservation efforts, it is important to understand its reproductive ecology to identify reproductive constraints for effective mitigation. However, our knowledge on the reproductive ecology of *Aquilaria* is meagre (Soehartono and Newton 2001, Uma Shankar 2012). Studies were carried out during 2011 and 2012 fruiting seasons on natural populations in Nagaland and also on those growing at Rain Forest Institute, Jorhat, and several homestead gardens in Nagaland and Assam.

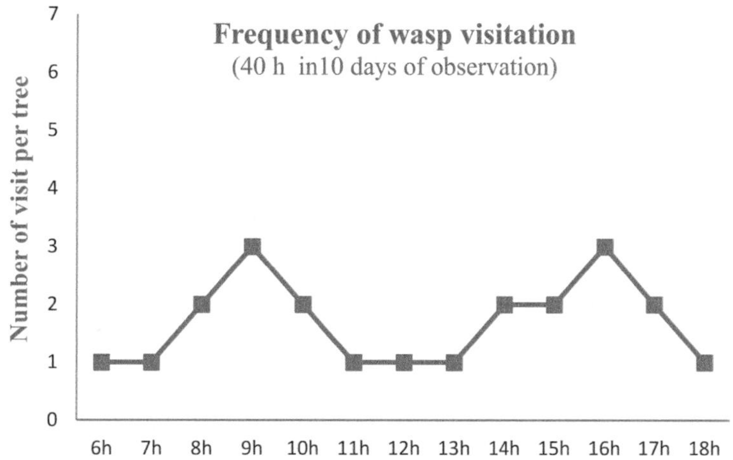

Fig. 13.2. Frequency of wasp visitation to the fruits of *Aquilaria malaccensis*

Trees of *A. malaccensis* start flowering and fruiting after they reach four to five years of age (Figure 13.1a). Flowering and fruiting occur in the dry season – flowering in April to May and the fruiting in June-July. The yellowish-green flowers are produced in umbels (Figure 13.1b). Flowers open (anthesis) during the night between 22.00 and 04.00 h. Flowering and fruiting are profuse in the populations studied; 4500-6000 flowers and 900-1200 fruits were recorded in six to eight-year-old trees. Nearly 20 to 25 per cent of the fruits were found to be infested by larvae of thrips and other insects. The fruit is a bilocular leathery dehiscent capsule measuring about 3.5 to 4.0 cm x 2.0 to 2.5 cm (Figure 13.1c,d) bearing a single seed in each locule. Mature fruits split longitudinally while they are still on the tree; the seeds remain hanging through thread-like structure (Figure13.1c) for a few days after dehiscence. Each seed bears a conspicuous crimson red, fleshy caruncle at the tip (Figure 13.1c,d). Detailed observations on dehisced fruits indicated that a yellow banded wasp, *Vespa affinis* L. (Hymenoptera: Vespidae) (Das and Gupta 1989, Kimsey and Carpenter 2012) is the only animal involved in seed dispersal. The brightly coloured caruncle seems to attract the wasp (Figure 13.1e). As soon as the wasp perches on the seed, the thread-like structure connecting the seed to the fruit breaks and the wasp carries the seed over varying distances before it lands on the leaves or branches of the same or any other tree. It feeds on the caruncle leaving behind an undamaged seed that falls to the ground. The flight of some of the seed-carrying wasp could be followed up to 500 m away from the parent tree. The wasps feed only on caruncle of fresh seeds in those one to two days after dehiscence, in which the caruncle remains fresh. Seeds that are more than three days old by which time the caruncle becomes dry, as well as naturally fallen seeds and those experimentally placed on the ground remained untouched by wasps. On a bright sunny day, three to four wasps were found visiting each tree. Wasp visitation starts by 06.00 h and continues until 18.00 h. The frequency of visits show two peaks one between 8 to 10 h in the morning and the other between 14 to 17 h in the afternoon (Figure 13.2). Although there were two or three other wasp species observed in the vicinity of the agarwood trees, only *V. affinis* visited the fruits and foraged the caruncle. During the peak of the fruiting season, 10-15 fruits per plant dehisce on each day and 20-30 fresh seeds are available on any given day. To determine the proportion of dispersed seeds, 278 seeds from six to eight-year-old marked trees (*N*=15) were kept under observation; of these 69 seeds (amounting to about one fourth of the observed seeds) were dispersed by wasps. The remaining seeds fell on the ground below the canopy by the action of wind and gravity. The viability of seeds dispersed by *V. affinis* (after consuming the caruncle) and fresh seeds harvested from the trees were tested by germinating them in Petri plates lined with moist filter paper. Both wasp-dispersed and fresh seeds germinated in about two weeks and in both the cases, 87 per cent germination was observed.

13.3 Concluding Remarks

As pointed out earlier, studies on seed dispersal have so far been vertebrate-centric. Wasps have been shown to be effective dispersers at least in six species including *Aquilaria malaccensis*. Five of the species dispersed by wasps are herbs/shrubs and their seeds produce elaiosomes as the reward for dispersal agents. The seeds of some of these species dispersed by wasps are also dispersed by ants. Wasps are the better dispersers when compared to ants as the distance of dispersal by wasps is markedly more than that of ants. Seeds of *A. malaccensis* contain caruncle as the reward for the wasp and wasp was the only animal involved in seed dispersal of this species. *A. malaccensis* is the first report of wasp dispersal in a tree species and also in tropical forests. More species of wasp dispersal are likely to be added in the coming years by carefully looking for non-vertebrate seed dispersals under varied habitats.

Acknowledgements

The financial assistance from ICFRE is gratefully acknowledged. I sincerely thank the Director, Rain Forest Research Institute, Jorhat, for providing necessary facilities and encouragement and Prof KR Shivanna, INSA Honorary Scientist, Ashoka Trust for Research in Ecology and Environment (ATREE), Bangalore for going through the manuscript.

References

Barden A 2000. Heart of the Matter. Agarwood Use and Trade and CITES Implementation for *Aquilaria malaccensis*. TRAFFIC International, Cambridge, pp 52.

Broad S 1995. Agarwood harvesting in Vietnam. TRAFFIC Bulletin 15: 96.

Bronstein JL, Alarcon R, Geber M 2006. Tansley Review: Evolution of insect/plant mutualisms. New Phytol 172:412–428.

Corlett RT 2007. Pollination or seed dispersal: which should worry about most? In: Dennis AJ, Schupp EW, Green RJ, and Wescott DA (Eds), Seed Dispersal: Theory and its Application in a Changing World. CAB International, Wallingford, UK, pp 523–544.

Crous PW et al. 1996. *Phaeoacremonium* gen nov associated with wilt and decline diseases of woody hosts and human infections. Mycologia 88:786–796.

Das BP, Gupta VK 1989. The Social Wasps of India and the Adjacent Countries. Gainesville, The Association for the Study of Oriental Insects, pp.292.

Dennis AJ, Schupp EW, Green RJ, Wescott DA 2007. Seed Dispersal: Theory and its application in a changing world. CAI International, Wallingford, UK.

Faegri K, van der Pijl L 1971. The Principles of Pollination Ecology. 2nd Edition, Pergamon Press, New York, pp 291.

Fenner M, Thompson K 2005. The Ecology of Seeds. Cambridge Univ Press, Cambridge, UK.

Handel SN, Beattie AJ 1990. Seed dispersal by ants. Sci Am 262:76–83.

Howe HF, Smallwood J 1982. Ecology of seed dispersal. Annu Rev Ecol Evol S 13:201–228.

IUCN, IUCN Red List of Threatened Species, Version 2012.2; www.iucnredlist.org. Retrieved on 22 March 2013.

Jules ES 1996. Yellow jackets (*Vespula vulgaris*) as a second seed disperser for the myrmecochorous plant, *Trillium ovatum*. Am Midland Nat 135:367–369.

Khan ML 2012. *Aquilaria malaccensis* Lam., a Red–listed and highly exploited tree species in the Assamese home garden. Curr Sci 102:546–547.

Kimsey L, Carpenter J 2012. The Vespinae of North America (Vespidae, Hymenoptera). J Hymenopt Res 28:37–65.

Manohara TN 2013. Wasp-mediated seed dispersal in agarwood plant (*Aquilaria malaccensis*), a critically endangered and overexploited species of North East India. Curr Sci 105:298–299.

Ng LT, Chang YS, Kadir AA 1997. A review on agar (gaharu) producing *Aquilaria* species. J Trop For Products 2:272–285.

Oldfield S, Lusty C, Mackinven A 1998. The Word List of Threatened Trees. World Conservation Press, Cambridge, pp 650.

Pellmyr O 1985. Yellow jackets disperse *Vancouveria* seeds (Berberidaceae). Madroño 32: pp 56.

Schaefer HM, Ruxton GD 2011. Plant–animal communication. Oxford Univ Press, New York.

Soehartono T, Newton AC 2001. Reproductive ecology of *Aquilaria* spp. In Indonesia. For Ecol Management 152:59–71.

Sühs RB, Somavilla A, Putzke J, Köhler A 2009. Pollen vector wasps (Hymenoptera, Vespidae) of *Schinus terebinthifolius* Raddi (Anacardiaceae), Santa Cruz do Sul, RS, Brazil. Braz J Biosci 7:138–143.

Tiffney BH 2004. Vertebrate dispersal of seed plants through time. Annu Rev Eco Evol S35:1–29.

Turner IM 2001. The Ecology of Seeds in Tropical Rain Forests. Cambridge Univ Press, Cambridge, UK.

Uma Shankar U 2012. Effect of seed abortion and seed storage on germination and seedling growth in *Aquilaria malaccensis* Lamk. (Thymelaeaceae). Curr Sci 102:596–604.

Zettler JA, Spira TP, Allen CR 2001. Yellow Jackets (*Vespula* spp.) disperse *Trillium* (spp.) seeds in Eastern North America. Am Midland Nat 146:444–446.

Chapter 14

Ants as Seed Predators and Dispersers of a Non-myrmecochorous tree *Lagerstroemia speciosa* in tropical moist-deciduous forest

Deepu Sivadas and A G Pandurangan*
deepusivadas01@gmail.com

Abstract

Ants have been traditionally considered either as predators or dispersers of seeds, but not both (Retana et al. 2004). However, harvesting ants might be simultaneously antagonistic and mutualistic towards seeds. This study analyses the predation-dispersal relationship between ants and seeds of *Lagerstroemia speciosa* (L.) Pers., a non-myrmecochorous tree, in a moist deciduous forest in Southern Western Ghats. The seeds are explosively dispersed and the mean ballistic dispersal distance for a tree was 3.7 m. Seven ant species were found to carry seeds but only one species (*Diacamma* sp.) regularly dispersed seeds, up to a distance of three to five metres. The ants usually deposit the seeds in the nest. The seed removal rate was 85 per cent (N = 200) over the first 24 h of exposure. The results obtained from the present study confirmed that ants acted as both predators and dispersers of seeds of *L. speciosa*.

14.1 Introduction

Ant-plant interactions are ubiquitous in nature and provide countless examples of mutualism (Bronstein et al. 2006). Ants may play an important role in the dynamics of communities by acting either as seed dispersal agents or as seed predators or both. Often, the services provided to the plant by the ants fall into two categories: protection from herbivory or seed dispersal. In return, the ants receive a nutritional benefits and/or nest space. There are two main mechanisms through which ants disperse seeds – myrmecochory (seed dispersal in which elaiosome acts as reward) and diszoochory (seed dispersal performed by seed-harvesting/granivorous ants) (Arnan et al. 2012). Ants can serve as secondary seed dispersers, collecting seeds on the ground and moving them away from the parent plant (Beaumont et al. 2009). Often, collected seeds are brought to the

DOI: 10.1201/9781003505471-14

nest where the elaiosome is consumed. The seeds then either remain in the nest or are deposited outside the nest in refuse piles (Beattie 1985, Renard et al. 2010, Servigne and Detrain 2008).

Several mechanisms have been proposed to explain the advantage of myrmecochory from the plant's perspective (Beattie 1985). The dispersal-for-distance hypothesis proposes that ants may decrease parent-offspring conflict or sibling competition simply by moving the seeds away from the parent's seed shadow (Andersen 1988, Higashi et al. 1989, Kalisz et al. 1999). Other hypotheses consider the advantage of seed burial by ants, including predator-avoidance (Bond and Breytenbach 1985, Boulay et al. 2009, Heithaus 1981), protection from fire (Berg 1966, Bond and Slingsby 1984), and directed dispersal (Howe and Smallwood, 1982). The directed dispersal hypothesis proposes that ant nests serve as favourable microhabitats that increase plant growth, survival, and/or fitness relative to random locations (Howe and Smallwood 1982, Wenny 2001).

In this study, we tested whether native ants confer benefits consistent with predictions under the directed dispersal hypothesis to a non-myrmecochorous plant (*Lagerstroemia speciosa* (L.) Pers., Queen's Crape-Myrtle, Lythraceae) in Southern Western Ghats, India. The tree is native to tropical Asia and distributed in Cambodia, China, India, Indonesia, Laos, Malaysia, Myanmar, Philippines, Thailand and Vietnam (Anonymous 1996, De Guzman et al. 1986, Krithikar and Basu 1987). It attains 7-21 m tall with wide spreading crown. It is usually found in the moist deciduous and semi evergreen forests. The tree is a potential source of corosolic acid, which is known for its anti-diabetic and anti-obesity activity (Klein et al. 2007). The plant also possesses compounds like α and β PGG, possessing many other health-beneficial bioactivities such as anticancer, anti-HIV, anti-SARS, anti-inflammation and antioxidant activity (Chen et al. 2005, Feldman et al. 1999, 2001, Klein et al. 2007, Nakashima et al. 1996). Due to its high medicinal and economic value, the tree is highly exploited from wild and this shows why its populationis in a declining trend. The recruitment is also found to be poor and hence we studied the seed dispersal of this species in the natural populations.

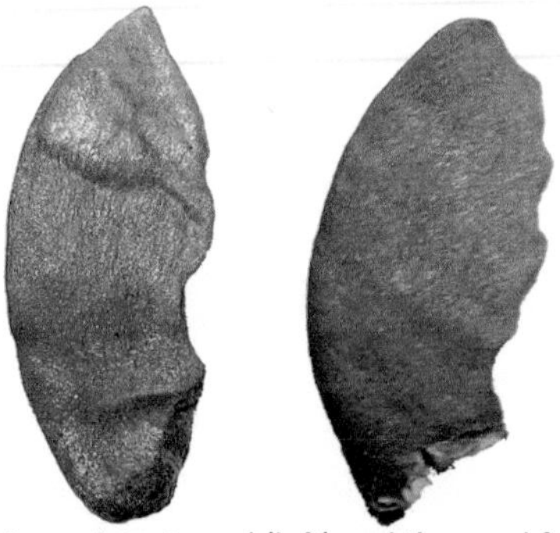

Fig. 14.1. *Lagerstroemia speciosa* – intact seed (left) and the seed from the ant refuse pile (right)

The flowering was observed during March to June and fruit setting initiated in April with a maturation period of 125–180 days. The fruit is a loculicidal capsule with 119 ± 8.09 seeds. A medium sized tree is capable of producing on an average 1,75,545 ± 11,934 seeds (persobs). The seeds are albuminous, about 71 ± 11 x 141 ± 42 mm in dimensions, weighing about 6.16 ± 1.14 mg (Figure 14.1). The primary dispersal of seeds occurs by explosive dehiscence of fruits.

Here we document a complex relationship of seed predation and secondary dispersal by ants and survival of seeds in the focal species. In light of these studies, we examine how native ants benefit *L. speciosa* through secondary seed dispersal and test whether the benefits are consistent with predictions under the directed dispersal hypothesis.

14.2 Methods

Studies were carried out at Chaliarmukku, a deciduous forest, located in the Nilambur forest range of Western Ghats (11°18'24.7"N; 76°14'43.8"E; ± 30 m a.s.l.) during March 2013. The study area is dominated by *L. speciosa* with associated species like *Terminalia elliptica*, *Diospyros ovalifolia*, *Terminalia paniculata*, *Bambusa bamboos* etc., without a prominent understory. The seed shadow (Wang and Smith 2002) was examined in the study site for 5 trees without overlapping canopy. To examine seed ant interactions, a seed removal experiment was conducted, 200 ink marked seeds were arranged on a 1 x 1 m plot about three metres from the fruiting tree. Interactions were observed between 0700 and 1800 h on 7 and 8 March 2013 at a time interval of 15 min. Ants were followed to determine the distance for which the seeds are carried and the site of deposition. Five replicates were performed on five different sites. In order to understand the seeds removal by other organisms like rodents or other vertebrates, the sites were examined for foot prints. The specimens of ants interacting with the seeds were collected, photographed and identified with the help of experts. The seeds collected from refuse piles were subjected to germination trials under *in situ* conditions for determining the germination percentage.

14.3 Results and Discussion

At the study site, *L. speciosa* fruits start dehiscing by January and continue up to March and the seeds were dispersed by the ballistic mechanism of the loculicidal capsules and then wind acting as a medium for the dispersal. However, not all the seeds are dispersed during this process and a certain amount of seeds will remain inside the capsule. The seeds are generally dispersed to a distance of three to five metres from the infructescence during the bursting of capsules. It was observed that the period for which the locules remain open depends on the atmospheric moisture, if an intermittent rain occurs, the locules will close.

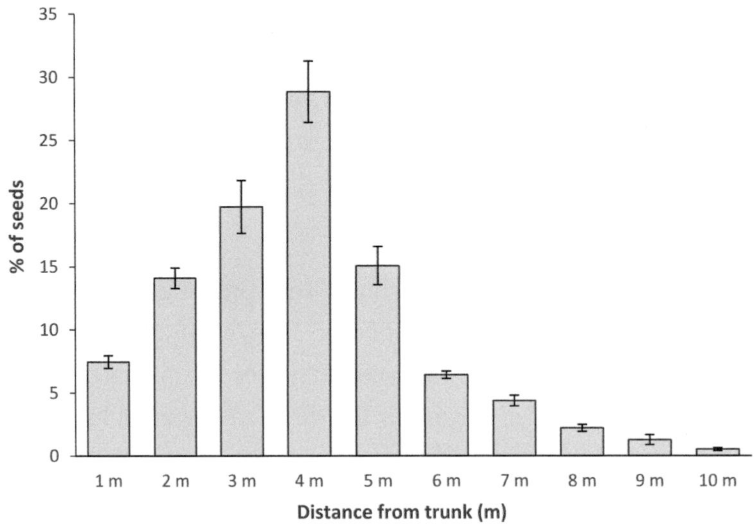

Fig. 14.2. Percent (mean ± SD of 5 plants sampled) of *L. speciosa* seed distribution by explosive mechanism

The seed dispersal in the absence of ants occurred over short distances from the parent plant (Figure 14.2). Five trees with non-overlapping canopy were studied to understand the seed shadow, considering that the mean radius of canopy is ca. three metres the explosive dispersal alone throws only a part of seeds beyond the maternal patch. Around 41 per cent seeds were dispersed with in the seed shadow and 28 per cent (N = 8357 seeds collected) were found at four metres from the plant, while the mean distance was 3.7 m (range: 0–10 m). Observations showed that seven ant species interacted with *L. speciosa* seeds at the study site, supporting the generalization that interactions between plants and ants are often not highly specialized or obligate to the particular species (Beattie et al. 1979, Davidson and Morton 1984, Passos and Ferreira 1996). Ant species, their behaviour and dispersal of seeds in the experimental plot is given in Table 14.1. Three ant species removed seeds to distances shorter than one metre and thus did not act as dispersers. Most frequently the species such as *Camponotus parius*, *Camponotus* sp.1, *Diacamma* sp. and *Cataulacus* sp. regularly removed the seeds up to 8.3 m. All ant species recruited a large numbers of workers for the seeds, and the individual ants usually returned to the seed depots several times to obtain more seeds. A single individual of *Diacamma* sp. was able to carry the collected seed into the nest. Three ant species viz., *Polyrhachis* sp., *Technomyrmex* sp. and *Oecophylla smaragdina* that only eat the seeds and reduce the chance of *L. speciosa* seeds being removed. The seeds of *L. speciosa* are oil rich, which contained several essential fatty acids, and therefore, have potential nutritional value (Tarandjiiska et al. 1996, Wei et al. 2006). The weaver ant *Oecophylla smaragdina*, *Polyrhachis* sp. and *Technomyrmex* sp., nesting on trees were found to take away non-dispersed seeds left on the dehisced fruits which are persistent on the trees.

Table. 14.1: Ant behaviour and dispersal of *L. speciosa* seeds in seed removal experiments.

Ant species	Behaviour	Distance seed moved (m) mean (N = 200)
Camponotus parius	d, p	4.50 ± 1.43 (n = 12)
Camponotus sp.1	d, p	5.21 ± 1.72 (n = 15)
Diacamma sp.	d, p	8.29 ± 3.12 (n = 132)
Cataulacus sp.	d, p	5.76 ± 2.22 (n = 7)
Polyrhachis sp.	p, t	—
Oecophylla smaragdina	p	—
Technomyrmex sp.	p	—

Behaviour: p - predator; d - disperser; t – try to remove seeds (distance < 1 m).

During the transport of the seeds to the nest through arbitrary trails by *Diacamma* sp., 13.1 ± 1% (mean + SD of three nests sampled) were lost (dispersal by seed drops on their foraging trail) before reaching the nest. Foragers of the same nest were not found to collect the seeds dropped along the trail to the nest again, at least during the current daily activity period. The seed removal rate was 85 per cent (N = 200), being similar to those reported for other Neotropical areas and higher than myrmecochores from temperate forest (Horvitz and Schemske 1986). Jungle fowl was the only vertebrate found to feed the seeds of *L. speciosa*, fallen on forest floor. Termites were also seen to collect the seeds of *L. speciosa*. Since our study was conducted in a forest fragment over a scheduled time, the local composition of seed predators may thus be altered and hence escape hypothesis cannot be excluded (Fonseca and Robinson 1990). The seeds without embryo and endosperm were frequent on refuse piles of ant nest in the study area (95.2%, N = 132), suggesting that transport of the seeds to the refuse zone does not benefit this plant. The intact seeds collected from the refuse piles (N = 3 nests), which contained 4.2 ± 1% ink marked seeds, were subjected to germination studies and the germination percentage was found to be five per cent (N = 100).

Seeds of *L. speciosa* are dispersed passively by gravity following ballistic dispersal. The results of this study confirm the role of seed harvesting ants as both predators and dispersers of the non-myrmecochorous *L. speciosa*. Granivorous ants are not usually very large in number, but they have a considerable impact on seeds by predation without playing any role in dispersal (Auld and Denham 1999, Brown et al. 1979, Diaz 1992). In the study area, seven ant species predated seeds, but only four of them were involved in seed dispersal. Nevertheless, not all seeds collected by the ants are consumed; some of them escape predation, either because they are dropped on the way to the nest, or because they are mistakenly rejected on the refuse pile. From the observations, *Diacamma* sp. acts as an effective dispersal agent for *L. speciosa* seeds as it displaces the seeds up to a distance of 8.29 ± 3.12 m from the mother plant.

Removal of *L. speciosa* seeds by native ants occurred rapidly with most seeds being removed within 24 hours. Three ant species acted as secondary dispersers and deposited the seeds in their nests. Considering that the mean radius of the canopy of *L. speciosa* is ca. three metres, the explosive dispersal alone throws only part of the seeds beyond the maternal patch. According to Berg (1966), removal of some seeds beyond the ballistic limits of the parent plants may be an effective method of local dissemination of seeds. As per the directed dispersal hypothesis the seeds dispersed by the ants should establish as they are deposited in a favourable microclimate.

However, the dual role of ants, such as *Diacamma* sp. in the present study, as both predators and dispersers of seeds, confirms that the relationship between seeds and ants is not always evident, i.e., whether it acts as a predator or disperser. Not all *L. speciosa* seeds removed by harvesting ants are taken into nests, and not all seeds taken into nests are consumed; that is, some of the seeds collected by ants might be dispersed as suggested by Retana et al. (2004).

Acknowledgements

We thank the Director, JNTBGRI for the facilities provided. Dr. Santhosh Shreevihar, Department of Zoology, Malabar Christian College, Kerala, India and Dr. Thressiamma, Centre for Ecological Sciences, IISC, Bangaluru are duly acknowledged for their support in the identification of ants. This work was under the financial assistance of the Department of Biotechnology, Govt of India. The authors also like to extend their thanks to the Forest Department, Kerala, for their logistic support.

References

Andersen A 1988. Dispersal distance as a benefit of myrmecochory. Oecologia 75: 507–511.

Anonymous 1996. The Wealth of India. National Institute of Science Communication and Information Resources, CSIR, India.

Arnan X, Molowny-Horas R, Rodrigo A, Retana J 2012. Uncoupling the Effects of Seed Predation and Seed Dispersal by Granivorous Ants on Plant Population Dynamics. PLoS ONE 7: e42869.

Auld TD, Denham AJ 1999. The role of ants and mammals in dispersal and post-dispersal seed predation of the shrubs *Grevillea* (Proteaceae). Plant Ecol 144: 201–213.

Beattie AJ 1985. The evolutionary ecology of ant-plant mutualisms. Cambridge University Press, New York.

Beattie AJ, Culver DC, Pudlo RJ 1979. Interactions between ants and the diaspores of some common spring flowering herbs in West Virginia. Castanea 44: 177–186.

Beaumont KP, Mackay DA, Whalen MA 2009. Combining distances of ballistic and myrmecochorous seed dispersal in *Adriana quadripartita* (Euphorbiaceae). Acta Oecol 35: 429–436.

Berg RY 1966. Seed dispersal of Dendromecon: its ecology, evolutionary and taxonomic significance. Am J Bot 53: 61–73.

Bond W, Slingsby P 1984. Collapse of an ant-plant mutualism: the Argentine ant (*Iridomyrmex humilis*) and myrmecochorous Proteaceae. Ecology 65: 1031–1037.

Bond WJ, Breytenbach GJ 1985. Ants, rodents and seed predation in Proteaceae. South African J Zool 20: 150–154.

Boulay R, Carro F, Soriguer R, Cerdá X 2009. Small-scale indirect effects determine the outcome of a tripartite plant-disperser-granivore interaction. Oecologia 161: 529–537.

Bronstein JL, Alarcón R, Geber M 2006. The evolution of plant-insect mutualisms. New Phytol 172: 412–428.

Brown JH, Reichman OJ, Davidson DH 1979. Granivory in desert ecosystems. Annu Rev Ecol Evol S 10: 201–227.

Chen CN, Lin CP, Huang KK, Chen WC, Hsieh HP, Liang PH, Hsu JT 2005. Inhibition of SARS-CoV 3C-like protease activity by Theaflavin-3, 30-digallate (TF3). Evidence-Based Complementary and Alternative Medicine 2(2): 209–215.

Davidson DW, Morton SR 1984. Dispersal adaptations of some *Acacia* species in the Australian arid zone. Ecology 65: 1038–1051.

De Guzman E, Umali R, Sotalbo E 1986. Guide to Philippine Flora and Fauna, III: Dipterocarps and Non-Dipterocarps. Quezon City: Natural Resources Management Centre and University of the Philippines, Diliman Quezon City, Philippines.

Diaz M 1992. Spatial and temporal patterns of granivorous ant seed predation in patchy cereal crop areas of central Spain. Oecologia 91: 561–568.

Feldman KS, Sahasrabudhe K, Laulor MD, Wilson SL, Lang CH, Scheuchenzuber WJ 2001. In vitro and in vivo inhibition of LPS stimulated tumour necrosis factor-a secretion by the gallotannin b-D-pentagalloylglucose. BMC Letters 11: 1813–1815.

Feldman KS, Sahasrabudhe K, Smith RS, Scheuchenzuber WJ 1999. Immunostimulation by plant polyphenols: a relationship between tumour necrosis factor a production and tannin structure. BMC Letters 9: 985–990.

Fonseca GAB, Robinson JR 1990. Forest size and structure: competitive and predatory effects on small mammal communities. Biol Conserv 53: 265–294.

Heithaus ER 1981. Seed predation by rodents on three ant-dispersed plants. Ecology 62: 136–145.

Higashi S, Tsuyuzaki S, Ohara M, Ito F 1989. Adaptive advantages of ant-dispersed seeds in the myrmecochorous plant *Trillium tschonoskii* (Liliaceae). Oikos 54: 389–394.

Horvitz CC, Schemske DW 1986. Seed dispersal of a neotropicalmyrmecochore: variation in removal rates and dispersal distance. Biotropica 18: 319–323.

Howe HF, Smallwood J 1982. Ecology of seed dispersal. Annu Rev Ecol Evol S 13: 201–228.

Kalisz S, Hanzawa FM, Tonsor SJ, Thiede DA, Voigt SJ 1999. Ant mediated seed dispersal alters pattern of relatedness in a population of *Trillium grandiflorum*. Ecology 80: 2620–2634.

Klein G, Kim J, Himmeldirk K, Cao Y, Chen X 2007. Antidiabetes and Anti-obesity Activity of *Lagerstroemia* speciosa. Evidence-Based Complementary and Alternative Medicine 4: 401–407.

Krithikar KR, Basu BD 1987. Indian Medicinal Plants vol 1-4. International book distributors. Dehradun, India.

Nakashima H, Ichiyama K, Hirayama, F, Uthino K, Ito M, Saitoh T, Ueki M, Yamamoto N, Ogawara H 1996. Sulfated penta galloylglucose (Y-ART-3) inhibits HIV replication and cytopathic effects in vitro, and reduces HIV infection in hu-PBL-SCID mice. Antiviral Res 30: 95–108.

Passos L, Ferreira SO 1996. Ant Dispersal of *Croton priscus* (Euphorbiceae) Seeds in an Tropical Semideciduous Forest in South eastern Brazil. Biotropica 28: 697–700.

Renard D, Schatz B, McKey DB 2010. Ant nest architecture and seed burial depth: implications for seed fate and germination success in a myrmecochorous savanna shrub. Ecoscience 17: 194–202.

Retana J, Pico FX, Rodrigo A 2004. Dual Role of Harvesting Ants as Seed Predators and Dispersers of a Non-Myrmechorous Mediterranean Perennial Herb. Oikos 105: 377–385.

Servigne P, Detrain C 2008. Ant-seed interactions: combined effects of ant and plant species on seed removal patterns. Insectes Soc 55: 220–230.

Tarandjiiska RB, Marekov IN, Nikolova-Damyanova BM, Amidzhin, BS 1996. Determination of triacylglycerol classes and molecular species in seed oils with high content of linoleic and linolenic fatty acids. J Sci Food Agr 72: 403–410.

Wang BC, Smith TB 2002. Closing the seed dispersal loop. Trends Ecol Evol 17: 379–385.

Wei Z, Guangyuan Z, Wenye Z 2006. Supercritical Fluid Extraction of *Lagerstroemia speciosa* Seed Oil and GC/MS Analysis. Chinese Agr Sci Bull 3: 621–625.

Wenny DG 2001. Advantages of seed dispersal: A re-evaluation of directed dispersal. Evol Ecol Res 3: 51–74.

Chapter 15

Plants Cry for Help upon Insect Attack and Parasitoids Come to their Rescue!!!

Sujata A Deshpande
d.sujata@gmail.com

Abstract

Plants are able to activate defence mechanisms against various forms of insect attacks. Insects produce various elicitors along with other components that induce defensive response in plants. The defence mechanisms could be direct in which the plants produce certain chemicals and/or structures to drive away or kill the insects attacking them. The defence could be indirect in which the plants call for help from other organisms by releasing some volatiles or other chemicals; the other organisms may be predators or parasitoids of the attacking insects. The following chapter elucidates tri-trophic interactions among plants; herbivores i.e., plant attacking insects and parasitoids of these herbivores. The emphasis is on indirect defence of plants in response to oviposition by herbivorous insects.

15.1 Introduction

Plants form the first trophic level in the food web. From microbes to man, many organisms feed on plants. Among these feeders insects are the most prominent ones. Approximately half a million species of insects are herbivorous (Dicke 2009). In tropical rain forests alone, insects devour 12 to 15 per cent of the total leaf area. All over the world, one sixth of the crops grown are eaten up by insects or fall prey to the diseases transmitted by them (Chapman et al. 2013). As a defence mechanism, the plants have evolved various means of perceiving and combating insect attacks on them. Once plants perceive insect attack they activate defence mechanisms which may be direct or indirect.

15.2 Forms of Insect Attack and Plant Responses

Plants not only sense the overt form of attack such as feeding damage by herbivore insects but also perceive subtle attack by herbivores even in the form of just touch and

DOI: 10.1201/9781003505471-15

egg deposition. Thus touching, egg laying and feeding could be considered as three forms of herbivorous insects' contact with plants. The following section briefly introduces these forms and plants' response to them. For more details refer to the review by Hilker and Meiners (2010).

15.2.1 Touch

Touch is the first contact of an herbivorous insect with its host plant. A plant may be touched by an adult insect to lay eggs and be fed on by the adult or its larval or nymphal stage. As the insects land and walk on the plants, they can be perceived by the plants as a result of increased pressure. Bristles present on insect tarsi may inflict scratches on the plants and such injuries are perceived by plants. Insects also leave behind their chemical footprints which can be sensed by the plants. The plants respond to these tactile sensations by rapidly changing cytosolic Ca^{++} signature (Haswell et al. 2008, Legué et al. 1997, Nakagawa et al. 2007) or by increasing the concentration of γ amino butyric acid (GABA) in the leaves (Bown et al. 2002; Hall et al. 2004). The leaves with high levels of GABA, when ingested by insects, are speculated to have detrimental effects on insect physiology by affecting their growth rate and survival. There is great scope to explore further the details of how plants perceive the touch by insects, and details of plants' responses and consequences of these responses on the ecology of plant and herbivore interactions.

15.2.2 Herbivore Oviposition

Egg deposition by herbivorous insects on plants is called oviposition. Herbivores usually glue their eggs to plant leaves with or without injury to the plant. Many studies have shown that oviposition-associated chemicals and/or injuries play a major role in plants' perception of egg deposition by herbivores. Once induced, the plants may defend themselves directly in the following ways. 1. Plants produce chemicals that repel the ovipositing insect females thus reducing chances of further egg deposition. The cabbage plant *Brassica oleracea* is known to produce oviposition deterrents in response to egg laying by the butterfly *Peiris brassicas* (Blaakmeer et al. 1994, Hilker et al. 2002b); 2. Plants produce ovicidal chemicals to kill herbivore eggs. Rice plants are known to produce chemical that kills eggs of plant-hoppers (Seino et al. 1996, Suzuki et al. 1996, Yamasaki et al. 2003); 3. Plants produce hypersensitive responses which cause eggs to detach from the leaves and fall off onto the ground. The larvae that hatch out from these fallen eggs rarely find their way back to the host plants and thus suffer an extremely high rate of mortality (Balbyshev and Lorenzen 1997, Hilker and Meiners 2006, Shapiro and DeVay 1987); 4. The plants produce neoplasms which raise the eggs from leaf surface and eventually detach them. Certain strains of the pea plant *Pisum sativum* are known to produce neoplasms in response to egg-laying by the pea weevil *Bruchus pisorum* and the cowpea weevil *Callosobruchus maculate*

(Doss et al. 1995; 2000, Hilker et al. 2002b). Neoplasm is basically a small protuberance consisting of undifferentiated plant cells. This protuberance raises the eggs up, and then the eggs detach and fall off. If the eggs are not detached, the newly hatching weevil larvae, which usually crawl directly from the egg to the pod surface underneath it, will now have to crawl down the neoplasm on the outside and then penetrate the pod. This increases exposure of the larvae to their predators and to desiccation.

Insect attack may induce some plants to produce indirect defences by calling for help of mobile 'bodyguards'. These bodyguards include parasitoids and predators of the herbivorous insects such as entamophagous nematodes and insectivorous birds (Bruinsma and Dicke 2008, D'Alessandro and Turlings 2006, Dicke 2009, Mantyla et al. 2008, Rasmann et al. 2005). The plants may attract these bodyguards by providing shelter or food to them or by emitting herbivore induced plant volatiles (HIPVs). This phenomenon of indirect plant defence in which the plants respond to egg deposition (or other forms of herbivory such as feeding, see below) by producing certain chemicals to attract carnivorous enemies of herbivores is referred to as 'plants crying for help' (Dicke and Sabelis 1988; Dicke et al. 1990a; Dicke 2009). One of the bodyguards that respond to plants' crying is insect parasitoids especially the egg and egg-larval parasitoids. This aspect is discussed further later in the chapter.

15.2.3 Feeding

Feeding by herbivores also induces plants to produce direct and indirect defences. Direct defences involve 1.Growth of thorns and trichomes to reduce access of herbivores to plant nutrients; 2. Production of toxins to kill the herbivores, and 3. Production of anti-nutritive and anti-digestive compounds to reduce digestibility of nutrients (Hilker and Meiners 2002 and references therein). The indirect defence involves producing chemicals to attract parasitoids and predators of the herbivores. As mentioned above this phenomenon is called 'plants crying for help'. It was first demonstrated in leaf eating mites and caterpillars (Dicke and Sabelis 1988, Dicke et al. 1990b; Turlings et al. 1990). Plants belonging to at least 13 different families have been shown to cry for help against feeding by a wide variety of insects and mites (Dicke 1999). Plants cry for help not only in response to leaf eating insects but also in response to the insects that bore stems, eat seeds and feed on roots (Dicke 2009 and references therein). Feeding leads to physical damage to the plant. Thus it was thought that plant defences can be induced by mimicking physical damage to the plants experimentally i.e., artificially damaging the plants. However, several studies have revealed that it is not the case; wounding alone does not induce plant defence. In addition to the wound, regurgitants of herbivores are necessary to induce plant defence (Hilker and Meiners 2002).

15.3 Elicitors of Plant Response

Different forms of insect attack induce plants to produce direct or indirect defences as mentioned above. The chemicals that bring out induction are called elicitors. In the case of insect touch, the probable elicitor is thought to be a chemical cocktail of insect footprint. However, this field requires deeper investigation.

A few oviposition associated elicitors are known with their chemistry fully identified. One of the well-studied elicitors is 'bruchin' that induces pea pods to form neoplasm, a form of direct plant defence. Bruchins are 3-hydroxypropanoic acid mono- or diesterified with monounsaturated C22-diols or mono- and diunsaturated C24-diols. These are found in egg laying females of Bruchid beetles (such as pea weevil and cowpea weevil, see above) and hence are called bruchins. Bruchins stimulate cell division which leads to formation of neoplasm. This stimulation occurs in those pea plants that carry *Neoplastic pod* (*Np*) gene. Bruchins also lead to up-regulation of several other genes involved in plant defence (Doss 2005, Doss et al. 1995 and 2000, Hilker and Meiners 2010). The other known elicitors that induce indirect plant defences are covered in the later part of the article.

There are several feeding-associated elicitors known (Hilker and Meiners 2010 and references therein). Fatty acid-amino acid conjugates (FACs), caeliferins and inceptins are some of the well-studied elicitors. FACs are found mainly in the regurgitants of lepidopteran larvae. FACs induce both direct and indirect plant defences. Caeliferins are saturated or monounsaturated sulphated α-hydroxy fatty acids. These are found in regurgitants of grasshoppers. Inceptins are peptides found in regurgitants of lepidopteran larvae. When FACs induce direct defence they do so by causing cell membrane depolarization which creates Ca^{++} influx resulting in activating biosynthesis of phytohormones and defensive chemicals. The indirect plant defence is brought about by changing plants' volatile signature, by triggering increase in phytohormones and defensive chemicals or by releasing plant volatiles to attract parasitoids.

15.4 Rescuers of the Plants Crying for Help

The main focus of this article is on indirect defence of plants in response to oviposition by herbivorous insect. Plants cry for help when herbivores deposit their eggs on the former. Parasitoids answer to the plants' cry and come to their rescue. The following part of the article deals with these rescuers.

15.4.1 Parasitoids

Parasitoids are insects that follow holometabolous or complete life cycle i.e., their life cycle goes through egg – larva – pupa – adult stages. As Godfray (1994) stated, a parasitoid is defined by the feeding habits of its larvae. Parasitoid larvae feed exclusively on the body of its host, eventually killing it. A single host is required for a parasitoid to complete its development although many parasitoids may develop gregariously in a single host. Parasitoids appear to be intermediates between predators and parasites. Like predators they always kill the host they attack and like many parasites they need a single host to mature. Parasitoids are also different from close relatives, the solitary wasps. Unlike solitary wasps, the parasitoids do not take the host to their nest. Before the term 'parasitoids' was introduced by Reuter in 1913, these insects were referred as 'insect parasites'.When the host is attacked by the parasitoids it is referred as the host being 'parasitized' and the phenomenon is called 'parasitism'.

15.4.1.1 Taxonomy of parasitoids

Parasitoids belong to the following orders of the class Insecta: Hymenoptera, Diptera, Coleoptera, Lepidoptera and Neuroptera (for details see Godfray 1994). Majority of parasitoids belong to the orders Hymenptera and Diptera. As many as about 50,000 described species of parasitoids belong to Hymenoptera. Order Diptera comes second with about 15,000 described species of parasitoids. Coleoptera, Lepidoptera and Neuroptera collectively have about 3,000 described species of parasitoids.

Fig. 15.1. a) An Eupelmid egg parasitoid wasp is emerging out of an egg case of an unknown insect; b) an Ichneumonid larval parasitoid wasp is piercing its ovipositor through the leaf case in an attempt to oviposit her eggs into a tea roller pest larva; c) a Chalcid pupal parasitoid wasp is emerging out from a pupa of rice-swift, a minor pest of rice; d) larvae of a gregarious braconid parasitoid wasp emerging out from the host larva of rice-swift by rupturing through the body wall.

15.4.1.2 Natural history of parasitoids

Natural history of parasitoids is varied (Godfray 1994). Adult female parasitoids lay their eggs directly on/in their host or near its vicinity. The parasitoids that lay their eggs directly on/in their hosts often have special adaptations for egg-laying. Hymenopteran parasitoids have specialized ovipositors. With these ovipositors the parasitoids manipulate eggs and also sting the hosts. Stinging may lead to permanent or temporary paralysis of the host. The parasitoids that attack well-concealed hosts possess ovipositors which are several times the length of the parasitoids' body. When not in use, such long ovipositors that extend well beyond abdomen are enclosed in protective valves or are coiled inside the abdomen. The ends of ovipositors may have cutting ridges so that the parasitoid can drill through the plant or wood tissue to reach the hosts hidden inside. In some parasitoids, the entire abdomen is flattened so that it can be inserted into the gills of fungi in which the hosts are concealed. In some cases parasitoids lay their eggs not directly on the host but in its vicinity. There are two ways by which the hosts get parasitized. One way is that the hosts eat the parasitoid eggs during feeding. The other way is that first instars of parasitoid larvae hatch out and actively locate the hosts. Adult parasitoids feed from flowers, plant sap and sometimes on the host itself.

15.4.1.3 Different types of parasitoids

The description of different types of parasitoids is based on Godfray (1994). Parasitoids mostly parasitize the developing or juvenile stages of their hosts although there are some parasitoids that attack adult hosts. Depending on the host stage that is attacked and killed parasitoids are classified into the following types:

Egg parasitoids: Parasitoids that lay their eggs in the host eggs (Figure 15.1a)

Larval parasitoids: Parasitoids that lay their eggs in the host larvae (Figure 15.1b)

Pupal parasitoids and **adult parasitoids:** They lay their eggs in host pupae or adult hosts respectively (Figure 15.1c)

Some parasitoids lay their eggs in one of the host stages but do not kill the host until it develops into the next stage. For example **egg-larval parasitoids** lay their eggs in the host eggs but allow the host eggs to hatch into host larvae which are then eaten up by the parasitoid larvae. Similarly **larval-pupal** parasitoids lay their eggs in the host larvae, allow the larvae to pupate and eventually consume the pupae. Such distinction among parasitoids based on the host stages that they attack and the stages they kill is more apparent in holometabolous hosts compared to the hemimetabolous hosts.

Koinobiont parasitoids allow the host to grow in size after parasitism. For example egg-larval parasitoids allow the host to develop from egg to larva; they also allow the host larvae to develop up to a certain extent before killing it. On the other hand, **idiobiont parasitoids** do not allow growth of the host. For example egg parasitoids lay their eggs in host eggs and consume the eggs before the host larva hatches out.

Based on whether the parasitoid larvae feed from inside or from outside the hosts, the parasitoids can be divided into two types. Larvae of **endoparasitoids** live and feed from inside the host. Larvae of **ectoparasitoids** live external to the host stages and feed from outside the host. Mouthparts of ectoparasitoids are usually buried in the host body. A vast majority of parasitoids are either ecto- or endoparasitoids. However, a small number of parasitoids live as ecto- as well as endoparasitoids.

When a single parasitoid develops from a single host, the parasitoid is called a **solitary parasitoid**. However when many parasitoids develop from a single host, they are called gregarious **parasitoids** (Figure 15.1d).

Sometimes parasitoids lay their eggs in an already parasitized host. If the eggs of the second parasitoid deposited happens to be the same as the first species of parasitoid, then **superparasitism** occurs. If eggs of the second parasitoid happens to be of a different species of parasitoids and the larvae of the two species compete over the resources (host body), then **multiparasitism** occurs. However, when the larvae of the second parasitoid feed, not on the host, but on the larvae of the first parasitoid, then **hyperparasitism** occurs. To complicate the matter further, there are two types of hyperparasitism. **Facultative hyperparasitoids** are able to attack unparasitized as well as parasitized eggs. On the other hand **obligatory hyperparasitoids** need previously parasitized hosts to lay their eggs. Finally, there are **cleptoparasitoids** which require previously parasitized eggs. Cleptoparasitoids feed on the host and not on the stages of the first parasitoid. Cleptoparasitoids often lack an ovipositor capable of drilling through the wood. They depend on the holes drilled by other parasitoids to reach the host located in the dead wood.

15.5 Tri-trophic Systems Involving Egg and Egg-larval Parasitoids

As the name indicates tri-trophic systems involve three trophic levels. Plants are at the first trophic level. Herbivorous insects which attack the plants by laying their eggs or by feeding on them represent the second trophic level. The third trophic level is represented by the parasitoids that come to the plant in response to plants' crying for help and attack herbivores. The focus here is on the egg and egg-larval parasitoids. When plants perceive oviposition by herbivores and raise their defence against it, the plants are at an advantage as they are defending themselves before any or little feeding damage has occurred. Egg parasitoids parasitize and kill the herbivore eggs before these eggs develop into the host larvae, thus the plant escapes feeding damage. Egg-larval parasitoids parasitize herbivore eggs but allow the egg to develop into the larva and eventually kill the herbivore larvae. The parasitized herbivore larvae usually die before they reach the final stage and thus they cause less damage to the plant while feeding. For example the egg-larval parasitoid *Ascogaster reticulata* oviposits in moth eggs. The parasitoid larvae hatch inside host eggs and develop inside as the host larvae grow, eventually feed on them. Parasitized host larvae develop up to fourth instar (Kawakami 1985) while healthy host larvae develop up to fifth instar. The final body weight of the parasitized host larvae was 27 per cent of the weight of healthy host larvae and consumes only 28 per cent of diet compared to that of healthy larvae (Ohnuma and Kainoh 1992). Details of some well-known tri-trophic interactions involving egg and egg-larval parasitoids are described below.

15.5.1 Case study systems

15.5.1.1 Elm system

The elm leaf beetle *Xanthogaleruca luteola* lays eggs on the leaves of the elm tree *Ulmus minor*. Their oviposition induces the elm leaves to emit volatiles or synomones that attract the eulophid wasp *Oomyzus gallerucae*. This parasitoid is a specialist of elm leaf beetle. Plants cry for help in response to herbivore oviposition was first demonstrated in the elm system. Before that indirect defence of plants was known only in cases of feeding damage. Females of elm leaf beetle remove the outer layers of the leaf surface just before gluing their eggs. This wounding, in addition to the elicitor that is present in oviduct secretion of the elm leaf beetle, is necessary for induction of plant's defensive response. Elm leaves with artificial oviposition wound do not attract the egg-parasitoid. Similarly intact elm leaves with oviduct secretion applied to them do not attract the egg-parasitoid. Only the elm leaves with artificial oviposition wound on which oviduct secretion is applied attract the egg-parasitoids. The exact chemical nature of the elicitor present in the oviduct secretion is yet to be known. Interestingly, not only the oviposited leaves but also the nearby leaves (without deposition of eggs) are induced to emit attractive volatiles. Thus induction of plant's defence is local (oviposited leaves) as well as systemic (un-oviposited leaves). The systemic induction extends at least up to 20 cm along the twig. For more details on the elm system see Meiners and Hilker (1997, 2000) and Meiners et al. (2000).

15.5.1.2 Pine system

The diprionid pine sawfly *Diprion pini* lays eggs in the needles of the pine tree *Pinus sylvestris*. The sawfly slits the pine needles from the distal end to the base and lays her eggs into this slit. The oviposition induces pine leaves to produce attractive volatiles that call the eulophid wasp *Chrysonotomyia ruforum* towards oviposited leaves. This wasp is a specialist on the eggs of *D. pini*. Similar to the elm system, the slit made in the leaf by *D.*

pini and the application of oviduct secretion induces the pine leaves to emit volatiles that attract the egg parasitoid. The chemistry of the elicitor is yet to be known. Similar to the elm system, induction of volatiles in pine is also local as well as systemic (Hilker et al. 2002a; 2005; Hilker and Meiners 2002).

15.5.1.3 Bean system

The pentatomid southern green stinckbug *Nezara viridula* lays eggs on the leaves of the broad bean *Vicia faba*. Oviposition induces the leaves to emit volatiles that attract the scelionid wasp *Trissolcus basalis*. Unlike the elm and pine systems, there is no oviposition associated wounding. However, the adult bugs feed on the leaves. Only those leaves that suffer feeding damage and also oviposited emit volatiles to attract the egg parasitoid. The parasitoid wasp (*Trissolcus basalis)* attacks different species of pentatomid bugs. The exact nature of elicitor is not known in this case. The leaves of French bean *Phaseolus vulgaris* also responds similarly to the same herbivore and parasitoid (Colazza et al. 2004a; b).

15.5.1.4 Brussels sprout system

This system shows varied results with different herbivores and parasitoids. Harlequin cabbage bugs lay eggs on leaves of Brussels sprout (Colazza et al. 2010 and references therein). As mentioned in the above systems, physical damage to the leaves in addition to oviposition is required to induce the plant defence. The leaf volatiles attract the egg parasitoid *Trissolcus brochymenae.*

There is another system in which the butterflies *Pieris brassicae* and *Pieris rapae* oviposit on the leaves of Brussels sprout without any wounding to the leaves. The egg parasitoid *Trichogramma brassicae* gets arrested on such leaves (Fatouros et al.2005). There is a difference between leaves attracting the parasitoids and leaves arresting the parasitoids. The term 'attracting' is used when leaves emit volatiles and these are detected by the parasitoids over a distance. It is thought that the parasitoids perceive these volatile through olfactory receptors. The term 'arresting' is used when leaves produce chemicals that are not volatile or are mildly volatile. Such chemicals are perceived over a closer range, such as by contacting the leaf. It is thought that the parasitoids perceive such chemicals through gustatory receptors. In the system of Brussels sprout – *P.brassicae* – *T. brassicae*, the elicitor of plant defence is well identified. It is Benzyl cyanide (BC). Females of *P. brassicae* receive this chemical through male ejaculate as an anti-aphrodisiac. The females release this chemical on to the plant while egg laying. BC induces the leaves emit chemicals that attract the egg parasitoid (Fatouros et al. 2008).

15.5.1.5 Tea Plant System

The smaller tea tortrix moth *Adoxophyes honmai* is a major pest of the tea plant *Camelia sinesis* in Japan. The brachonid wasp *Ascogaster reticulata* is an egg-larval parasitoid that parasitizes several tortricid species. The moth female lays eggs on underside of tea leaves without wounding the leaves. The leaves are induced in such a way that they arrest the parasitoids on their surface (Deshpande and Kainoh 2012). No volatile chemicals have been detected from the leaves. Therefore it is likely that tea leaves produce non-volatile or mildly volatile chemicals that could be detected once the parasitoids land on leaves and use these chemicals to distinguish leaves with eggs from the ones without eggs. The elicitor of plant defence appears to be present in the reproductive system of the female moth, however the exact nature of elicitor is yet to be identified.

15.6 Concluding Remarks

Insect parasitoids are considered as 'keystone' species (Colazza et al. 2010 and references therein). Insect parasitoids play an important role not only on their own ecosystems but also for the scientific community. They are used as model systems by behavioural ecologists to understand interspecific communications as well as by practitioners of applied biological control programs to manage pests of economically important crops. From the ecological point of view future studies should focus on the following areas: 1. What are the costs and benefits of defence against herbivore eggs to the plants? The obvious benefit is that by producing direct or indirect defence against herbivores eggs, the plants are reducing future feeding damage that would be caused by the larvae hatching out of these eggs. However, to produce these defences, the plants have to synthesize and emit certain chemicals or produce extra cell growth as in the case of neoplasm formation. It is not yet known as to what is the 'cost' of production of defensive mechanisms to the plants and how do plants pay for it. It is also not known if all plants, at all times, give defensive response against herbivore oviposition. 2. At present, the interactions among plants, herbivore eggs and parasitoids appear to be very diverse; it ranges from being highly species-specific to being generalists. It will be interesting to know common trends, if there are any, with respect to elicitors of plant response, their chemistry and behaviour of parasitoids in species-specific and generalist scenarios. 3. What is the exact mechanism of signalling cascade in plants, especially in case of oviposition induced plant defence? How plants produce local and systemic induction? 4. The herbivores are known to have evolved counter-adaptations to plant defence (for review see Hilker and Meiners 2010). Such kind of arms race is expected and it would be interesting to know the details of counter adaptations in different species and their effects on herbivory. For example, it is known that herbivorous arthropods increase biosynthesis of detoxifying enzymes in response to high amount plant toxins (Hilker and Meiners 2010 and references therein). 5. Recent studies (Hilker and Meiners 2011 and references therein) have shown that plants' response to early stages of herbivore attack, such as touch and oviposition, may affect plants' response to later stages such as larvae hatching out from surviving eggs. It is also speculated that feeding by adult herbivores prior to oviposition may enhance plants' defence against the eggs that will be deposited in the near future. More investigations are needed to know how plants 'remember' these previous attacks and how they use that memory against subsequent attacks. I hope that future investigations in this interesting field of tri-trophic interactions would provide information on some of the points raised above.

Acknowledgements

I thank Dr Pallaty Allesh Sinu for inviting me to write this chapter and for giving ample time and encouragement to do so. I am grateful to Prof KR Shivanna for reviewing the chapter.

References

Balbyshev NF, Lorenzen JH 1997. Hypersensitivity and egg drop, a novel mechanism of host-plant resistance to Colorado potato beetle (Coleoptera: Chrysomelidae). J Econ Entomol 90: 652–657.

Blaakmeer A, Hagenbeek D, van Beek TA, de Groot AE, Schoonhoven LM and van Loon JJA 1994. Plant response to eggs vs. host marking pheromone as factors inhibiting oviposition by *Pieris brassicae*. J Chem Ecol 20: 1657–1665.

Bruinsma M, Dicke M 2008. Herbivore-induced indirect defence: from induction mechanisms to community ecology. In A Schaller (Ed) Induced Plant Resistance to Herbivory, Springer Publishers, Berlin, Germany, pp 31–60.

Bown AW, Hall DE, MacGregor KB 2002. Insect footsteps on leaves stimulate the accumulation of 4-aminobutyrate and can be visualized through increased chlorophyll fluorescence and superoxide production. Plant Physiol 129: 1430–1434.

Chapman RF, Simpson SJ, Douglas AE 2013. The Insects: Structure and Function, fifth edn Cambridge University Press, Cambridge, UK.

Colazza S, Fucarino A, Peri E, Salerno G, Conti E, Bin F 2004a. Insect oviposition induces volatile emission in herbaceous plants that attracts egg parasitoids. J Exp Biol 207: 47–53.

Colazza S, Mcelfresh JS, Millar JG 2004b. Identification of volatile synomones, induced by *Nezara viridula* feeding and oviposition on bean spp, that attract the egg parasitoid *Trissolcus basalis*. J Chem Ecol 30: 945–964.

Colazza S, Peri E, Salerno G, Conti E 2010. Host searching byegg parasitoids: Exploitation of host chemical cues. In FL Consoli, JRP Parra and RA Zucchi (Eds) Egg Parasitoidsin Agroecosystems with Emphasis on Trichogramma. Springer, New York, NY, USA, pp 97–147.

D'Alessandro M, Turlings TCJ 2006. Advances and challenges in the identification of volatiles that mediate interactions among plants and arthropods. Analyst 131: 24–32.

Deshpande SA, Kainoh Y 2012. Herbivore egg deposition induces tea leaves to arrest the egg-larval parasitoid *Ascogaster reticulata*. Entomol Exp et Appl 144: 1–9.

Dicke M 1999. Evolution of induced indirect defence of plants.In R Tollrian, CD Harvell (Eds). The Ecology and Evolution of Inducible Defences. Princeton University Press, Princeton, NJ, USA, pp 62–88.

Dicke M 2009. Behavioural and community ecology of plants thatcry for help. Plant Cell Environ 32: 654–665.

Dicke M, Sabelis MW 1988. How plants obtain predatory mitesas bodyguards. Neth J Zool 38: 148–165.

Dicke M, Sabelis MW, Takabayashi J 1990a. Do plants cry for help? Evidence related to a tritrophic system of predatory mites, spider mites and their host plants. Symp Biol Hung 39: 127–134.

Dicke M, Van Beek TA, Posthumus MA, Ben Dom N, VanBokhoven H, De Groot AE 1990b. Isolation and identification of volatile kairomone that affects acarine predator–prey interactions. Involvement of host plant in its production. J Chem Ecol 16: 381–396.

Doss RP 2005. Treatment of pea pods with bruchin B results in up-regulation of a gene similar to *MtN19*. Plant Physiol Biochem 43: 225–231.

Doss RP, Proebsting WM, Potter SW, Clement SL, Williamson RT 1995. Response of Np mutant of pea (*Pisum sativum* L.) to pea weevil (*Bruchus pisorum* L.) oviposition and extracts. J Chem Ecol 21: 97–106.

Doss RP, Oliver JE, Proebsting WM, Potter SW, Kuy SR, Clement SL, Williamson RT, Carney JR, Devilbiss ED 2000. Bruchins-insect-derived plantregulators that stimulate neoplasm formation. Proc Natl Acad Sci USA 97:6218–6223.

Fatouros NE, Bukovinszkine'Kiss G, Kalkers LA, Gamborena RS, Dicke M, Hilker M 2005. Oviposition-induced plant cues: Do they arrest *Trichogramma* wasps during host location? Entomol Exp et Appl 115: 207–215.

Fatouros NE, Broekgaarden C, Bukovinszkine'Kiss G, Van Loon JJA, Mumm R, Huigens ME, Dicke M, Hilker M 2008. Male-derived butterfly antiaphrodisiac mediates induced indirect plant defence. Proc Natl Acad Sci 105: 10033–10038.

Godfray HCJ 1994. Parasitoids: Behavioral and evolutionary ecology, first edn. Princeton University Press, Princeton, New Jersey.

Hall DE, MacGregor KB, Nijsse J, Bown AW 2004. Footsteps from insect larvae damage leaf surfaces and initiaterapid responses. Eur J Plant Pathol 110: 441–447.

Haswell ES, Peyronnet R, Barbier-Brygoo H, Meyerowitz EM, Frachisse J-M 2008. Two MscS homologs provide mechanosensitive channel activities in the Arabidopsis root. Curr Biol 18: 730–734.

Hilker M, Meiners T (2002). Chemoecology of Insect Eggs and Egg Deposition. Blackwell Publishing, Berlin, Germany.

Hilker M, Meiners T 2006. Early herbivore alert: insect eggs induce plant defence. J Chem Ecol 32: 1379–1397.

Hilker M, Meiners T 2010. How do plants "notice" attack by herbivorous arthropods? Biol Rev 85: 267–280.

Hilker M, Meiners T 2011. Plants and insect eggs: How do they affect each other? Phytochemistry 72: 1612–1623.

Hilker M, Kobs C, Varama M, Schrank K 2002a. Insect egg deposition induces *Pinus sylvestris* to attract egg parasitoids. J Exp Biol 205: 455–461.

Hilker M, Rohfritsch O, Meiners T 2002b. The plant's response towards insect egg deposition. In Hilker M and Meiners T (Eds) *Chemoecology of insect eggs and egg deposition,* first edn. Blackwell Publishing, 205–233.

Hilker M, Stein C, Schröder R, Varama M, Mumm R 2005. Insect egg deposition induced defence response in *Pinus sylvestris*: Characterization of the elicitor. J Exp Biol 208: 1849–1854.

Kawakami T 1985. Development of immature stages of *Ascogaster reticulatus* Watanabe (Hymenoptera: Braconidae), an egg-larval parasitoid of the smaller tea tortrix moth, *Adoxophyes* sp. (Lepidoptera: Tortricidae). Appl Entomol Zool 20: 380–386.

Leguè V, Blancaflor E, Wymer C, Perbal G, Fantin D, Gilroy S 1997. Cytoplasmic free Ca 2+ in Arabidopsis roots changes in response to touch but not gravity. Plant Physiol 114: 789–800.

Mantyla E, Alessio GA, Blande JD, Heijari J, Holopainen JK, Laaksonen T, Piirtola P, Klemola T 2008. From plants to birds: higher avian predation rates in trees responding to insect herbivory. PLoS ONE 3: e2832.

Meiners T, Hilker M 1997. Host location in *Oomyzus gallerucae* (Hymenoptera: Eulophidae), an egg parasitoid of the elm leaf beetle *Xanthogaleruca luteola* (Coleoptera, Chrysomelidae). Oecologia 112:87–93.

Meiners T, Hilker M 2000. Induction of plant synomones by oviposition of a phytophagous insect. J Chem Ecol 26: 221–232.

Meiners T, Westerhaus C, Hilker M 2000. Specificity of chemical cues used by a specialist egg parasitoid during host location. Entomol Exp et Appl 95: 151–159.

Nakagawa Y, Katagiri T, Shinozaki K, Qi Z et al. 2007. Arabidopsis plasma membrane protein crucial for Ca 2+ influx and touch sensing in roots. Proc Natl Acad Sci 104: 3639–3644.

Ohnuma Y, Kainoh Y 1992. Effect of parasitism by *Ascogaster reticulatus* [Hym: Braconidae] on growth of the host *Adoxophyes* sp (Lep: Tortricidae). Entomophaga 37: 327-332.

Rasmann S, Kollner TG, Degenhardt J, Hiltpold I, Toepfer S, Kuhlmann U, Gershenzon J, Turlings TCJ 2005. Recruitmentof entomopathogenic nematodes by insect-damaged maizeroots. Nature 434: 732–737.

Reuter OM 1913. Lebensgewohnheiten und Instinkte der Insekten, Friedlander, Berlin.

Shapiro AM, DeVay JE 1987. Hypersensitivity reaction of *Brassica nigra* L. (Cruciferae) kills eggs of *Pieris* butterflies (Lepidoptera: Pieridae). Oecologia 71: 631–632.

Seino Y, Suzuki Y, Sogawa K 1996. An ovicidal substance produced by rice plants in response to ovipositionby the whitebacked planthopper, *Sogatella furcifera* (Horvath)(Homoptera: Delphacidae). Appl Entomol Zool 31: 467–473.

Suzuki Y, Sogawa K, Seino Y 1996. Ovicidal reaction of rice plants against the whitebacked planthopper, *Sogatella furcifera* Horvath (Homoptera:Delphacidae). Appl Entomol Zool 31: 111–118.

Turlings TCJ, Tumlinson JH, LewisWJ 1990. Exploitation of herbivore-induced plant odors by host-seeking parasitic wasps. Science 250: 1251–1253.

Yamasaki M, Yoshimura A, Yasui H 2003. Genetic basic of ovicidal response to Whitebacked planthopper (*Sogatella furcifera* Horvath) in rice (*Oryza sativa* L.). Mol Breed 12: 133–143.

Chapter 16

The Promise of Genomics in Plant-Animal Interaction Studies

M Nagarajan, K Nimisha, AK Nimisha and PA Sinu*
nagarajan.spkces@gmail.com

Abstract

Community dynamics and ecosystem processes depend on two major kinds of species interactions, positive interaction and negative interaction. Positive interactions in the form of mutualisms and negative interactions in the form of predations are the two most common interactions in natural communities. Plant-pollinator interactions and predator-prey interactions though two contrasting forms of associations, are unavoidable with respect to proper community functioning. Studies related to pollinator-plant interactions are important, as they play important roles in floral evolution, inflorescence design and also a plant's adaptation to various environmental conditions, ultimately leading to speciation. Likewise, predator-prey interactions influence the balance of organisms in a particular ecosystem. Any alteration in the balance may cause drastic changes in the population density of predators as well as other organisms linked to the food web. A large number of studies have been carried out to understand these complex interactions. Earlier, indirect methods such as field observation, radio tracking of organisms along with numerous biochemical and microhistological studies were used for observing various ecological interactions. These conventional methods and techniques are now being replaced by new improved methods and technologies such as PCR based DNA barcoding and high throughput DNA sequencing. Application of these techniques in the field of pollination biology and prey detection of the predators has improved the precision and accuracy in interpreting such interactions in a more accurate way.

16.1 Introduction

The interaction between plants and animals, one of the important factors in community ecology, contributes immensely to the phenotypic diversity and evolution of species. A

DOI: 10.1201/9781003505471-16

better understanding of these relationships, and the plant and animal species involved would help to forecast species abundance and food web structure of ecosystems. But it is an extremely diverse and complex relationship to study thoroughly using traditional methods. Genomics is providing links between various sub disciplines of biology. It has grown remarkably over the last two decades. All branches of natural science research are permeated by genomics including transgenic plants, animal production, pharmaceuticals and enzyme production. Extensive participation of researchers across the world has enabled complete sequencing of entire genomes of important plant and animal species. It has resulted in analysing millions of nucleotides of DNA sequences and several thousands of different types of genetic markers. Genomic research on both animals and plants has provided new methods and tools for studying plant and animal interactions in detail, without altering the natural habitat/environmental conditions. The use of genomic techniques in addressing ecological questions like plant animal interactions has provided new insights into fundamental processes such as diversity, evolution, ecology, functionality of communities, food web dynamics and nutrient cycling. This chapter seeks to provide information on the usefulness of genomic approaches to study the interaction between plant and animal species with special emphasis on predator-prey and plant-pollinator interactions.

16.2 Predator-prey Interaction

The predator–prey interaction is one of the important components of community ecology which controls the changes in animal populations. Therefore, understanding the predator–prey interaction is not only an essential requirement but also a major challenge. For elucidating complex food web structures, precise identification of predator-prey interaction at different trophic levels of the community is essential. Various techniques including video capturing are being frequently used for this purpose. But these methods, fail in measuring trophic level connections between species particularly, those that are highly mobile and cryptic in nature (Symondson 2002, Sheppard and Harwood 2005). In the case of generalist predators, identification and quantification of the preys become more challenging as they feed on a wide range of species. In such situations, choice of the prey may be decided based on various factors of the prey such as nutritional qualities, toxins, defence mechanism, escape strategies and previous feeding history of the predator.

Visual identification of predatory interactions through cameras may provide only a limited knowledge of the predatory events. The correct validation of any quantitative data on predation can be obtained only if it is allowed to occur naturally with minimum disturbance without any interference or bias towards certain types of prey (Sunderland 1988, Harwood and Obrycki 2005). Most of the time researchers fail to recreate and maintain the environment similar to that of the field as the methods used may affect the ability of the predator to find and capture its prey. To overcome these difficulties, a number of techniques have been developed to identify the prey being consumed by predators in the field. One widely used method is the collection of predators from the field and carrying out post-mortem analysis of their gut or faecal samples. Morphological identification of prey remains can be difficult in some situations where a thorough mastication and digestion processes are involved (degradation of soft parts and high fragmentation of hard parts). For example, the insectivorous bats, preying on great diversity of species chew and digest the prey thoroughly there by making the prey identification a difficult task (Zeale et al. 2011, Zarzsos-Lacoste et al. 2013). Most of the invertebrate predators such as spiders and hemipteran bugs are basically fluid feeders. Therefore, they don't produce any identifiable remains of the prey. In such cases the digestive processes render many samples visually unrecognizable. Despite these difficulties there are several successful studies with regard to gut dissection and faecal sample analyses for identification of prey remains passing through a predator's stomach (Burger et al. 1999, Webb et al. 2000).

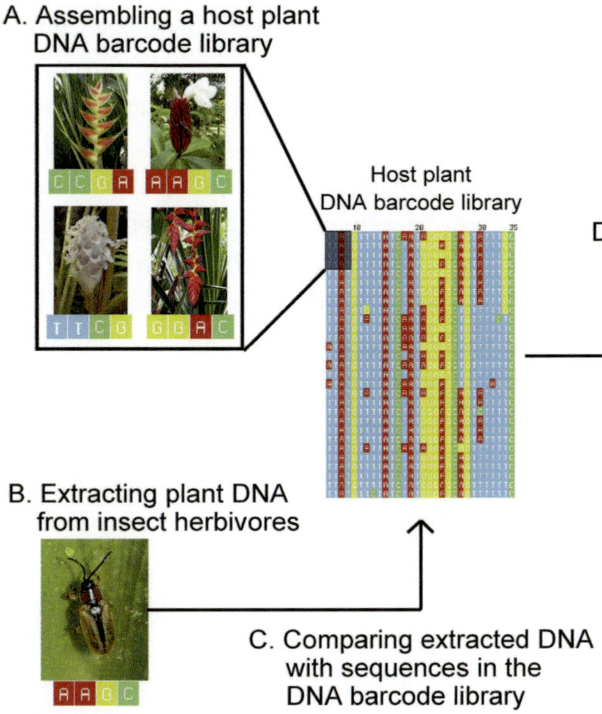

Fig.16.1. Diagrammatic representation of process involved in diet identification of insect herbivore using DNA barcode technique. A) Assembling of the entire DNA barcode library containing sequence of all potential host plants. B) Collection of insect herbivore from felid followed by DNA extraction from insect's gut content, PCR (of *rbcL*, nuclear ribosomal *ITS2* and plastid *trnH-psbA* intergenic spacer) and sequencing C) Sequence comparison between sequence of insect with sequences in DNA library D) Host plant identification. (*Adapted from García-Robledo et al. 2013*).

16.2.1 Conventional Prey-detection Techniques

Till date, a wide range of techniques have been developed for studying the prey-predator interactions. These include the application of stable isotope, radioisotope labelling, electrophoretic and chromatographic techniques for identifying prey proteins and pigments. Apart from these techniques ELISA, polyclonal and monoclonal antibodies are also being used for studying predator-prey interactions. ELISA was the most often used technique for identifying insect-specific proteins in the gut content of predator (Sunderland 1988, Pierce and Boyle 1991, Greenstone 1996, Symondson 2002).

16.2.2 PCR- based Techniques

For the past few years, the conventional techniques have been superseded by DNA based techniques especially PCR which makes prey identification easier even from degraded and semi-digested prey remains. Polymerase chain reaction (PCR)-based techniques are being used for detecting prey remains in the stomach, scats or regurgitates of predators. PCR has the ability to differentiate between unique pieces of DNA from predator and prey species. Another unique feature of PCR technique is that multiple prey items can be screened simultaneously (Harper et al. 2005). DNA-based prey assays have also been proven as more sensitive and often provide greater taxonomic resolution of prey. Several recent studies on molecular detection of prey remains in the gut of terrestrial arthropods have shown the usefulness of molecular level of prey detection. For example, Zaidi et al. (1999) conducted experiments on *Pterostichus cupreus* preying on *Culex quinquefasciatus*, using the primers designed to amplify a variety of DNA fragments of diverse lengths, targeting fully characterized multiple-copy esterase genes. Chen et al. (2000) performed studies on invertebrate systems using mitochondrial gene, cytochrome oxidase subunit II (COII) to amplify insects such as aphids present inside the gut of Coccinellidae. Microsatellite markers have also proved useful for identifying predator diets. For instance, DeWoody et al. (2001) carried out predation study of filial cannibalisms by the *Ethiostoma olmstedi* using microsatellite markers.

DNA technology is also being widely utilized to detect plant species involved in herbivory as it is the most extensive feeding habit of insects. In addition, there exists a tight ecological link between herbivores and their food plants. Short–sequence fragments, 'DNA barcodes' used for identification of plant species in the insect gut provide the host information and there by provide a starting point for analysing co-evolutionary relationships and trophic interactions of plants and herbivores. Jurado-rivera et al. (2009) studied gut content of leaf beetles (subfamily Chrysomelinae) from Australia using the chloroplast *trnL* (UAA) intron as barcode. The phylogenetic tree constructed using *trnL* intron sequences, included 13 major plant families out of which, species belonging to *Myrtaceae* (eucalypts) and *Fabaceae* (acacia) were found to be dominant. This study provided a good physical evidence for understanding the plant-herbivore associations using plant DNA barcode. There are several other ecological studies reported on this aspect. Matheson et al. (2008) studied the ecological interactions existing between the insects and plant species. Here, a 157bp region of ribulose bisphosphate carboxylase gene large subunit (*rbc*L) was used as marker for plant species identification in the insect gut. Further, Junnila et al. (2010) have used PCR-based approach for diet analysis of mosquito (*Anopheles sergenti*) feeding on a variety of plant species. Similarly, using *rbc*L gene fragment as potential barcode, interaction between soil-living insect herbivore and its associated plant has been studied by Staudacher et al. (2011). This is one of the first such studies based on DNA based approach to analyse the dietary choices of soil-living insects. Other than insects and beetles, PCR-based methods have been used successfully to study the diet of a wide range of organisms such as marine vertebrates (Jarman et al. 2002, Deagle et al. 2009), sea birds (Deagle et al. 2007, Ohem et al. 2011), herbivorous terrestrial vertebrates (Hibert et al.2011), insectivorous bats (Zeale et al. 2011) and many marine and terrestrial invertebrates (Deagle et al. 2005, Harper et al. 2005, Juen and Traugott 2005). Dietary analysis of ruminants such as cattle using PCR for amplifying plant specific DNA in semi-digested feed has been reported by Perez-Amaro et al. (2010). This approach proved to be more useful in species-specific identification of plants than the microhistological techniques used previously for cattle diet analysis. Other than gut analysis, predator scat analysis are also becoming popular methods used in elucidating prey-predator relationship as gut analysis often requires sacrifice of the organism especially when larger vertebrates are involved in the process and it is sometimes considered as unethical. In studies using scats, epithelial cells sloughed off from the walls of the lower intestine of predator are targeted for DNA extraction (Reed et al. 1997). The scats of predator identified in this way are subjected to physical examination for solid prey remains but when prey remains are very much unrecognizable, the prey DNA is determined using PCR. Purcell et al. (2004) used DNA extracted from bones found in Pacific harbour seal scats to obtain species-level identification of salmonids. Successful analysis of DNA from the plant components in the scats of Brown bears and extinct ground sloths have been reported by Hoss et al. (1992) and Poinar et al. (1998).

16.2.3 NGS: Revolutionizing Community Genomics

With the advancement in technology, new techniques such as Next generation sequencing (NGS) have come up with the potential to identify huge number of targets (prey species) present in predator's diet on one singlehanded experiment. Even predation monitoring at single haplotype level is possible using NGS. Therefore, in future this new technology would possibly help to carry out direct genetic selection, by comparing the genotypes present in the field with those existing in predator's stomach. Using NGS, now it is possible to sequence all the prey species present in predator gut or faecal samples simultaneously. Initial studies using NGS reported by Deagle et al. (2009, 2010) used the pyro-sequencing technology on GS-FLX platform (Roche), also known as 454 sequencing. It generates about 100 million base pairs with approximate read length of 250bp over 7.5h analytical run. The plant-herbivore interaction has been studied using this same technology. For instance, Soininen et al. (2009) used the high-throughput sequencing technique to identify the diet of small subarctic species of voles, *Microtus oeconomus* and *Myodes rufocanus* using the P6 loop region of chloroplast gene (*trnL*- UAA intron) as 'barcode'. In this particular study, the authors have tried to validate the usefulness of DNA barcoding based on NGS over microhistological analysis of the species diet. It was found that NGS technique provided far-more better taxonomical resolution in the diet compared to the conventional techniques. Similar study was done by Raye et al. (2011) in wild herbivore, *Rupicapra rupicapra* (the alpine chamois). Recently, the dietary behaviour of South Asian wild cat (*Prionailurus bengalensis*) has been studied from the faeces DNA using IIumumina genome analyser IIx (Shehzad et al. 2012). Other than vertebrate predators, both marine and terrestrial, NGS has been tested for the diet analysis of invertebrates. In addition, NGS technology has found usefulness in the dietary analysis of a number of other organisms like bats (Bohmann et al. 2011), puffins (Bowser et al. 2013), slow worm (Brown et al. 2012) and wild ruminants such as bison (Kowalczyk et al. 2011).

All the above mentioned studies employed a common technique before going for NGS that is the use of modified predator specific oligonucleotide or blocking primers other than universal primers and species-specific or group-specific primers (used to amplify the prey DNA). These blocking primers could prevent the co-amplification of predator DNA along with the target prey DNA (Vestheim and Jarman, 2008). As in many of the studies non-specific DNAs such as predator DNA, DNA of gut parasites and symbionts may get abundantly amplified along with target prey DNA (Deagle et al. 2006, Pompanon et al. 2012), which can lead to biases while analysing and interpreting the prey data.

A completely new approach for the analysis of diets of generalist predators using next-generation sequencing technology without blocking probes have been reported recently by Pinol et al. (2014). The sequencing technology used here was Ion Torrent, which produced 10Mb to 100Mb sequences with 200bp read length. This technology is faster when compared to previous NGS sequencers. This particular study has been conducted in Linyphiid spider (*Oedothorax fuscus*) to demonstrate that NGS technology using general primers without predator-specific blocking primers could save time and money. All other earlier techniques required predator-specific blocking primers to be designed. Also, the cost per read is less compared to techniques involving cloning and sequencing. This method is mainly applicable for Arthropods such as spiders as in this case predator and prey species are phylogenetically closer (closer spider groups were detected in *Oedothorax fuscus* diet). In terms of precision and information delivery, Next-generation sequencing technology has been far more successful as it could analyse a large number of samples simultaneously with a little additional cost. In the years to come the cost of NGS analysis is expected to get lower and it is expected to be widely used.

16.3 Frugivore-Plant Interactions

Frugivore-plant interaction is another important type of mutualistic interaction, which plays a crucial role in ecological networks. Furgivores act as seed disseminators of plants. In an evolutionary point of view, this interaction is considered as an example of 'diffuse co-evolution' in which evolutionary changes in traits occur from interactions between groups of species. Diet investigation studies are needed for elucidating various kinds of trophic interactions. Similar to pollinator-plant interactions, frugivore-plant interactions could be studied using PCR-based approaches. Marrero et al. (2009) developed a protocol for DNA extraction from faecal sample and regurgitated seeds of two sympatric frugivorous pigeons. This study showed, high PCR efficiency compared to DNA extracted using other protocols, and identified bird species that ejected the seeds. In order to have an accurate knowledge of the role of birds as seed disperser's or predators, proper identification of bird as well as plant species holds much importance.

In recent years, DNA barcoding and microsatellite genotyping has proved to be very useful in studying seed dispersal by fruit-eating organisms. For example, Godoy and Jordano (2001) reported a new approach to identify source tree *Prunus mahaleb* seeds dispersed by frugivorous birds by using microsatellite makers. This approach proved useful in unambiguous identification of the maternal source tree and also analysis of spatial distance between the tree source and dispersed seed. DNA barcode technique has been extensively used in studies related to plant-frugivore interactions. For instance, Hayward (2013) conducted dietary studies using faecal samples of Jamaican fruit bats where he studied dietary constituent of three morphologically dissimilar bat species (*Artibeus jamaicensis, Ariteus flavescens* and *Glossophaga soricina*). DNA barcoding analysis using *rbcL* DNA marker resulted in identification of a total of 11 fruit taxa in the fruit bat's diet, out of which seven were new, and they were not able to be determined through traditional methods. Yoshihiro et al. (2010) used cytochrome *b* gene as marker for identifying different species of Civets from the faecal DNA. This study revealed that palm civet can disperse large seeds and plays a significant role as long-distance dispersers.

16.4 Plant-pollinator Interactions

Mutual beneficial relationships are seen between pollinators and plants. The pollinators visit flowers to obtain food, mainly in the form of pollen or nectar which are, in fact, the rewards paid in return for the services of the pollinators for carrying pollen from one flower to another. Pollinator-mediated selection is regarded as a common phenomenon that contributes to the evolution of floral and inflorescence design resulting in adaptation and speciation in plants (Schemske et al. 1999). It is important to identify the source of pollen collected by the visitors to understand the type of pollination and its efficacy. Traditionally, pollen collected by the pollinators are identified by examining its morphological characters using a microscope and then comparing it with the reference pollen of the focal plant. But this approach is laborious and time consuming. Also, it may have limited taxonomic precision as the pollen of related species may not show readily recognizable morphological differences (Richardson et al. 2015).

Fig. 16.2. Schematic view of genomic approach to study plant-pollinator interactions: A) Visitation of pollinators to different flowers for nectar collection; B) Pollen dispersal by the pollinators; C) Laboratory process for pollen identification using genomic techniques; D) DNA sequence analysis and DNA barcode development.

16.4.1 Flow Cytometry

Flow cytometry has been widely used in pollen identification which uses fluorescence to analyse thousands of particles at a time. Large numbers of pollen grains collected by pollinators can be tested for their DNA content and thus the species composition also can be estimated. Kron et al. (2014) used this technique for estimating species composition of pollen by collecting pollen grains from honey bees and bumble bees that usually pollinate *Monarda, Centaurea* and *Solidago*. The nuclei of pollen grains were isolated, stained and passed on through flow cytometer. Using laser based technique fluorescence was detected by electronic detector. The data generated were plotted on logarithmic scales to produce histogram. Similarly, the fluorescence of sampled pollen or leaf collected directly from the plant was used as reference. By comparing fluorescence peak of reference species with that of pollen collected on bees, identification of their corresponding plant species was possible.

16.4.2 DNA Barcoding

DNA barcode is a short sequence of DNA from nuclear or chloroplast region that can be used for identifying the organisms through barcoding. In the case of plants, chloroplast genes such as *rbcL, matK* and *trnH-psbA* or nuclear ITS region are considered as potential barcode regions. Using PCR, these plant markers are amplified primarily from a mixture of pollen samples collected by the pollinator and the fragments are sequenced. It can thus help ecologists to identify pollinator which was previously unknown to them. In addition, this approach can indirectly help in measuring ecologically and evolutionarily important events such as selective pressure of one species on another, relative specialization and generalization of specific plant-animal pairs and the economic impact of pollination of a particular crop of interest (Clare et al. 2013). Zhou et al. (2007) carried out DNA barcoding studies on pollen grains of *Chenopodiaceae* plants. Here they sequenced internal transcribed spacer region (ITS) of nuclear ribosomal DNA from the plant leaves and established a reference database for ITS. The ITS1 region of pollen grains of chenopod plant was then amplified using nested PCR and sequenced. The pollen grains were identified by comparing with Chenopodiaceous reference database. Similar kind of study was done by Richardson et al. (2015) where they applied the DNA metabarcoding technique using internal transcribed spacer ITS2 in which they compared their result with those obtained from traditional microscopic melissopalynology to determine the pollen wandering habits of honey bees in an agricultural area. It further suggests that DNA metabarcoding was far better in providing highly sensitive results than the traditional method and also a genus level identification of plant was possible.

16.4.3 Next Generation Sequencing

It is possible to assess the multiple pollen grains at a stretch originating from many plant species using ITS2 region of nuclear DNA through NGS technique. The whole procedure involves DNA extraction using standard protocol followed by amplification of chloroplast gene *trnL* by PCR. During PCR amplification the forward primer (*TrnL-C)* and reverse primer (*trnL-h)* is fitted with different adapters, subsequently the amplified products are sequenced. For instance, Kraaijeveld et al. (2015) used NGS for identification of airborne pollen. The study revealed that NGS technology was efficient and sensitive in determining the pollen genera in mixed samples. Keller et al. (2015) conducted an experiment for species identification from mixed pollen samples of honey bee using the ITS2 region. They also concluded that NGS is an efficient method for identification of pollen at the species and genus level.

16.4.4 Transcriptome Sequencing

The RNA-Seq also provides a new opportunity for *de novo* genome sequencing in non-model organisms. In recent years, high-throughput RNA sequencing has been shown as a highly efficient method for analysis of gene expression, discovery of novel transcripts and identification of differentially expressed genes. Pollen grains are composed of RNAs-encoding proteins involved in germination, tube growth and degradation of the cell wall. RNA sequence in these protein coding genes expressed in pollen grains are detected using high throughput RNA sequencing. Loraine et al. (2013) used this technique on *Arabdiopsis* pollen genome, to detect 4,172 protein-coding genes, which are expressed in pollen.

16.5 Concluding Remarks

It is now clear that advent of genomics in the field of community ecology has brought tremendous improvement in this field of research. From the ecological point of view, it is important to learn the predator-prey relationships and plant pollinator co-evolution as they are used to define food chains, which provide the background from which we can understand how energy flows through an ecosystem. All sort of phenotypes, floral patterns, evolution and adaptations, floral traits, have their origin at genomic level. Using the most advanced technologies available currently, it is now possible to conduct experiments at whole genome level, which has made it possible to elucidate the correlations in plant-animal interactions. These advancements, thus may allow researchers to quantify the diversity of plant-animal interactions by comparative approaches. In future, genomic technologies are expected to effectively improve molecular detection systems not only with respect to predator-prey and plant–pollinator interactions but also on analogous researches which are concerned with parasitoid-host, pathogen-host and host-herbivore interactions.

Acknowledgements

We are grateful to Prof KR Shivanna and Prof KG Sivaramakrishnan for reviewing the chapter. The first author is also grateful to the Science and Engineering Research Board (SERB), Department of Science and technology, Government of India, New Delhi, for financial Assistance (SR/So/AS-84/2012).

References

Bohmann K, Monadjem A, Noer CL et al. 2011. Molecular dietary analysis of two African free-tailed bats (*Molossidae*) using high throughput sequencing. PLoS One 6:e21441. doi:10.1371/journal.pone.0021441

Bowser AK, Diamond AW, Addison JA 2013. From puffins to plankton: a DNA-based analysis of a seabird food chain in the northern gulf of maine. PLoS One 8:e83152. doi:10.1371/journal.pone.0083152

Brown DS, Jarman SN, Symondson WOC 2012. Pyrosequencing of prey DNA in reptile faeces: analysis of earthworm consumption by slow worms. Mol Ecol Resour 12: 259–266.

Burger JC, Patten MA, Rotenberry JT, Redak RA 1999. Foraging ecology of the California gnatcatcher deduced from fecal samples. Oecologia 120: 304 –310.

Chen Y, Giles KL, Payton ME, Greenstone MH 2000. Identifying key cereal aphid predators by molecular gut analysis. Mol Ecol 9: 1887–1898.

Clare EL, Schiestl FP, Leitch AR and Chittika L 2013. The promise of genomics in the study of plant pollinator interactions. Genome Biol 14:207

Deagle BE, Kirkwood R, Jarman SN 2009. Analysis of Australian fur seal diet by pyrosequencing prey DNA in feces. Mol Ecol 18: 2022–2038.

Deagle BE, Chiaradia A, McInnes J, Jarman SN 2010. Pyrosequencing faecal DNA to determine diet of little penguins: is what goes in what comes out. Conserv Genet 11: 2039–2048.

Deagle BE, Eveson JP, Jarman SN 2006. Quantification of damage in DNA recovered from highly degraded samples — a case study on DNA in faeces. Front Zool 3: e11. doi:10.1186/1742-9994-3-11.

Deagle BE, Gales NJ, Evans K et al. 2007. Studying seabird diet through genetic analysis of faeces: a case study on Macaroni penguins *Eudyptes chrysolophus*. PLoS One 2: e831.

Deagle BE, Jarman SN, Pemberton D, Gales NJ 2005.Genetic screening for prey in the gut contents from a giant squid (*Architeuthis* sp.). J Hered 96:417–423.

DeWoody JA, Fletcher DE, Wilkins SD, Avise JC 2001. Genetic documentation of filial cannibalism in nature. Proc Natl Acad Sci USA 98: 5090– 5092.

García-Robledo C, Erickson DL, Staines CL, Erwin TL, Kress WJ 2013. Tropical Plant–Herbivore Networks: Reconstructing Species Interactions Using DNA Barcodes. PLoS ONE 8(1): e52967. doi:10.1371/journal.pone.0052967

Godoy JA and Jordano P 2001. Seed dispersal by animals: exact tracking of the source trees with endocarp DNA microsatellites. Mol Ecol 10: 2275-2283.

Greenstone MH 1996. Serological analysis of arthropod predation:past, present and future. In: The Ecology of Agricultural Pests: Biochem Approaches (eds Symondson WOC, Liddell JE), pp 265 –300. Chapman and Hall, London.

Harper GL, King RA, Dodd CS et al. 2005. Rapid screening of invertebrate predators for multiple prey DNA targets. Mol Ecol 14: 819 –827.

Harwood JD, Wallin WG, Obrycki JJ 2005. Uptake of Bt endotoxins by non-target herbivores and higher order arthropod predators: molecular evidence from a transgenic corn agro ecosystem. Mol Ecol 14: 2815–2823.

Hayward C E 2013. "DNA barcoding expands dietary identification and reveals dietary similarity in Jamaican frugivorous bats" University of Western Ontario - Electronic Thesis and Dissertation Repository. Paper 1697.

Hibert F, Sabatier D, Andrivot J et al. 2011. Botany, genetics and ethnobotany: a crossed investigation on the elusive tapir's diet in French Guiana. PLoS One 6: e25850.doi: 10.1371/journal.pone.0025850

Höss M, Kohn M, Pääbo S, Knauer F, Schroeder W 1992.Excrement analysis by PCR. Nature 359: 199.

Jarman SN, Gales NJ, Tierney M, Gill PC, Elliott NG 2002. A DNA-based method for identification of krill species and its application to analysing the diet of marine vertebrate predators. Mol Ecol 11: 2679–2690.

Juen A, Traugott M 2005. Detecting predation and scavenging by DNA gut-content analysis: a case study using a soil insect predator-prey system. Oecologia 142: 344–352.

Junnila A, Müller GC, Schlein Y 2010. Species identification of plant tissues from the gut of An. sergentiiby DNA analysis. Acta Trop 115:227–233.

Jurado-Rivera JA, Vogler AP, Reid CAM,Petitpierre E, Go´mez-Zurita J 2009. DNA barcoding insect–host plant associations. Proc R Soc B 276: 639–648.

Keller A, Danner N, Grimmer G, Ankenbrand M, von der Ohe K, von der Ohe W, Rost S, Härtel S, Steffan-Dewenter I 2015. Evaluating multiplexed next-generation sequencing as a method in palynology for mixed pollen samples. Plant Biol (Stuttg) 2:558-566.

Kowalczyk R, Taberlet P, Coissac E et al. 2011. Influence of management practices on large herbivore diet—Case of European bison in Bialowieza Primeval Forest (Poland). For Ecol Manage 261: 821–828.

Kraaijeveld K, Letty A. Weger D, Marina Ventayol Garcia, Buermans H, Frank J, Hiemstra PS, den Dunner JT 2015. Efficient and sensitive identification and quantification of airborne pollen using next-generation DNA sequencing. Mol Ecol Resour 15:8–16

Kron P, Kwok A and Husband BC 2014. Flow cytometric analysis of pollen grains collected from individual bees provides information about pollen load composition and foraging behavior. Ann Bot 113: 191–197

Loraine AE, McCormick S, Estrada A, Patel K, and Qin P 2013. RNA-seq of arabidopsis pollen uncovers novel transcription and alternative splicing. Plant Physiol 162: 1092–1109.

Matheson CD, Muller GC, Junnila A, Vernon A, Hausmann A, Millere MA, Greenblatt C, Schlein Y 2008. A PCR method for detection of plant meals from the guts of insects. Org Divers Evol 7: 294–303.

Marrero P, Fregel R, Cabrera VM and Nogales M 2009. Extraction of high-quality host DNA from feces and regurgitated seeds: a useful tool for vertebrate ecological studies. Biol Res 42: 147-151.

Ohem J, Juen A, Nagiller K, Neuhauser S, Traugott M 2011. Molecular scatology: how to improve prey DNA detection success in avian faeces. Mol Ecol Resour, 11: 620–628.

Perez-Amaro JA, Mata-Gonzalez R, DelCurto T, Bohnert D, Riera-Lizarazu O, Boyd C, Leonard J 2010. The potential of DNA analysis for cattle diet determination in sagebrush rangelands. Oregon Beef Council Report. http://beefcattle.ans.oregonstate.edu.

Pierce GJ, Boyle PR 1991. A review of methods for diet analysis in piscivorous marine mammals. Oceanogr Mar Biol Annu Rev 29: 409 – 486.

Pinol J, San andres, Clare EL, Mir G and Symondson WOC 2014. A pragmatic approach to the analysis of diets of generalist predators: the use of next-generation sequencing with no blocking probes. Mol Ecol Resour 14: 18–26.

Poinar HN, Hofreiter M, Spaulding WG et al. 1998. Molecular coproscopy: dung and diet of the extinct ground sloth Nothrotheriops shastensis. Science 281: 402–406.

Pompanon F, Deagle BE, Symondson WOC, Brown DS, Jarman SN, Taberlet P 2012.Who is eating what: diet assessment using next generation sequencing. Mol Ecol 21: 1931–1950.

Purcell M, Mackey G, LaHood E 2004. Molecular methods for the genetic identification of salmonid prey from the Pacific harbor seal (*Phoca vitulina richardsi*) scat. Fish Bull 102:213–220.

Raye G, Miquel C, Coissac E, Redjadj C, Loison A, Taberlet P2011. New insights on diet variability revealed by DNA barcoding and high-throughput pyrosequencing: chamois diet in autumn as a case study. Ecol Res 26: 265–276.

Reed JZ, Tollit DJ, Thompson PM, Amos W 1997. Molecular scatology: the use of molecular genetic analysis to assign species, sex and individual identity to seal faeces. Mol Ecol 6: 225–234.

Richardson RT, Lin CH, Sponsler DB, Quijia JO, Goodell K, Johnson RM 2015. Application of ITS2 metabarcoding to determine the provenance of pollen collected by honey bees in an agro ecosystem. Appl plant Appl Plant Sci 3: 1400066

Schemske DW, Bradshaw HD 1999. Pollinator preference and the evolution of floral traits in monkey flowers (*Mimulus*). Proc Natl Acad Sci USA 96: 11910-11915

Shehzad W, Riaz T, Nawaz MA et al. 2012. Carnivore diet analysis based on next-generation sequencing: application to the leopard cat (*Prionailurus bengalensis*) in Pakistan. Mol Ecol 8: 1951–1965.

Sheppard SK, Harwood JD 2005. Advances in molecular ecology: tracking trophic links through predator–prey food-webs. Funct Ecol 19: 751–762.

Soininen EM: Valentini A, Coissac E, Miquel C, Gielly L, Brochmann C, Brysting AK, Sonstebo JH, Rolf A Ims1, Yoccoz NG and Taberlet P 2009. Analysing diet of small herbivores: the efficiency of DNA barcodingcoupled with high-throughput pyrosequencing for deciphering thecomposition of complex plant mixtures. Front Zool 6:16.

Staudacher K, Wallinger C, Schallhart N, Traugott M 2011.Detecting ingested plant DNA in soil-living insect larvae. Soil Biol Biochem 43: 346-350.

Sunderland KD 1988. Quantitative methods of detecting invertebrate predation occurring in the field. Ann Appl Biol 112: 201–224.

Symondson WOC 2002. Molecular identification of prey in predator diets. Mol Ecol 11: 627–641.

Vestheim H, Jarman SN 2008. Blocking primers to enhance PCR amplification of rare sequences in mixed samples–a case study on prey DNA in Antarctic krill stomachs. Front Zool 5:12.

Webb JK, Shine R, Branch WR and Harlow PS 2000. Life underground: food habits and reproductive biology of two amphisbaenian species from southern Africa. J Herpetol 34: 510–516.

Yoshihiro N, Eiji I, Miho Inoue-Murayama and Jumrafiha Abd. Sukor 2010. High potential of a disturbance-tolerant frugivore, the common palm civet *Paradoxurus hermaphrodites* (Viverridae), as a seed disperser for large-seeded plants. Mammal Study 35: 209–215.

Zaidi RH, Jaal Z, Hawkes NJ, Hemingway J, Symondson WOC 1999. Can multiple-copy sequences of prey DNA be detected amongst the gut contents of invertebrate predators. Mol Ecol 8: 2081– 2087.

Zarzoso-lacoste D, Corse E, Vidal E 2012. Improving PCR detection of prey in molecular diet studies: importance of group-specific primer set selection and extraction protocol performances. Mol Ecol Resour. 13: 117-127.

Zeale MRK, Butlin RK, Barker GLA, Lees DC, Jones G 2011. Taxonspecific PCR for DNA barcoding arthropod prey in bat faeces. Mol Ecol Resour 11: 236–244.

Zhou LJ, Pei KQ, Zhou B, and Ma KP 2007. A molecular approach to species identification of Chenopodiaceae pollen grains in surface soil. Am J Bot 94: 477-481.

Chapter 17

A key to the divisions of Apoidea, families of Apiformes and genera of Apidae of India

T Jobiraj
jobibee@gmail.com

17.1 Introduction

Bees occupy a unique role in the world of arthropods. They have an important role in agriculture, economy, ecology etc. Bees serve as significant pollinators of flowering plants, helping to sustain ecological systems worldwide. In the Order Hymenoptera, there is a section called Aculeata (females have a sting, which is a modification of ovipositor of ancestors). The Aculeata includes the wasps, ants and bees. Bees and sphecoid wasps together constitute the super family Apoidea (Michener 1986). The super family Apoidea is divisible into two informal series, the Sphecoid wasps or Spheciformes and the bees or Apiformes (Brothers 1975). In this chapter a key for the identification of the members of Apoidea is provided.

17.2 Key to the Divisions of Apoidea

1. Body setae simple; metabasitarsus not broader than remaining tarsal segments, frequently with stringil; female T7 not divided into hemitergites .... **Spheciformes**
   - At least some setae, particularly those of legs and surrounding propodeal spiracle, branched or plumose; metabasitarsus flattened, wider than remaining tarsal segments, without stringil; female T7 divided into hemitergites..............**Apiformes**

17.3 Key to the Families of Bees, based on Adults

(Modified from Michener 2000)

1. Labial palpus with first two segments elongated (Figure 17.1), flattened, the last two segments small, usually diverging laterally from axis of first two, not flattened, rarely absent; galeal comb absent or rarely weakly indicated; stipial comb and concavity commonly present (Figure 17.2); galeal blade elongate, commonly as long as or longer than stipes (Figure 17.3) [long-tongued bees] **2**

DOI: 10.1201/9781003505471-17

– Labial palpus with the first four segments similar to one another (Figure 17.4), or first or rarely first two elongate but not much flattened; galeal comb commonly present; stipial comb and concavity absent (Figure 17.5); galeal blade usually shorter than stipes (Figure 17.5) [short-tongued bees] **3**

2. Labrum with basolateral angles enlarged, base forming broad articulation with clypeus, labrum thus widest at base (Figure 17.6); labrum at least 0.8 times as long as broad and usually as long as broad or longer; scopa, when present, restricted to metosomal sterna .... **Megachilidae**

– Labrum with basolateral angles little developed, articulation with clypeus thus narrower than full width of labrum (Figure 17.7); labrum usually broader than long, but in some parasitic forms (where scopa is absent) labrum elongate; scopa when present, on hind leg, particularly the tibia, and usually absent on metasomal sterna .... **Apidae**

3. Glossa pointed at apex, sometimes with flabellum **4**

– Glossa bluntly rounded, truncate, or bilobed at apex; flabellum absent **7**

4. Lacinia represented by scale like lobe with hairs near base of galea (Figure 17.4); mentum and lorum forming proboscidial lobe (Figure 17.8, 17.9, 17.11); both at least partly sclerotized; lorum not flat..**5**

– Lacinia inconspicuous or displaced, not a scale like lobe at base of galea (Figure 17.38, 17.39); mentum and lorum not forming proboscidial lobe (Figure 17.10), mentum sometimes membranous; lorum membranous or nearly flat sclerotized membranes between cardines (Figure 17.10, 17.12) **6**

5. Lorum more or less plate like but produced in middle for attachment to base of mentum; facial fovea present in females (Figure 17.15) and some males, fovea sometimes a groove rather than broad as in figure; subantennal area almost always defined by two subantennal sutures below each antennal socket (Figure 17.15) ... [in part] ....
Andrenidae (Andreninaea and Panurginae)

– Lorum slender, V-shaped or Y-shaped, (Figure 17.1); facial fovea absent; a single subantennal suture below each antennal socket (Figure 17.16) .... **Melittidae**

6. Lacinia a small, hairless sclerite hidden between expanded stipites; sub antennal area defined by two subantennal sutures below each antennal socket (Figure 17.15); stigma nearly absent; first flagellar segment as long as scape or longer ...... [In part] ....
Andrenidae (Oxaeinae)

– Lacinia represented by small, hairy lobe on anterior surface of labio-maxillary tube above rest of maxilla (Figure 17.16); a single subantennal suture below each antennal socket (Figure 17.28); stigma well developed; first flagellar segment much shorter than scape ....**Halictidae**

7. Apex of glossa bluntly rounded, without preapical fringe or apical glossal lobes; episternal groove absent below scrobal groove; scopa present on hind tibia, but absent on femur ....**Stenotritidae**

   Apex of glossa truncate to bilobed; episternal groove usually present below scrobal groove; scopa, when present, well developed on hind femur as well as tibia ....**Colletidae**

17.4 Key to the Indian Genera of Apidae (With Subfamilies and Tribes)

1. Submarginal cross veins and second recurrent vein weak compared with other veins, commonly absent (Figure 17.17), marginal cell open at apex or closed by weakened vein; hind tibial spur absent ... (Apinae, Meliponini) **2**

   – All veins well developed, conspicuous, marginal cell closed by strong vein (Figure 17.18) 3

2. Forewing commonly over 4 mm; wing venation usually not greatly reduced (Figure 17.19), but if minute and with some of the wing reduction characters listed below, then upper margin of hind tibia with plumose hairs intermixed with simple ones (Figure 17.19); hind wing commonly with cells R and Cu closed by at least weakly brownish veins (Figure 17.19); forewing with one or two submarginal cross veins usually indicated, first submarginal cell usually recognizable; second Cu of forewing completely indicated at least by faint veins ....***Trigona*** Jurine

   – Forewing length less than 3 mm, wing venation greatly reduced (Figure 17.20); hindwing without closed cells (Figure 17.20), veins closing cells R and Cu, if visible at all, clear and unpigmented; upper margin of hind tibia without plumose hairs (Figure 17.20); forewing with submarginal cross veins almost always completely absent, thus without indication of submarginal cells ....***Lisotrigona*** Moure

3. Scopa of female, when present, forming corbicula on posterior tibia (Figure 17.21); inner apical margin of posterior tibia of non parasitic females with row of stiff bristles (rastellum Figure 17.22); pygidial and basitibial plates absent **4**

   – Scopa of female not forming a tibial corbicula, and scopa sometimes absent; inner apical margin of posterior tibiae bare or hairy, without comb; pygidial and basitibial plates frequently present **5**

4. Eyes hairy (Figure 17.23); jugal lobe of hindwing present (Figure 17.23); hind tibial spurs absent; arolia present (Figure 17.21) .....(Apinae, Apini) ....***Apis*** Linnaeus

 – Eyes bare; jugal lobe of hind wing absent (Figure 17.29); hind tibial spurs present; arolia absent .... (Apinae, Bombini) .... ***Bombus*** Latreille

5. Stigma absent (Figure 17.30); middle and hind basitarsi longer than tibiae; clypeus nearly flat, lower lateral areas not curved backward (Figure 17.24, 17.25) .... (Xylocopinae, Xylocopini) ....***Xylocopa*** Latreille

 – Stigma present (Figure 17.26); middle and hind basitarsi usually shorter than tibiae; clypeus variable but usually more convex with lower lateral parts bent backward (Figure 17.27, 17.28) **6**

6. Pygidial plate absent, sometimes represented by spine on pygidial fimbriae of female; scopa present (Figure17.22, 17.31); epistomal suture between lateral extremity and subantennal suture usually bent mesad such that upper part of clypeus is almost parallel-sided; clypeus not or weakly protuberant 7

 – Pygidial plate present in females and most males, or if absent in female, then scopa also absent; epistomal suture not or rarely bent mesad such that upper part of clypeus is nearly parallel-sided, clypeus commonly strongly protuberant, lower lateral parts bent strongly posteriorly (Figure17.27,17.28) 8

7. Submarginal cells two (Figure 17.32); clypeus typically slightly constricted at level of tentorial pits (Figure 17.33), this level near middle of clypeus .... (Xylocopinae, Allodapini) ....***Braunsapis*** Michener

 – Submarginal cells three (Figure 17.34); clypeus not constricted at level of tentorial pits, this level above middle of clypeus (Figure17.35) (Xylocopinae, Ceratinini)...***Ceratina*** Latreille

8. Labrum longer than broad (Figure 17.36) or rarely about as long as broad; scopa absent; body rather short-haired, or if with long hairs, then metasoma usually with pattern of white, appressed hair spot or bands **9**

 – Labrum broader than median length (Figure 17.37), longer than broad, in which case scopa present, body with long hairs, and metasoma without spots or bands of appressed white hairs **13**

9. Distal parts of wings, beyond venation, hairless and coarsely papillate (Figure 17.38) (Apinae, Melectini) **10**

 – Distal parts of wings, beyond venation with hairs (Figure 17.45), not or weakly papillate (Figure 39) .... (Nomadinae, Ammobatini) **11**

10. Scutellum flat or nearly so, produced posteriorly over metanotum, propodeum, and in some positions, base of T1, as sharply margined plate, bidentate with broad V or U-shape emargination and posterior part of scutellar surface on underside of plate, facing downward; body with areas of appressed plumose hairs forming white, blue, or green spots or broken bands (Figure17.40) .... ***Thyreus*** Panzer

– Scutellum convex, biconvex, bifurculate, or bispinose, surface thus downward posteriorly, and posterior part of scutellar surface thus declivous, facing posteriorly; body with or without areas of appressed plumose hairs, but if present, then these hairs white, not at all bluish or greenish .... ***Tetralonioidella*** Strand

11. Mandibles in repose directed posteromesally, so that, they cross one another well before their apices, their anterior margins forming angle of 90° to 145° .... ***Parammobatodes*** Popov

– Mandibles in repose directed mesally, so that, they overlap one another and cross, if at all, at extremely obtuse angle **12**

12. S5 of female with posterior margin unmodified, covered with free hairs; labrum not extending under or posterior to mandibles in repose .... ***Pasites*** Jurine

– S5 of female with posterior part not so strongly bent upward, visible from below, preapical concavity more U-shaped and defined by hairy ridge; labrum commonly extending under and sometimes beyond mandibles in repose .... ***Ammobates*** Latreille

13. Slender, sparsely haired, wasp like bees; the apex of marginal cell sharply pointed on wing margin (Figure 17.41); scopa not developed; three submarginal cells, the first nearly as long as second and third taken together (Figure 17.41) .... (Nomadinae, Nomadini) .... ***Nomada*** Scopoli

– More hairy and robust bees; the apex of marginal cell not sharply pointed (Figure 17.42); scopa usually more or less well developed **14**

14. Paraglossa as long as first two segments of labial palpus taken together; jugal lobe of hindwing about half as long as vanal lobe (Figure 17.43); antennae of males commonly elongate .... (Apinae, Eucerini) **15**

– Paraglossa much shorter than first segment of labial palpus; jugal lobe of hindwing usually less than half as long as vanal lobe; antennae of males not usually elongate **17**

15. Maxillary palpus, three to four segmented or with minute fifth segment, often not longer than width of galeal blade; keirotrichiate area of female apically at least half as wide as hind tibia; scopa consists of sparse, strongly plumose hairs on outer surface of tibia, each arising from tubercle separated from its neighbours by full width of tubercle ....***Tetralonia*** Spinola

– Maxillary palpus five or six segmented, often larger than maximum width of galeal blade; keirotrichiate area and associated finely and closely punctate surface of female usually less than half as wide as hind tibia; scopa consists of denser hairs, often simple, if plumose, then arising from tubercles less widely separated from one another..**16**

16. Clypeus usually protuberant for more than width of eye or more times length of eye; middle femur of male on undersurface commonly with area of dense, appressed, red-brown hair hiding surface ....***Eucera*** Scopoli

– Clypeus usually protuberant for less than width of eye as seen in lateral view; middle femur of male sometimes with area of sparse, appressed whitish hair not hiding surface, more commonly with unspecialized sloping or erect hair .... ***Tetraloniella*** Ashmead

17. Stigma small, usually shorter than prestigma, parallel-sided (Figure 17.44), vein arising near it apex; distal parts of wings usually strongly papillate, basal parts usually with large bare area (Figure 17.44) .... (Apinae, Anthophorini) **18**

– Stigma larger, longer than prestigma, tapering beyond vein R, which arises near the middle of stigma (Figure 17.47); distal parts of wings usually only weakly papillate; second abscissa of vein M + Cu of hind wing half as long as vein M or less (Figure 17.45) .... (Apinae, Ancylini) ....***Tarsalia*** Morawitz

18. First recurrent vein joining second submarginal cell near middle; third submarginal cell subquadrate, with front and rear margins of about equal length, and basal and distal margins of about equal length (Figure 17.46) **19**

– First recurrent vein terminating at or near apex of second submarginal cell (Figure 17.26); third submarginal cell with front margin much shorter than rear margin, because of curvature of third submarginal cross vein, the distal margin of the cell is longer than basal margin (Figure 17.26) ....***Habropoda*** Smith

19. Arolia present, well developed; hair never metallic ....***Anthophora*** Latreille

– Arolia absent (Figure17.48); hair sometimes metallic .... ***Amegilla*** Friese

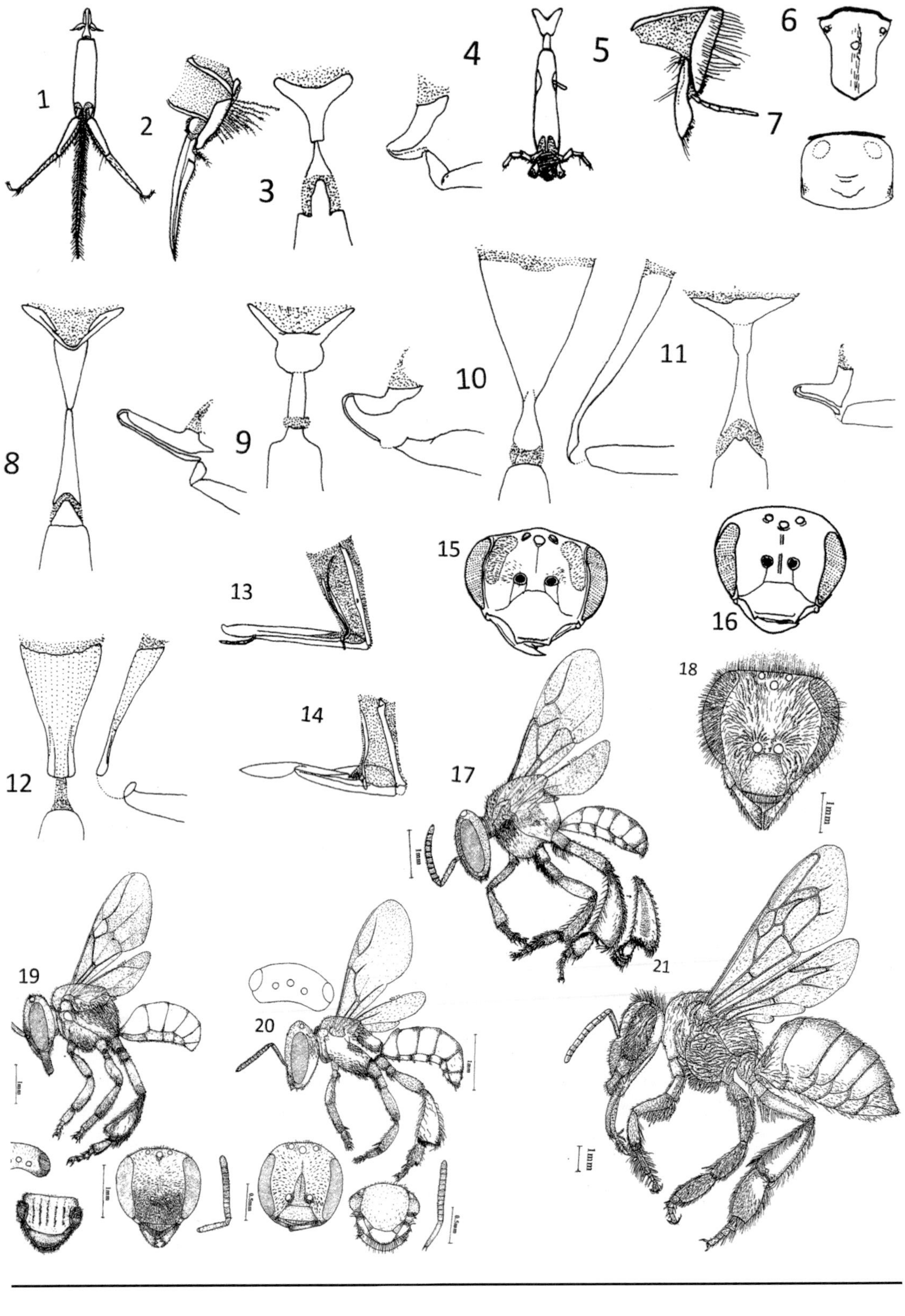
1
2
3
4
5
6
7
8
9
10
11
12
13
14
15
16
17
18
19
20
21

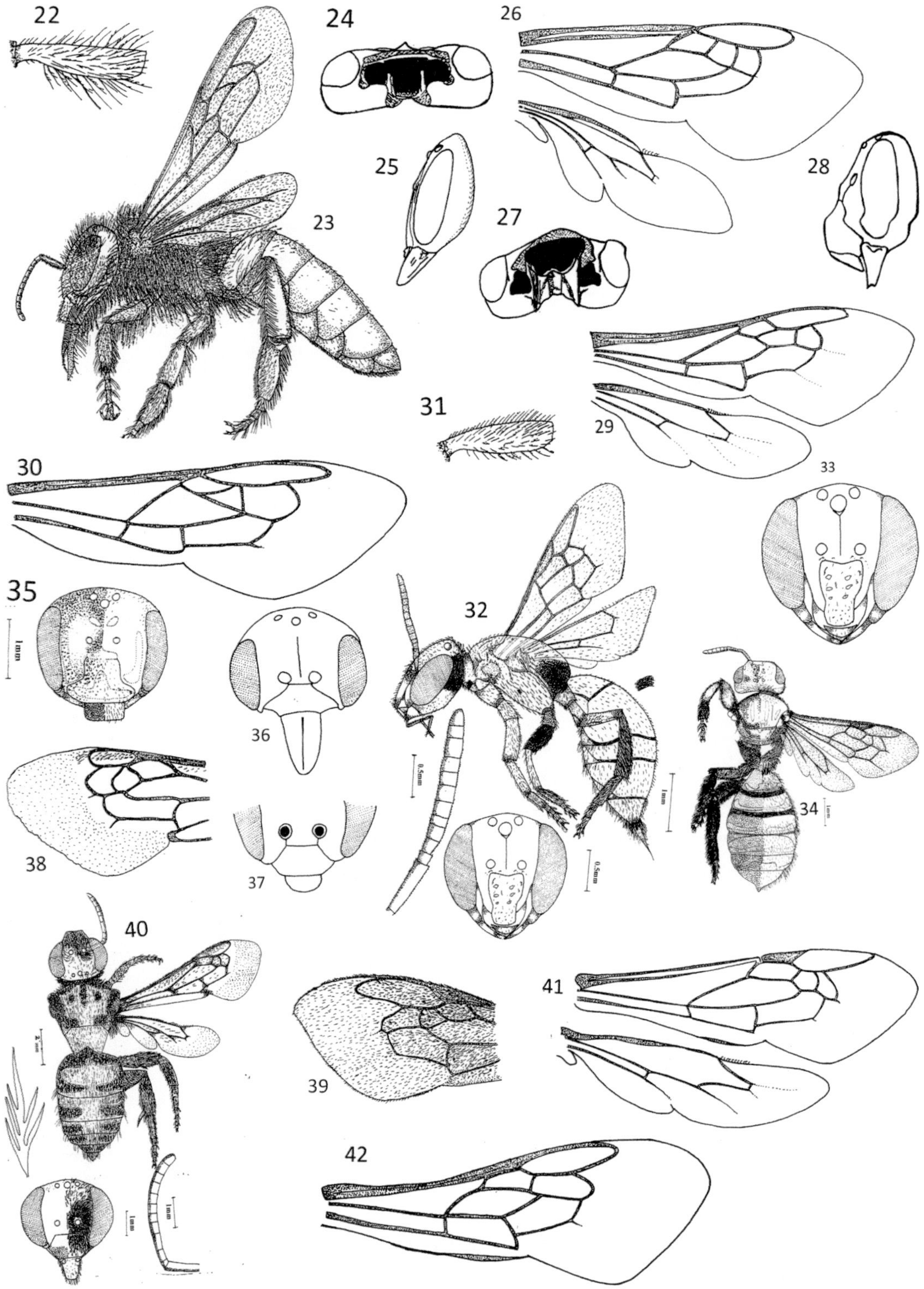
22
24
26
25
28
23
27
31
29
30
33
35
1mm
32
36
0.5mm
1mm
0.5mm
34
1mm
38
37
40
1mm
1mm
41
39
42

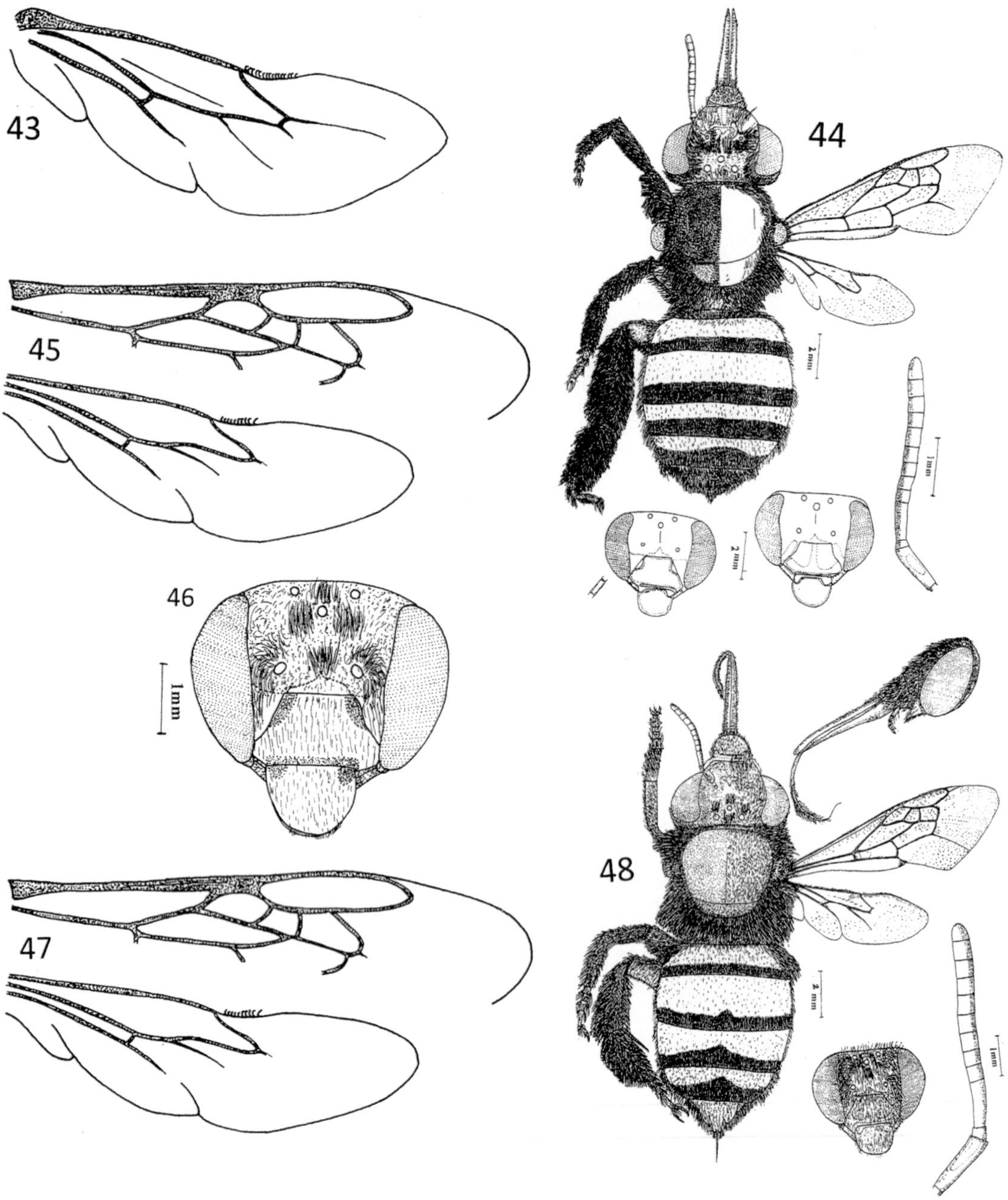

Fig. 17.1-17.48. Please refer the text for the details

Acknowledgements

The author is grateful to the authorities of Government College, Kodenchery, for the facilities provided.

References

Bingham CT 1897. The Fauna of British India Including Ceylon and Burma, Hymenoptera, Vol.1. Wasps andBees. XXiX + 577pp., 4 Pls. London: Taylor and Francis. [Reprinted, 1975. New Delhi: Today and Tomorrow's.]

Brothers DJ 1975.Phylogeny and classification of Aculeate Hymenoptera, with special reference to Muttilidae. Uni Kan Sci Bull 50: 483–648.

Engel MS 1998. Fossil Honey bees and evolution in the Genus Apis (Hymenoptera: Apidae). *Apidologie* 29: 265–281.

Michener CD 1986. Family group names among bees. J Kan Entomol Soc 59: 219–234.

Michener CD 2000. *The Bees of the world.* Johns Hopkins University press, Baltimore Maryland, USA.

For Product Safety Concerns and Information please contact our
EU representative GPSR@taylorandfrancis.com Taylor & Francis
Verlag GmbH, Kaufingerstraße 24, 80331 München, Germany